Pulse Improvement

Shabir Hussain Wani • Mukesh Jain
Editors

Pulse Improvement

Physiological, Molecular and Genetic Perspectives

Springer

Editors
Shabir Hussain Wani
Mountain Research Centre for Field Crops
Khudwani, Sher-e-Kashmir University
of Agricultural Sciences and Technology
of Kashmir
Srinagar, Jammu and Kashmir, India

Mukesh Jain
School of Computational & Integrative
Sciences
Jawaharlal Nehru University
New Delhi, India

ISBN 978-3-030-01742-2 ISBN 978-3-030-01743-9 (eBook)
https://doi.org/10.1007/978-3-030-01743-9

Library of Congress Control Number: 2018962000

This Springer imprint is published by the registered company Springer Nature Switzerland AG
The registered company address is: Gewerbestrasse 11, 6330 Cham, Switzerland

Dedication

Dr. R. S. Paroda

Dr. Raj Paroda, former Director General, Indian Council of Agricultural Research (ICAR), and Secretary, Department of Agricultural Research and Education (DARE), Government of India, is an accomplished plant breeder and geneticist by profession and an able research administrator. He has made significant contributions in the field of crop science. He is known for the modernization and strengthening of Indian National Agricultural Research System (NARS). He is the main architect of one of the world's largest and most modern National Gene Banks in New Delhi. He has received numerous prestigious awards and recognitions, namely, Padma Bhushan by the

Government of India, Rafi Ahmed Kidwai Prize, ICAR Team Research Award, FICCI Award, Om Prakash Bhasin Award, BP Pal Gold Medal, Borlaug Award, Mahindra Shiromani Award, Asia-Pacific Seed Association Special Award, CGIAR Award for Outstanding Partnership, Life Time Achievement Award by Association of Agricultural Scientists in America, Dr. Harbhajan Singh Memorial Award, ISCA Gold Medal for Excellence in Science, Gold Medals from Ministry of Agriculture of Armenia and Vietnam, Life Time Achievement Award of "Agriculture Today," Dr. A. B. Joshi Memorial Award, Prof. S. Kannaiyan Memorial Award and Awasthi-IFFCO Award, for Life Time Achievement in agricultural science and development.

Dr. Paroda had been the Founder Chairman of Global Forum on Agricultural Research (GFAR) based at FAO, Rome. He also served for more than two decades as the Executive Secretary of Asia-Pacific Association of Agricultural Research Institutions (APAARI), Bangkok, a well-known regional organization fostered by him to strengthen regional collaboration. He had served as Chairman as well as Vice-Chairman of ICRISAT Board, Member of Board of Trustees of IRRI, Member of WMO High Level Task Force on Climate Services, Member of Advisory Council of Australian Centre for International Agricultural Research (ACIAR), Member of Finance Committee of CGIAR and Member of the Governing Board of the Commonwealth Agriculture Bureau International (CABI). Currently, he is a member of high-level Strategic Impact, Monitoring and Evaluation Committee (SIMEC) of CGIAR.

He also was the President of Indian Science Congress in 2000–2001 and President of the National Academy of Agricultural Sciences, besides being president of a dozen national scientific societies in agricultural sciences. He has been conferred the Fellowship of several National Science Academies like INSA, NAAS and NASI and was elected as the General President of the prestigious Indian Science Congress in 2000–2001. Among international recognitions, he was elected as a Fellow of Agricultural Academies of Russia, Georgia, Armenia, Tajikistan and the Third World Academy of Sciences (TWAS).

Seventeen universities have awarded him D.Sc. (Honoris Causa) degree including Ohio State University, Columbus, and Indian Agricultural Research Institute, New Delhi. Dr. Paroda also worked as Chairman, Farmers Commission of Haryana, Chairman of Working Group on Agriculture and Member of Rajasthan Planning Board. Currently, he is the Chairman of the Trust for Advancement of Agricultural Sciences (TAAS).

Preface

With the increase in human population which is believed to exceed 9.7 billion by 2050, the growing demand for food and nutritional requirements is increasing at an alarming rate, and pulse crops act as a proficient spring of plant-derived proteins which involve negligible inputs. Also, pulses are principal and inexpensive source of proteins and minerals, which play a major part in improving the protein calorie malnutrition. However, pulses are grown in low-input and rain-fed conditions which result in lower yield and poor nutritional quality. In addition to the rising pulse requirement, the varying climate scenario has led to the increase in innumerable production limitations, including abiotic and biotic stresses. Pulse researchers globally and particularly in India have made strenuous efforts to defy the above challenges, and also with the progression of biotechnological tools, novel avenues for increased pulse production have come up in the last decade. Recent advancements in plant molecular biology and genomics in the form of the whole genome sequence, physical maps, genetic and functional genomic tools, integrated approaches using molecular breeding and genetic engineering put forward innovative prospects for improving production and productivity of many pulse crops. Many genes have been discovered which include regulatory genes that regulate stress response (e.g. transcription factors and protein kinases) and functional genes, which guard the cell (e.g. enzymes for generating protective metabolites and proteins). These genes are used to enhance stress tolerance in pulses. Advances in genomic tools have resulted in availability of genome-wide scattered molecular markers and the transcriptome and whole-genome assemblies. Through this book *Pulse Improvement: Physiological, Molecular and Genetic Perspectives*, effort has been made to include chapters unravelling the molecular and genomic mechanisms behind improved yield, quality traits and tolerance to biotic and abiotic stress tolerance in pulse crops using molecular breeding, and modern genomic and genome editing tools.

This book provides a detailed and novel reference material for researchers, teachers and graduate students involved in pulse crop improvement using recent advanced molecular and genomic tools. The chapters are written by world-class reputed researchers and academicians in the field of pulse crop improvement. We express sincere thanks and gratefulness to our esteemed authors; without their strenuous efforts, this book project would not have been possible. We are also thankful to Springer Nature for providing such opportunity to complete this book. We are also thankful to all our family members for their support during the entire book project completion.

Srinagar, India Shabir Hussain Wani
New Delhi, India Mukesh Jain

Contents

Chapter 1
Pulses for Human Nutritional Security

Manisha Goyal, Jitender Singh, Pankaj Kumr, and Anil Sirohi

1.1 Introduction

Pulses belong to the family Leguminosae and are the second most economically important—as well as nutritionally important—crop (International Legume Database and Information Service 2006–2013). It is estimated that pulses have been consumed for at least 10,000 years, and they are among the most widely consumed foods in the world. Pulses play important roles in contributing to food and nutritional security and replenishing soil nutrients, with a huge potential to address needs such as future global food security, nutrition, and environmental sustainability.

The higher financial returns from the production of cereals have contributed to pulses being grown on marginal land and to their low levels of cultivation in general. Adding pulses to crop rotations produces environmental, social, and economic benefits of pulse production as it fulfils needs for protein, minimizes soil degradation, and supports diversification in food production and consumption. The livelihood and development impacts of increased pulse production and consumption need to be understood by pulse producers and all stakeholders. The legume research program conducted by CGIAR [formerly known as the Consultative Group for International Agricultural Research] (2012) reported that the share of arable land is 36% in Myanmar, 30.6% in Niger, 27% in Kenya, 22% in Mozambique, 20% in Burkina Faso, and around 10–18% in Uganda, Nigeria, Tanzania, Ethiopia, India, Mexico, and Pakistan. Increasing population is a leading cause of penury in terms of scarcity of nutritional food worldwide. The greatest shares of all forms of malnutrition are borne by Africa and Asia (WHO 2016). According to World Bank estimates, India is one of the highest-ranking countries in the world for the number of children suffering from malnutrition. In 2017 the Global Hunger Index (GHI) reported that India ranked 97th out of 118 countries with a serious hunger situation. With a GHI

M. Goyal · J. Singh (✉) · P. Kumr · A. Sirohi
College of Biotechnology, S. V. Patel University of Agriculture & Technology,
Meerut, 250110, Uttar Pradesh, India

S. H. Wani, M. Jain (eds.), *Pulse Improvement*,
https://doi.org/10.1007/978-3-030-01743-9_1

score of 29.0, India ranked third after Afghanistan and Pakistan among all South Asian nations. From the above estimated data emerges an immediate need to discuss various measures to deal with human nutritional security for human beings. To eradicate the world food problem, extensive breeding work is going on to increase the nutritional values and yields of several edible seeds worldwide. Moreover, developed countries account for only about an eighth of global pulse cultivation areas but more than a fifth of global production, because the yields in developed countries are almost twice those in developing countries. In developing economies, to achieve food and nutritional security, more efforts are needed to harness extrusion technology for producing cost-effective food, utilizing locally grown pulses with more inputs. Moreover, by improving cropping patterns using pulses, farmers can improve their yields, limit the long-term threat to food security, and improve the availability of pulses. This review considers the environmental friendliness of pulses as a crop, as well as the extensive uses and demands for pulses as a nutritional food source.

1.2 Benefits of Pulse Production

Pulse crops play a leading role in the global nitrogen cycle by fixing atmospheric nitrogen in soils by symbiotic association with rhizobia, making them self-sufficient in nitrogen and enabling them to grow in almost any soil without any or with much less fertilizer input. From 1960 to 2000, nitrogen fertilizer use increased by roughly 800%, with half of that being utilized for wheat, rice, and maize production (Canfield et al. 2010). Cereal crops such as wheat, rice, and maize utilize 40% of the fertilizer that is applied, leading to significant waste and environmental impacts such as eutrophication of coastal waters and creation of hypoxic zones (Canfield et al. 2010). Incorporation of pulses into crop rotation reduces the fertilizer requirement for self crop, as well as following cereals or any other crops. Systematic crop rotation based on incorporating pulses into cereal-based systems reduces synthetic fertilizer use and optimizes the timing and amounts of fertilizer application to crops, which are the two most important interventions to decrease nitrogen application (Canfield et al. 2010). Biological nitrogen fixation is an alternative source of nitrogen, which can be enhanced along with other integrated nutrient management strategies such as use of animal manure, as well as recycling of the nutrients contained in crop residues (Lal 2004). The greater availability of nitrogen for subsequent cereal crops also benefits the yields of those cereal crops.

1.3 Conservation Tillage

Conservation tillage is defined as physical, chemical, or biological soil manipulation to optimize conditions for germination, seedling establishment, and crop growth (Lal 1985). Conventional plow-based farming leaves soil vulnerable to water and wind erosion, increases agricultural runoff, degrades soil productivity, and releases greenhouse gases (GHGs) through both soil disturbance and fossil fuel use. No-tillage or direct seeding under a mulch layer from the previous crop reverses this process by implementing a package of practices, which includes minimum mechanical soil disturbance, permanent organic soil cover, and diversification of crop species grown in sequences and/or associations (FAO 2013). Conservation tillage involves the introduction of pulses and oilseeds into cereal-based crop rotation. With implementation of multiple years of no-tillage, it has been demonstrated that the nitrogen fixation benefits of conservation tillage or no-tillage with pulse and oilseed bean nodulation improves and nitrogen fixation rates also increase to a large extent (Van Kessel and Hartley 2000).

1.4 Reduced Greenhouse Gas Emissions

Pulse crop cultivation also plays an important role in reducing GHG emissions in agricultural production through lower fertilizer requirements, as pulses supply their own nitrogen, as well as contributing nitrogen to succeeding crops (Lemke et al. 2007).

A study conducted by Lal (1985) depicts the considerable effects of no-tillage in comparison with conventional tillage on soil chemical properties after 6 years of tillage imposition (Table 1.1).

1.5 Social Benefits

1.5.1 Nutritional Benefits

Pulses have several health benefits, which vary according to the species and cultivar. Pulses contain approximately 20–30% protein content (Iqbal et al. 2006), which is about twice the amount of protein present in wheat, oats, barley, and rice (GPC 2016). Pulses are a very rich source of lysine but are relatively low in sulfur-containing amino acids such as cysteine, methionine, and tryptophan; this can be balanced by intake of pulses together with cereals, which are abundant sources of these essential amino acids (Boye et al. 2010). Pulses are relatively low in energy density (i.e., 1.3 kcal/g) but possess a high carbohydrate content (McCrory et al.

Table 1.1 Chemical properties of soil after implementation of no-tillage

Soil property	Conventional tillage	No-tillage
pH (1:1 in water)	4.7	5.3
Organic carbon (%)	1.35	1.48
Total nitrogen (%)	0.195	0.191
Bray – phosphorus (ppm)	42.8	25.0

ppm parts per million

2010) and are excellent sources of fiber, containing mostly insoluble fiber, as well as soluble fiber (Tosh and Yada 2010). Moreover, pulses are characterized by densely packed storage of micronutrients such as folate, zinc, iron, calcium, potassium, and magnesium (Patterson et al. 2009; Kearney 2010). Pulses are also well known for containing other healthy components such as vitamin E, vitamin A, riboflavin, niacin, pyridoxine, mono- and polyunsaturated fats, and plant sterols (Lovejoy 2010; Iqbal et al. 2006; Patterson et al. 2009; USDA 2012). In contrast to dried pulses, sprouted pulses are a significant source of vitamin C (Raatz 2010).

All around the world, pulses are primarily used to feed animals, but they are also emerging as an ethanol alternative fuel (Tigunova et al. 2013). Among several benefits of pulse crops such as peas and lentils are nitrogen fixation, with little or no requirement for nitrogen fertilizer (Burgess et al. 2012) along with chickpea, pigeon pea, urad, and mung beans. Limiting the amount of GHG release helps in lowering the carbon footprint of other crops grown in rotation (Gan et al. 2011; Harrison 2011). Thus, farming of pulses is highly beneficial for sustainability of the environment, and inclusion of pulses in the daily diet is a healthy way to reduce the risks of several chronic diseases. The human body requires a daily intake of about 50 g of protein, whereas in India the per capita daily intake is only about 10 g, which has direct adverse effects on the health and work efficiency of the people. Only 14 essential amino acids are supplied by the human body itself; the remaining six have to come from food. If all six amino acids are present in a single food item, it is called a complete protein food (Table 1.2).

By combining pulses with other nutritious food items, the nutritional value of pulses can be further enhanced, as other foods enhance the absorption of all nutrients found in pulses. Studies suggest that when beans are eaten with grains the body is better able to absorb iron and other minerals found in pulses. Therefore, for the vegetarian population, combinations of two or more pulses or other crops are needed in the diet.

Table 1.2 Nutritional content of various pulses

Bean variety	Calories	Protein (g)	Carbohydrates (g)	Fiber (g)	Calcium (mg)	Iron (mg)	Potassium (mg)
Faba bean	187	12.92	33.40	9.2	61	2.55	456
Mung bean	15	17.05	2.67	0.16	18	0.54	34
Green bean	36	20.00	8.40	3.6	58	2.16	220
Lima bean	209	11.58	40.19	9.0	54	4.17	969
Adzuki bean	294	17.30	56.97	16.8	64	4.60	1224
Black bean	227	15.24	40.78	15.0	46	3.61	611

Source: http://www.scind.org/462/Health/nutritional-benefits-of-pulses.html

1.6 Economic Benefits

The availability and price of agricultural commodities are strongly affected by climate change, resource depletion, and demographics (MSCI 2012). Farmers involved in grain and oilseed production have found economic benefits with lower input costs and increased profits by including a pulse crop in their rotation; the benefits include enhanced efficiency of nitrogen fertilizer use, reduced tillage, and in some cases reduced pesticide use. It is estimated that farmers save between 30% and 40% of their time, labor, and fossil fuel use by using a no-tillage method compared with conventional tillage (FAO 2001; Lorenzatti 2006).

The skyrocketing prices of pulses since 2008 can be attributed to almost stagnant production, leading to a decline in the per capita availability of pulses. Presently about 25–26 million hectares of land are under pulse cultivation in India, producing about 17 million tons of pulses annually. To meet the demand, about 2–3 million tons of pulses need to be imported every year as the yield (around 700 kg per hectare) is less than the global average and the per capita availability is one fifth lower than what nutritionists recommend. The demand for pulses in India is bound to increase further with a growing population, as well as sustained and inclusive economic growth. About 12 million hectares that are under rice production during the rainy season in India remain fallow in the subsequent post-rainy (rabi) season. Efforts to introduce pulses in these rabi conditions could have significant economic and poverty alleviation benefits (Joshi et al. 2002). The sustainability of the Indian agricultural system as a whole in the long run is a major concern because of consistent reduction in soil fertility and loss of essential soil nutrients due to exhaustive cropping systems being implemented after the green revolutions. The first revolution provided self-sufficiency in cereal production for the country, but it pushed pulses into rain-fed environments, which led to their poor productivity, besides leading to an imbalance in soil micronutrients. Though substantial progress has been made in evolving techniques to obtain high yields of pulses, their production has stagnated for the last several decades, primarily because of the number of biotic and abiotic constraints in rain-fed environments. The current shortfall in pulse availability is mainly due to a lower seed replacement rate for improved varieties, poor adoption of improved technologies by farmers, abrupt climate change, insect pests, emergence of new biotypes and races of key pests and pathogens, and declining total factor productivity. The reduction in yield as a result of climate change will be more pronounced for rain-fed pulses, especially those cultivated in areas frequently prone to drought. The Intergovernmental Panel on Climate Change (IPCC) has projected a rise in temperature of 2–3 °C over current levels by 2050. The predicted changes in temperature and their associated impacts on water availability, pests, disease, and extreme weather events are likely to affect the potential of pulse production. The pulse requirement in India is projected to be 50 million metric tons by the year 2050, necessitating an annual growth rate of 4.2%.

1.7 Reasons Why Pulses Are Underestimated

The cooking time for pulses is much longer than that for other vegetables, which may be one of the biggest constraints for including pulses in daily diets. Pulses are difficult to digest because of the presence of oligosaccharides. Moreover, pulses contain high levels of antinutrients such as phytate (Sandberg 2004), tannin, and phenol, which can limit the body's absorption of minerals such as iron and zinc, and can lead to iron deficiency (Petry et al. 2010). However, soaking (particularly in a sodium bicarbonate solution), germination, and fermentation may effectively reduce the antinutrient content in pulses (Vadivel and Pugalenthi 2009).

1.8 Pulses Help to Control Chronic Diseases

Being low in fat, cholesterol free, rich in fiber and protein, and an abundant source of iron, folate, etc., a diet of pulses has been found to be highly effective in reducing the risks of a number of chronic disease (such as cancer, diabetes, and cardiovascular disease) and aging, as well as in maintaining the lipid profile in the human body (Dahl et al. 2012; Iqbal et al. 2006; Winham et al. 2008).

World Cancer Research Fund/American Institute for Cancer Research (WCRF/AICR) research (2010) demonstrated that the high fiber content of pulses is negatively associated with colorectal cancers. Another study showed that pulses possess antiproliferative activity, as well as inducing apoptosis in colon cancer cells through the presence of the undigestible fiber component of pulses (Haydé et al. 2012; Campos-Vega et al. 2013). Moreover, pulses contain selenium, which has the ability to inhibit tumor cell development in mice. Therefore, selenium has been suggested to play a preventive role in breast, esophageal, and stomach cancers (Greeder and Milner 1980). Additionally, pulses are a rich source of zinc, which is known to be associated with improved function in immune cells by reducing oxidative stress in cells (Ibs and Rink 2003; Eide 2011). Not only do the nutrients in pulses exhibit anticancer activity but also some of the antinutrients such as phytic acid and tannins suppress cancer by preventing oxidative DNA damage in cells (Midorikawa et al. 2001; Dai and Mumper 2010). In general, high total cholesterol levels and high low-density lipoprotein (LDL) cholesterol levels are recognized as major risk factors for heart disease. Consumption of pulses has been found to be associated with lower cholesterol levels because of the presence of high soluble fiber content, thereby reducing the risk of cardiovascular disease (Bazzano et al. 2011).

A high-fiber diet with low energy density, a low glycemic load, and moderate protein content is thought to be particularly important for weight control. Pulses are densely packed with fiber, contain limited protein content, and are well established as low–glycemic index (low-GI) foods, thus pulses can play an effective role in weight management activities (Foster-Powell et al. 2002; Albete et al. 2010). Inclusion of pulses in the diet may also reduce the risk of type 2 diabetes by lowering the GI (Jenkins et al. 2012). Recently, the Canadian Diabetes Association recommended eating greater amounts of high-fiber foods, such as wholegrain breads and cereals, lentils, dried beans and peas, brown rice, vegetables, and fruit. Similarly, the American Diabetes Association suggests that people with diabetes include dried beans (such as kidney or pinto beans) and lentils in meals (Bantle et al. 2008).

1.9 Global View of Pulse Security

The majority of the world's population suffers from qualitative and quantitative insufficiency of dietary protein and calorie intake. Admitting the fact that grain legumes have the potential to provide a sustainable solution to human food security worldwide, remarkable efforts are currently being made to increase the availability of genomic resources of legumes. Moreover, to improve the yield and nutritional quality, and to enhance the resilience of legume crops to climate change, various innovative breeding techniques should be applied. Production agronomy and crop rotation approaches could also be intensified to address the associated economic and environmental challenges. The International Year of Pulses in 2016 was facilitated by the Food and Agriculture Organization (FAO) of the United Nations, focusing on the contribution of pulses to production and dietary diversity to eradicate hunger and malnutrition. The major objectives of this initiative were (1) to promote the value and utilization of pulses throughout the food system; (2) to raise the awareness of their benefits; (3) to foster enhanced research; (4) to advocate for better utilization of pulses in crop rotations; and (5) to address challenges in trade (Considine et al. 2016). The International Institute of Tropical Agriculture (IITA) has developed more than 80% of cowpea varieties released to farmers in Nigeria through its genetic breeding programs. Cowpea is one of the most drought-tolerant crops adapted to the semiarid tropics in Africa and Asia and is also an alternative to expensive animal sources of protein (Inter Press Service 2016).

1.10 Pulse Security in the Indian Scenario

India is a world leader in pulse production, consumption, and importation. In the year 2010–2011, India imported 2–3 million tons (MT) of pulses but, by 2050, production will be as high, as India will have changed from being a net importer to being a net exporter of pulses. Measures that could be taken on a priority basis for improving pulse productivity include (1) encouraging accelerated adoption of current technology for bridging the yield gap; (2) support to boost the seed replacement rate and quality; (3) strengthening of lifesaving irrigation in pulse-growing areas; (4) guaranteeing availability of critical inputs (viz., seed fertilizer and pesticides); (5) gradual mechanization of pulse production; (6) public–private partnerships for sustaining the value chain and minimizing postharvest losses; and (7) policy support for the value chain for pulses (Singh et al. 2015).

The Farmer Welfare Ministry has also taken several steps to accelerate pulse production. According to a Government of India Press Information Bureau report (2016), only 482 districts in 16 states were included in the National Food Security Mission (NFSM) for 2013–2014. Now all 638 districts in 29 states—including Goa, Kerala, eight northeastern states, and three hilly states—are included in this mission. Out of the total sanction of INR 1700 crores (INR 17 billion) under the NFSM, the central government has been allocated INR 1100 crores (INR 11 billion), whereas INR 430 crores (INR 4.3 billion) have been allocated by state governments for pulses only. A major amount of this allocation goes into production of new varieties of pulses under the NFSM. For expansion of cultivation of new kinds of seeds, INR 7.85 lakhs (INR 785,000) worth of minikits were distributed to farmers free of cost through state governments in the year 2016–2017. Moreover, in the same year, demonstrations of new techniques for pulse production, with the allocation of INR 25.29 crores (INR 252.9 million) for this purpose, were carried out on 31,000 hectares by 534 Krishi Vigyan Kendra (KVKs) [farm science centers] through the Indian Council of Agricultural Research (ICAR) and state agricultural universities. The government is also encouraging farmers to adopt efficient technologies and to grow new types of breeder seeds of pulses, using intercropping with oil seeds, cotton, and other crops.

By utilizing the nutritional value of pulses, production of delicious and healthy pulse-based food could meet the demands of human nutritional security worldwide. Various food-processing techniques have the potential to increase the safety and nutritional value of food. These technologies used for preparation of weaning foods include roasting, germination, milling, baking, cooking, drying, fermentation, and extrusion. Extrusion is one of the most popular food-processing techniques worldwide. Extrusion cooking plays an important role in restricting the presence of antinutritional compounds, and the mechanical and chemical processes involved in it lead to an increase in protein digestion through denaturation of protein structures (Stojceska et al. 2008; Temba 2016).

In addition to human nutrition, pulses are used as a feedstuff for pigs and poultry. Crops are key players in food security. By consuming pulses, one can keep oneself healthy and energized. Because of the highly nutritive nature of pulses a large genetic diversity in the seeds of grain legumes has been studied and huge volumes of information have already been deposited in GenBank databases, but these are not fully used in potential breeding programs (Cowling et al. 2017). Moreover, the high nutrient quality of pulses paves the way to supporting future health issues and thereby has the potential to influence consumer demand and food industry interest in developing pulse-based food products. Thus, there is an urgent need for intensification of basic and applied research on pulses, which will form a remarkable cornerstone of future food and nutritional security globally.

To increase pulse cultivation and yield, an inclusive approach is needed that considers farmers' resource limitations, confronts the realities of diverse socioecological environments, and addresses external factors affecting production, such as competition of pulses with cereals. Increased investment in research and development is needed to improve productivity and make the information produced accessible to and understandable by farmers. There is also a need for increased investment in breeding underutilized, high-quality varieties that are pest, disease, and climate resilient. The necessary steps should be initiated with the help of seed support systems to improve the availability of and access to varieties that are suited to local conditions, thereby empowering smallholders. With the help of different extension programs regarding cropping systems, farmers should be aware of and take advantage of the beneficial impacts of pulses, such as crop rotation and intercropping. Farmers should have access to markets, and the government should establish cooperatives, minimum support prices, and weather-based price insurance. The value chain should be strengthened with the help of different production models, support networks, development of agribusiness services, and development of storage warehouses and logistics.

References

Albete I, Astrup A, Martinez JA et al (2010) Obesity and the metabolic syndrome: role of different dietary macronutrient distribution patterns and specific nutritional components on weight loss and maintenance. Nutr Rev 68:214–231

Bantle JP, Wylie-Rosett J, Albright AL, Apovian CM, Clark NG, Franz MJ et al (2008) Nutrition recommendations and interventions for diabetes: a position statement of the American Diabetes Association. Diabetes Care 31:61–78

Bazzano LA, Thompson AM, Tees MT et al (2011) Non-soy legume consumption lowers cholesterol levels: a meta-analysis of randomized controlled trials. Nutr Metab Cardiovasc Dis 21:94–103

Boye J, Zare F, Pletch A (2010) Pulse proteins: processing, characterization, functional properties and applications in food and feed. Food Res Int 43(2):414–431

Burgess MH, Miller PR, Jones CA (2012) Pulse crops improve energy intensity and productivity of cereal production in Montana, U.S.A. J Sustain Agr 36(6):699–718

Campos-Vega R, Oomah BD, Loarca-Piña G, Vergara-Castañeda HA (2013) Common beans and their non-digestible fraction: cancer inhibitory activity—an overview. Foods 2(3):374–392

Canadian Diabetes Association (2012) The benefits of eating fibre. http://www.diabetes.ca/diabetes-and-you/nutrition/fibre/

Canfield D, Glazer A, Falkowski P (2010). The Evolution and Future of Earth's Nitrogen Cycle. Science; 8 October 2010, Vol. 330

CGIAR research program report on grain legumes (2012) Grain legumes: leveraging legumes to combat poverty, hunger, malnutrition, and environmental degradation. https://www.icrisat.org/.

Considine MJ, Considine JA (2016) On the language and physiology of dormancy and quiescence in plants. J. Exp. Bot. 67, 3189–3203.

Cowling WA, Li L, Siddique KHM, Henryon M, Berg P, Banks RG, Kinghorn BP (2017) Evolving gene banks: improving diverse populations of crop and exotic germplasm with optimal contribution selection. J Exp Bot 68:1927–1939

Dahl WJ, Foster LM, Tyler RT (2012) Review of the health benefits of peas (*Pisum sativum* L). Br J Nutr 108(S1):3–10

Dai J, Mumper RJ (2010) Plant phenolics: extraction, analysis and their antioxidant and anticancer properties. Molecules 15(10):7313–7352

Eide DJ (2011) The oxidative stress of zinc deficiency. Metallomics 3(11):1124–1129

FAO (2001) The economics of conservation agriculture. Food and Agriculture Organization of the United Nations, Rome
FAO (2013) Conservation agriculture. Agriculture and Consumer Protection Department. Food and Agriculture Organization of the United Nations, Rome
Foster-Powell K, Holt SH, Brand-Miller JC (2002) International table of glycemic index and glycemic load values: 2002. Am J Clin Nutr 76:5–56
Gan Y, Liang C, Wang X, McConkey B (2011) Lowering carbon footprint of durum wheat by diversifying cropping systems. Field Crops Res 122(3):199–206
GHI (2015) 2015 global hunger index report. International Food Policy Research Institute (IFPRI), Washington, DC
Gore, Padmavati & Tripathi, Kuldeep & Upadhyay, Richa & Deepak Dubey, Swati. (2017). Nutritional Benefits of Pulses. Scientific India Magazine.
Greeder GA, Milner JA (1980) Factors influencing the inhibitory effect of selenium on mice inoculated with Ehrlich ascites tumor cells. Sci 209(4458):825–827
GPC (2016) Grain processing Corporation, Washington. https://www.grainprocessing.com/
Harrison M (2011) Pulse crops may reduce energy use and increase yields for farmers. http://www.physorg.com/news/2011-05-pulse-crops-energy-yields-farmers.html
Haydé VC, Ramón GG, Lorenzo GO, Dave OB, Rosalía RC, Paul W et al (2012) Non-digestible fraction of beans (*Phaseolus vulgaris* L.) modulates signalling pathway genes at an early stage of colon cancer in Sprague–Dawley rats. Br J Nutr 108:145–154
Ibs KH, Rink L (2003) Zinc-altered immune function. J Nutr 133(5):1452–1456
ILDIS. 2006-2013. Legume web from the International Legume Database & Information Service (ILDIS) World Database of Legumes, version 10.38 by Richard White (Cardiff University). School of Biological Sciences in the University of Southampton. (ILDIS).
Inter press service (2016) http://www.ipsnews.net/2016/10/the-beating-pulse-of-food-security-in-africa/.
Iqbal A, Khalil IA, Ateeq N, Sayyar Khan M (2006) Nutritional quality of important food legumes. Food Chem 97(2):331–335
Jenkins DJ, Kendall CW, Augustin LS, Mitchell S, Sahye-Pudaruth S, Mejia SB et al (2012) Effect of legumes as part of a low glycemic index diet on glycemic control and cardiovascular risk factors in type 2 diabetes mellitus: a randomized controlled trial effect of legumes on glycemic control. Arch Intern Med 172(21):1653–1660
Joshi PK, Birthal P, Bourai V (2002) Socioeconomic constraints and opportunities in rainfed rabi cropping in rice fallow areas of India. International Crops Research Institute for the Semi-Arid Tropics, National Centre for Agricultural Economics and Policy Research, New Delhi
Kearney J (2010) Food consumption trends and drivers. Phil Trans R Soc 365:2793–2807
Lal R (2004) Carbon Emissions from Farm Operations. Environment International. 30: 981–990.
Lal R (1985) Mechanized tillage systems effects on properties of a tropical alfisol in watersheds cropped to maize. Soil Tillage Res 6(2):149–161
Lemke RL, Zhong Z, Campbell CA, Zentner R (2007) Can pulse crops play a role in mitigating greenhouse gases from North American agriculture. Agron J 99:1719–1725
Lorenzatti S (2006) Factibilidad de implementación de un certificado de agricultura sustentable como herramienta de diferenciación del proceso productivo de Siembra Directa. Universidad de Buenos Aires, Buenos Aires
Lovejoy JC (2010) Fat: the good, the bad, and the ugly. In: Wilson T, Bray GA, Temple NJ, Struble MB (eds) Nutrition guide for physicians. Humana Press, Totowa, pp 1–11
McCrory MA, Hamaker BR, Lovejoy JC, Eichelsdoerfer PE (2010) Pulse consumption, satiety, and weight management. Adv Nutr Res 1(1):17–30
Midorikawa K, Murata M, Oikawa S, Hiraku Y, Kawanishi S (2001) Protective effect of phytic acid on oxidative DNA damage with reference to cancer chemoprevention. Biochem Biophys Res Commun 288(3):552–557
MSCI (2012) Morgan Stanley Capital. International https://www.msci.com/.
Patterson CA, Maskus H, Dupasquier C. (2009) Pulse crops for health. Pulse Canada. AACC International Inc

Petry N, Egli I, Zeder C, Walczyk T, Hurrell R (2010) Polyphenols and phytic acid contribute to the low iron bioavailability from common beans in young women. J Nutr 140(11):1977–1982

Press Information Bureau, Gov. of India report. (2016) http://www.pib.nic.in/AllRelease.aspx?MenuId=3#

Raatz S (2010) The Bean Institute. Nutritional values of dry beans. http://beaninstitute.com/health-benefits/nutritional-value-ofdry-beans/

Sandberg GJ, Miller BR, and Harper MJ (2004) A Qualitative Study of Marital Process and Depressionin Older Couples. 51, 256–264

Singh AK, Sigh SS, Prakash V, Kumar S, Dwivedi SK (2015) Pulses production in India: present status, bottleneck and way forward. J Agri Search 2(2):75–83

Stojceska V, Ainsworth P, Plunkett A, Ibanoglu E, Ibanoglu S (2008) Cauliflower by-products as a new source of dietary fibre, antioxidants and proteins in cereal based ready-to-eat expanded snacks. J Food Eng 87:554–563

Temba MC (2016) The role of compositing cereals with legumes to alleviate protein energy malnutrition in Africa. Int J Food Sci Tech 51(3):543–554

Tigunova OA, Shulga SM, Blume YB (2013) Biobutanol as an alternative type of fuel. Cyto and Gen 47(6):366–382

Tosh S, Yada S (2010) Dietary fibres in pulse seeds and fractions: characterization, functional attributes, and applications. Food Res Int 43(2):450–460

USDA, Agricultural Research Service (2012) USDA national nutrient database for standard reference, release 25. Nutrient data laboratory home page. www.ars.usda.gov/ba/bhnrc/ndl

Vadivel V, Pugalenthi M (2009) Effect of soaking in sodium bicarbonate solution followed by autoclaving on the nutritional and antinutritional properties of velvet bean seeds. J Food Proc Pres 33:60–73

Van Kessel C, Hartley C (2000) Agricultural management of grain legumes: has it led to an increase in nitrogen fixation. Field Crops Res 65:165–181

WCRF/AICR (2010) Food, nutrition, physical activity, and the prevention of cancer: a global perspective. AICR, Washington, DC http://www.dietandcancerreport.org/cancer_resource_center/downloads/Second_Expert_Report_full.pdf

Winham DM, Webb D, Barr A (2008) Beans and good health. Nutr Today 43(5):201–209

World Health Organization (2016) World health statistics 2016: monitoring health for the SDGs, sustainable development goals. World Health Organization. http://www.who.int/iris/handle/10665/206498.

Chapter 2
Genomic Resources and Omics-Assisted Breeding Approaches for Pulse Crop Improvement

Javaid Akhter Bhat, S. M. Shivaraj, Sajad Ali, Zahoor Ahmad Mir, Aminul Islam, and Rupesh Deshmukh

2.1 Introduction

The Fabaceae (legumes) is the second most important family after Poaceae as a cornerstone for human food/protein security and the third largest family in the plant kingdom with ~20,000 species (Smýkal et al. 2015). Legumes are known for their unique ability to fix biologically atmospheric nitrogen through symbiosis with the bacteria, which makes them an essential component of sustainable agricultural production systems (Zhu et al. 2005). As per the Food and Agriculture Organization (FAO), the term "pulse" is reserved for those legumes harvested solely for the dry seed. Furthermore, FAO considers only those legumes as *pulses* that are harvested exclusively for grain purpose, which comprises a total of 11 pulse crops (Akibode and Maredia 2011). FAOSTAT (2012) reported that a total of 70.41 million tons (mt) of pulses are harvested annually in the world from 77.5 million hectares (Mha) area with a productivity of 907 kg/ha. Out of which, 62.98 mt (i.e., 90%) of global pulse production is contributed by seven major pulse crops, viz., common bean, pea (dry peas), chickpea, pigeon pea, faba bean, and lentil. Moreover, in the semiarid

J. A. Bhat
School of Biotechnology, Sher-e-Kashmir University of Agricultural Sciences & Technology of Jammu, Jammu, Jammu and Kashmir, India

S. M. Shivaraj · Z. A. Mir
National Research Centre on Plant Biotechnology, New Delhi, India

S. Ali
Centre of Research for Development, University of Kashmir, Srinagar, India

A. Islam
National Institute of Plant Genome Research, New Delhi, India

R. Deshmukh (✉)
National Agri-Food Biotechnology Institute (NABI), Mohali, Punjab, India
e-mail: rupesh.deshmukh.1@ulaval.ca

S. H. Wani, M. Jain (eds.), *Pulse Improvement*,
https://doi.org/10.1007/978-3-030-01743-9_2

and sub-tropical farming systems of the world, pulse crops are the critical components in maintaining the sustainability of food production as well as food security and generating livelihood to millions of resource-poor people living in these areas (Broughton et al. 2003).

Over one billion people around the world are suffering from malnutrition due to a deficiency in protein and micronutrient (Godfray et al. 2010). In this regard, the pulses being a rich source of vitamins, several essential minerals, and lysine-rich protein, complementing the conventional cereal-based carbohydrate-rich diets, are important for human nutrition (Coles et al. 2016). Due to high protein, mineral, and vitamin content in their seeds, pulses play a crucial role in alleviating malnutrition in the developing and underdeveloped countries (Broughton et al. 2003). Considering immense value in global food security, conventional breeding approaches have been extensively used for pulse improvement, which led to development and release of several high-yielding varieties. Although the area under pulse crops has been increased globally from 64 to 77.5 Mha over the last 50 years (FAOSTAT 2012; Bohra et al. 2014), they are usually cultivated in marginal land resulting in a significant reduction in productivity (FAO 2012). The productivity of major pulse crops remains desolately low (1000 kg/ha), and also the gap between their actual and potential yield was observed to be very high (FAOSTAT 2012; Varshney et al. 2013a). FAO of the United Nations designated the year 2016 as the International Year of Pulses (IYP) to bring the attention of global human community toward this underestimated group of crops. Moreover, the role of global climate change on food security and agriculture production is being discussed throughout the world (Reynolds et al. 2010). By 2050, the population of the world is predicted to be twice as that of the current population (Tester and Langridge 2010), which requires an increase in food production by 70% (Furbank and Tester 2011). Therefore, to cope such challenges and to maintain food security, the development of high-yielding varieties is to be accelerated at a more excellent pace to increase food production as well as withstand environmental stresses. Some efforts have been made in this direction using conventional breeding methods but with insufficient understanding of the underlying genetical or molecular mechanisms conferring resistance/tolerance to biotic/abiotic stresses (Varshney et al. 2013a). In this context, advances in modern "omics" approaches like genomics, transcriptomics, phenomics, proteomics, metabolomics, and ionomics have improved our understanding toward genetic architecture, molecular network, and physiological basis of complex traits in legumes. This novel approaches also assist in identification of marker-trait associations for economically important traits to enhance selection efficiency in breeding (Chaudhary et al. 2015). Hence, the holistic interdisciplinary approach encompassing "omics" technologies, breeding, physiology, and bioinformatics will accelerate the development of stress tolerant genotypes in pulses to sustain their productivity in the face of global climate change. This chapter mainly focused on how the availability of genomic resources as well as high-throughput "omics" approaches will aid the crop improvement program in pulse crops.

2.2 Genomic Resources: Pulse Crops

The available genomic resources were insufficient for major pulse crops until 2005, but consequently, significant progress has been made in the development of large-scale genomic resources (Table 2.1). Recent advancement in pulse genomic resources has become possible due to coordinated efforts and financial support from various organizations such as Bill & Melinda Gates Foundation, CGIAR's Generation Challenge Programme, United States Agency for International Development (USAID), Australian Council of International Agricultural Research (ACIAR), International Development Research Centre (IDRC), Kirkhouse Trust, US Department of Agriculture (ARS-USDA), Agriculture and Agri-Food Canada (AAFC), Commonwealth Scientific and Industrial Research Organization (CSIRO), L'Institut national de la recherché agronomique (INRA), Department for Environment, Food, and Rural Affairs (DEFRA), Turkish General Directorate of Agricultural Research (GDAR), Brazilian Agricultural Research Corporation (EMBRAPA), Ethiopian Institute of Agricultural Research (EIAR), Indian Council of Agricultural Research (ICAR), Department of Biotechnology (DBT) of Government of India, US National Science Foundation (NSF), and The Peanut Foundation of the American Peanut Council. Furthermore, application of next-generation sequencing (NGS) techniques aided in generating genomic resources of pulse crops on a whole-genome scale at a much faster pace with considerably less cost and time. In brief, these efforts led to complete draft genome sequences, functional genomics resources, enormous numbers of molecular markers, and high-density genetic maps of many pulse crops. The genomic resources available for various pulse crops are presented in Table 2.1. These resources accelerated genetics and genomic research in these pulse crops species such as the establishment of various marker-trait associations and initiation of molecular breeding (Bohra et al. 2014, Table 2.1). The draft genome sequences based on next-generation sequencing platforms are now publicly available for many pulse crops such as pigeon pea (Singh et al. 2012; Varshney et al. 2012), chickpea (Varshney et al. 2013b), mung bean (Kang et al. 2014), Lupin (Yang et al. 2013), and common bean (Schmutz et al. 2014). Hence, these crops are graduated to "genomic resource-rich crops" from their "orphan" (less studied crops) state. Although the last decade has witnessed the development of genomic resources for major pulse crops, several other pulses such as, viz., lentils, lima bean, adzuki bean, tepary bean, faba bean, cowpea, vetches, etc., are still not being exploited. Moreover, in many lesser developed countries due to lack of resources, infrastructure, and research facilities, the development of genomic resources in many underutilized or less studied pulse crops grown in such countries is limited. The taxonomic relationships among different pulse crops based on their chromosome number and estimated genome size (Mbp) are represented in Fig. 2.1.

Table 2.1 Details of genomic architecture and available genomic resources in important pulse crops

S.No	Genetic resources	Chickpea	Pigeonpea	Common bean	Pea	Lentil	Faba bean	Lupin	Mungbean
1	Ploidy (2*n*)	16	22	22	14	14	12	40	22
2	Estimated genome size	~740 (Mbp)	833 (Mbp)	625 (Mbp)	4685 (Mbp)	4032 (Mbp)	13,032 Mbp	951 (Mbp)	~579 Mbp (Mbp)
3	BAC libraries	10× (Thudi et al. 2011)	11× (Bohra et al. 2011)	10× (de Faria Müller et al. 2014)	2 (Yu 2011)			12× (Gao et al. 2011)	3.5× (Miyagi et al. 2004)
4	BAC-end sequences	46,270 (Thudi et al. 2011)	88,860 (Bohra et al. 2011)	52,270 (de Faria Müller et al. 2014)		–	–	13,985 (Gao et al. 2011)	
5	TILLING population	5000 mutant M_2 lines	5000 mutant lines (Varshney et al. 2010a); Varshney et al. (2010b)	3000 mutant lines (Porch et al.(2009)	3027 (Dalmais et al. 2008); 4704 (Triques et al. 2007	–	–	–	22 (Barkley et al. 2008)
6	ESTs deposited at NCBI	53,333	24,177	94,450	256,174	10,389	20,704	9537	1005

7	SSRs	2000 (Varshney et al. 2013a)	4000 (Bohra et al. 2011; Dutta et al. 2011; Raju et al. 2010)	~2000 (Blair et al. 2009a, 2011, 2012; Buso et al. 2006)	2911 (Bohra et al. 2014)	360 (Andeden et al. 2013), ~75 SSRs (Durán et al. 2004; Hamwieh et al. 2005, 2009), 2393 (Kaur et al. 2011), 5673 (Verma et al. 2013)	73 (Zeid et al. 2009), 802 (Kaur et al. 2012), 336 (Kaur et al. 2014a; Kaur et al. 2014b)	1497 (Gao et al. 2011), 66 (Kamphuis et al. 2015)	1742 (Gupta et al. 2014), 13,134 (Chen et al. 2015b), 93 (Gwag et al. 2006)
8	DArT clones	5397 (Thudi et al. 2011)	15,360	2501 (Brinez et al. 2012)	–	–	–	3072 (Vipin et al. 2013)	7680 (Hang et al. 2012)
9	454/FLX reads	4,35,018 (Hiremath et al. (2011), 1,931,224 (Garg et al. 2011a), 969,132 (Jhanwar et al. 2012), 7.12 million (Kudapa et al. 2014)		16,92,972 (Kalavacharla et al. 2011), 160,036 (Hyten et al. 2010)	3,19,402 (Macas et al. 2007)	1.38 million (Kaur et al. 2011)	3,04680 (Kaur et al. 2012)	1.9 million (Parra-González et al. 2012)	–

(continued)

Table 2.1 (continued)

S.No	Genetic resources	Chickpea	Pigeonpea	Common bean	Pea	Lentil	Faba bean	Lupin	Mungbean
10	Transcriptome assemblies	103,215 contigs (Hiremath et al. 2011), 46,369contigs (Kudapa et al. 2014), 34,760,contigs (Garg et al. 2011b), 969,132contigs (Jhanwar et al. 2012)	48,476 contigs (Dubey et al. 2011), 21,434 contigs (Kudapa et al. 2012)	62828contigs (Wu et al. 2014)	324428contigs (Franssen et al. 2011)	15,354 contigs (Kaur et al. 2011);	58,962 contigs (Braich et al. 2017)	63,271 contigs (Kamphuis et al. 2015)	48,693 contigs (Chen et al. 2015a)
11	SNPs	519,066	10,000	44,875 (Ariani et al. 2016)	72,090 (Bohra et al. 2014)	44,879 (Sharpe et al. 2013); 1095 (Temel et al. 2014)	75 (Cottage et al. 2012), 14,522 (Kaur et al. 2014a; Kaur et al. 2014b)	9137 (Hane et al. 2017), 207,887 (Yang et al. 2015)	40,503 (Van et al. 2013), 6000 (Schafleitner et al. 2016)
12	Mapping populations	30 (Upadhyaya et al. 2011)	25 (Varshney et al. 2013a)	~20 (Galeano et al. 2011)	~25 (McPhee 2007; Rubiales et al. 2011)	~20 (Kumar et al. 2015a; Kumar et al. 2015b)	~ 20 (Arbaoui et al. 2008a; Ma et al. 2013; Torres et al. 2006)	13 (Phan et al. 2007; Cowley et al. 2014)	~19 (Chen et al. 2016a; Chen et al. 2016b)
13	Genetic maps	29 (Gaur et al. 2014)	10 (Saxena et al. 2017a)	~10 (Galeano et al. 2011)	~42 (Bohra et al. 2014)	~20 (Kumar et al. 2015a; Kumar et al. 2015b)	~10 (Gutiérrez et al. 2013; Ma et al. 2013; Torres et al. 2010)	~9 (Yang et al. 2013)	~12 (Chen et al. 2016a; Chen et al. 2016b)

14	Physical maps	Available (Zhang et al. 2010a; Zhang et al. 2010b)	Not available	Available (Córdoba et al. 2010)	Not available	Not available	Not available	Not available	Not available
15	Complete genome sequence	Available (Varshney et al. 2013b)	Available (Varshney et al. 2012; Singh et al. 2012)	Available (Schmutz et al. 2014)	Not available	Available	Not available	Available (Hane et al. 2017)	Available (Kang et al. 2014)

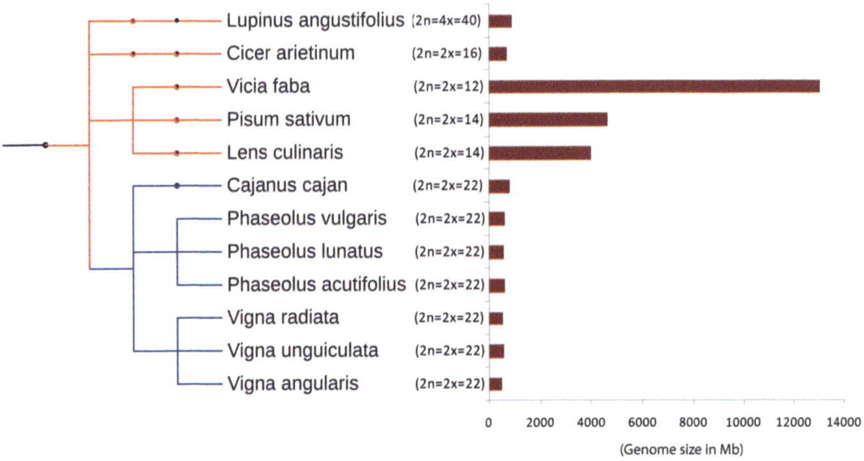

Fig. 2.1 Phylogenetic relationship among different pulse crops based on their chromosome number and estimated genome size

2.3 Gene and QTL Discovery: An Overview of Pulse Crops

The identification of a gene/QTL underlying the trait of interest is a prerequisite before initiation of any marker-assisted selection (MAS)/genomics-assisted breeding (GAB) program. By considering the importance of genomic resources, molecular markers are of direct utility in crop breeding, because they are heavily deployed for gene/QTL mapping studies using either linkage mapping (LM) or association mapping (AM) approaches (Das et al. 2017; Zargar et al. 2015). As mentioned above, the advent of NGS technologies has produced large-scale genomic resources in almost all the major pulse crops, which led to the development of either the first-generation or comprehensive genetic maps in these crop species (Table 2.1). Furthermore, the combined analysis of phenotypic data for the trait of interest and genetic maps of the respective segregating populations has facilitated identification of molecular markers associated with several agronomically important traits. The LM-based QTL discovery requires the appropriately developed experimental population with considerable size, which lies at the core of such studies (Mitchell-Olds 2010). In contrast, the linkage disequilibrium (LD)/AM analysis requires non-experimental populations consisting of genetically diverse genotypes for unraveling the genetic architecture of agronomic traits (Mackay and Powell 2007). The LM and AM analysis constitute the forward genetics approach where the known phenotypic variation is used to detect an underlying causal genetic polymorphism (McCallum et al. 2000). In this section, we briefly review the work carried for the discovery of gene(s)/QTLs in different pulse crops for agriculturally important traits using these two approaches (Table 2.2).

In most of the major pulses, QTLs controlling several agriculturally important traits have been identified through linkage mapping approach (Table 2.2). In certain

Table 2.2 Details of significant QTL/linkage mapping studies in important pulse crop species

Crop species	Name of population	Type of population	Population size	Number of QTL/genes identified	Trait	Types of markers	References
Chickpea	FLIP84-92C × PI 599072	RILs	142	2	*Ascochyta* blight resistance	Isozyme, RAPD, ISSR	Santra et al. (2000)
	'ICCV-2' × 'JG-62'	RILs	76	6	Double podding and other morphological traits	STMS, RAPD, ISSR	Cho et al. (2002)
	PI 359075(1) × FLIP84-92C(2)	RILs	133	5	*Ascochyta* blight resistance	STMS	Cho et al. (2004)
	Kabuli × desi	RILs	159	–	Fusarium wilt resistance	ISSR, RAPD, STMS	Cobos et al. (2005)
	C 214′ × 'WR 315'	F_2	188	–	Fusarium wilt (FW) and	SSR	Sabbavarapu et al. (2013)
	C 214′ × 'ILC 3279'	F_2	188	–	*Ascochyta* blight	SSR	Sabbavarapu et al. (2013)
	ICC3996 × S95362 & S95362 × cv. Howzat	RILs	113 & 128		Seed size traits	SSR	Hossain et al. (2010)
	SBD377 × BGD112	RILs	177		Seed traits	SNP	Verma et al. (2015a); Verma et al. 2015b)
	ICCV 96029 × CDC frontier	RILs	92	15	Early flowering and *Ascochyta* blight	SNP	Daba et al. (2016)
	ICC 4958 × ICC 1882	RILs	264	164	Drought tolerance	SNP	Jaganathan et al. (2015)
	ICCV 96029 × CDC frontier, ICC 5810 × CDC frontier, BGD 132 × CDC frontier and ICC 16641 × CDC frontier	F_2	190, 190, 190& 146	10	Flowering time	SSR	Mallikarjuna et al. (2017)
	Pusa 1103 × ILWC 46 & Pusa 256 × ILWC 46	RILs	102 & 98	3	Pod number and seed yield	InDels	Srivastava et al. (2016)

(continued)

Table 2.1 (continued)

Crop species	Name of population	Type of population	Population size	Number of QTL/genes identified	Trait	Types of markers	References
Pigeonpea	Pusa dwarf' × 'HDM04–1'	F_2	186	13	Plant type and earlierness	SNP and SSR	Kumawat et al. (2012)
	ICP 8863 × ICPL 20097 & TTB 7 × ICP 7035	F_2	190 & 130	6	Resistance to sterility mosaic disease	SSR	Gnanesh et al. (2011)
	ICPL 20096 × ICPL 332	RILs	188	8	Fusarium wilt and sterility mosaic disease	SNP	Singh et al. (2016)
	ICPB 2049 × ICPL 99050	F_2	188	_	Fusarium wilt	SSR	Bohra et al. (2012)
	ICPB 2049 × ICPL 99050	RILs	188	14	Fusarium wilt	SNP	Saxena et al. (2017b)
	ICPL 20096 × ICPL 332	RILs	188	10	Sterility mosaic disease	SNP	Saxena et al. (2017a)
	ICP-26 × ICPW-94	F_2	116	3	Blue butterfly and plume moth resistance	SSR, RAPD, ISSR	Mishra et al. (2016)
	ICP-26 × ICPW-94	F_2	116	7	Pod borer resistance	SCoT, RAPD, ISSR	Sahu et al. (2015)
	Bahar × KPL 43	F_2	271	3	Fusarium wilt resistance	SSR	Patil et al. (2017a); Patil et al. (2015b)

Common bean	DOR364 × G19833	RILs	86	17	Root hair, acid exudation, and phosphorus-uptake traits	RFLP, AFLP, RAPD, SSR, SCAR	Yan et al. (2004)
	PC-50′ × XAN-159	RILs	~100	16	White mold disease resistance	RAPD	Park et al. (2001)
	G19833 × DOR 364	RILs	86	26	Root architecture traits and phosphros accumulation	RFLP, AFLP, RAPD, SCAR	Beebe et al. (2006)
	Xana × Cornell 49,242	RILs	104	31	Morpho-agronomic and seed quality traits	AFLP, SSR, SCAR, ISSR, RAPD	Pérez-Vega et al. (2010)
	DOR364 × G19833	RILs	87	26	Iron and zinc accumulation	SSR	Blair et al. (2009b)
	Jalo EEP 558 × small white	F_2	142	6	Angular leaf spot	SSR	Teixeira et al. (2005)
	ICA Cerinza × G24404	$BC_2F_{3:5}$	157	14	Yield traits	SSR, SCAR	Blair et al. (2006)
	PMB0225 × PHA1037	RILs	185	26	Anthracnose disease resistance	SSR, SNP, SCAR, AFLP	González et al. (2015)
	G5686 × sprite	F_2	180	3	Angular leaf spot resistance	SSR	Keller et al. (2015)
	AND 277 × SEA 5	RILs	105	6	Angular leaf spot and powdery mildew	SSR, SNP	Bassi et al. (2017)
	DOR 364 × BAT 477	RILs	98	4	Yield and symbiotic nitrogen fixation	AFLP, RAPD, SSR, and gene-based markers	Diaz et al. (2017)

(continued)

Table 2.1 (continued)

Crop species	Name of population	Type of population	Population size	Number of QTL/genes identified	Trait	Types of markers	References
Pea	DP × JI296	RILs	135	17	*Ascochyta* blight resistance	RAPD, SSR, STS	Prioul et al. (2004)
	Carneval × MP1401	RILs	88	5	Lodging and *Ascochyta* blight resistance, plant height	AFLP, RAPD, STS	Tar'an et al. (2003)
	Kaspa × Parafield	RILs	134	4	Salinity tolerance	SNP, EST-SSRs	Leonforte et al. (2013)
	3148-A88 × Rovar	F_2	133	13	*Ascochyta* blight resistance	SSR, RAPD	Timmerman-Vaughan et al. (2002)
	Carneval' × 'MP1401	RILs	88	7	Grain yield, seed protein concentration and early maturity	AFLP, RAPD, STS	Tar'an et al. (2004)
	Shawnee × Bohatyr	RILs	187	2	Fusarium wilt resistance	SSR, RAPD, isozyme	Mc Phee et al. (2012)
	A26 × Rovar and A88 × Rovar.	F_2	148 & 133	14	*Ascochyta* blight resistance	SSR	Timmerman-Vaughan et al. (2004)
	Kaspa × PBA Oura	RILs	185	1	Metribuzin tolerance	SNP, SSR	Javid et al. (2017)
	P665 × cv. Messire	RILs	98	10	Drought tolerance	SSR, SNP	Iglesias-García et al. (2015)
	Aragorn × Kiflica	RILs	158	89	Seed mineral concentration and content	SSR, SNP	Ma et al. (2017)

Lentil	ILL5588 × ILL7537	F_2	150	5	*Ascochyta* blight resistance	RAPD, ISSR, AFLP	Rubeena Taylor et al. (2006)
	Lupa' × BG 16880	F_2	113	23	Plant structure, growth habit, and yield	RAPDs, ISSRs, AFLPs, SSRs	Fratini et al. (2007)
	ILL-6002 × ILL-5888	RILs	206	4	*Stemphylium* blight resistance	SSR, RAPD, SRAP	Saha et al. (2010)
	WA8649090 × Precoz	RILs	106	2	Leaf area and the association with winter hardiness	RAPD, ISSR, AFLP	Kahraman et al. (2010)
	Cassab × ILL2024	RILs	126	1	Boron tolerance	SSR, SNP	Kaur et al. (2014a); Kaur et al. (2014b)
	CDC Robin × 964a-46	RILs	139	13	Seed quality characteristics	SSR, SNP	Fedoruk et al. (2013)
	Precoz × L830	RILs	126	2	Seed weight and seed size	SSR	Verma et al. (2015a); Verma et al. (2015b)
	WA 8649090 × Precoz	RILs	94	18	Seed relevant traits	SSR, ISSR, RAPD	Jha et al. (2017)

(continued)

Table 2.1 (continued)

Crop species	Name of population	Type of population	Population size	Number of QTL/genes identified	Trait	Types of markers	References
Faba bean	Vf6 × Vf136	F_2	196	3	Broomrape resistance	Isozyme, RAPD, SSR	Román et al. (2002)
	Icarus × Ascot	RILs	95	4	*Ascochyta* blight resistance	SSR, SNP	Kaur et al. (2014a); Kaur et al. (2014b)
	Côte d'Or 1 × BeanPureLine 4628	RILs	101	5	Frost tolerance	RAPD	Arbaoui et al. (2008b)
	Vf6 × Vf136	RILs	144	2	Orobanche resistance	RAPD, STSs, ESTs, SSR, SCAR	Díaz-Ruiz et al. (2009)
	Côte d'Or 1 × Bean PureLine 4628	RILs	101	17	Frost tolerance	SNP	Sallam et al. (2016a); Sallam et al. (2016b)
	29H × Vf136	RILs	119	1	*Ascochyta* blight resistance	RAPD, SSR	Atienza et al. (2016)
	Icarus × Ascot	RILs	87	35	Flowering traits	SNP	Catt et al. (2017)

Mung bean	Berken × ACC41	RILs	227	15	Hard-seededness and seed weight	RFLP	Humphry et al. (2005)
	JP211874 × P229096 cv. Sukhothai	BC_1F_1	250	105	Domestication-related traits	SSR	Isemura et al. (2012)
	'Berken × ATF 3640'	RILs	147	1	Powdery mildew resistance	RFLP	Humphry et al. (2003)
	V1725BG × AusTRCF321925	F_2	170	23	Phytic acid and phosphorus contents	SSR	Sompong et al. (2012)
	"Kamphaeng Saen 1" × "VC6468–11-1A"	RILs	190	2	Powdery mildew resistance	SSR	Kasettranan et al. (2010)
	KPS1 × V4718	F_2	155	1	Cercospora leaf spot resistance	SSR	Chankaew et al. (2011)
	V6087AG × V5020BY	F_2	123	1	Seed starch content	SSR	Masari et al. (2017)
	VC2917 × ZL	RILs	256	58	Drought tolerance	SSR	Liu et al. (2017a); Liu et al. (2017b)

cases, where genetic linkage map is unavailable, the bulked segregant analysis (BSA) is a preferred method of choice to find linkage relationship of markers with the phenotypic trait of interest, especially resistance to biotic and abiotic stresses (Zhang et al. 2012). Furthermore, BSA has emerged as powerful mapping strategy for understanding marker-trait associations when near-isogenic lines (NILs) are available as genetic resources (Pérez-de-Castro et al. 2012). In pulse crops, notable examples involving BSA-based molecular tagging were employed for screening striga resistance in cowpea, *Ascochyta* blight resistance in lentil, growth habit in faba bean, and powdery mildew in pea using various types of molecular markers, viz., RAPD, AFLP, SCAR, and CAPS (Boukar et al. 2004; Chowdhury et al. 2001; Avila et al. 2007; Pereira et al. 2010). Till now genomics-assisted breeding (GAB) had limited application in pulse crop improvement, but the availability of relevant markers from ongoing mapping projects for important traits will effectively accelerate GAB in the pulse crops.

In contrast to linkage mapping, AM does not involve the laborious and costly development of large experimental population but uses a set of diverse and non-related individuals to test nonrandom association of alleles/LD (Mackay and Powell 2007). The AM depends on the rate of LD decay for establishing marker-trait associations (Andre et al. 2017; Tardivel et al. 2014). Although LD is not uniform across the whole genome but decays at a much faster rate in cross-pollinated species as compared to inbreeding crops (Yu and Buckler 2006). Before 2009, there was no significant evidence of AM reported in pulse crops, but with the increasing availability of large-scale genetic markers, SNP-based genotyping platform, and NGS data, it has emerged as a method of choice for high-resolution QTL discovery. For example, Murray et al. (2009) first reported AM to examine the associations of various candidate genes with yield/yield-relevant traits in a diverse collections from "USDA Pea Core," and consequently, the role of some pea homologues of APETALA2 (AP2) and GA3-oxidase (GA3ox) with regard to yield was revealed. Furthermore, use of AM for understating marker-trait association of agronomically important traits in various pulse crops has been well reported in the last 6–7 years (Table 2.3).

2.4 Genomics-Assisted Breeding (GAB)

The integration and use of genomic tools in breeding practices for development of superior cultivars with improved yield and quality as well as enhanced resistance to biotic and abiotic stresses are called as genomics-assisted breeding. To detect and utilize the association between genotype and phenotype for crop breeding is a core objective of GAB (Fig. 2.2). The GAB involves a range of approaches, viz., genomics, transcriptomics, and proteomics, to identify the molecular markers linked to traits of interest that assist breeders to predict the phenotype of plants from their marker genotype. The GAB approaches involve marker-assisted backcrossing (MABC), marker-assisted recurrent selection (MARS), and genomic selection (GS)

Table 2.3 Details of significant genome-wide association studies (GWAS) reported in important pulse crops

Crop species	Population size of core set	Trait	Type and number of marker used	Number of QTLs/marker trait associations	References
Chickpea	92	Seed-Fe and Zn concentrations	24,620 SNPs	16	Upadhyaya et al. (2016a)
	336	Seed-protein content	16,376 SNPs	7	Upadhyaya et al. (2016b)
	94	Seed-Fe and Zn concentrations	1536 SNP	8	Diapari et al. (2014)
	300	Drought and heat stresses	1072 DArTs, 764 SNPs, 36 SSR	312	Thudi et al. (2014b)
	187	Protein content	33 SSR	9	Jadhav et al. (2015)
	187	Seed weight	23 SSR	4	Kulwal et al. (2016)
	69	*Ascochyta* blight resistance	144,000 SNPs	1	Li et al. (2017)
Pigeonpea	292	Domestication and agronomic traits	446,568 SNPs	241	Varshney et al. (2017)
	89	Fusarium wilt resistance	65 SSRs	6	Patil et al. (2017a); Patil et al. (2017b)
	94	Determinacy	6144 DArT, 768 SNPs	25	Mir et al. (2013)
Common bean	504	Halo blight resistance	17,750 SNPs	1	Tock et al. (2017)
	206	Cooking time	~5000 SNP	3	Cichy et al. (2015)
	180	Anthracnose and angular leaf spot resistance	103 SSRs, 384 SNPs	66	Perseguini et al. (2016)
	237	Agronomic traits	5398 SNPs	20	Kamfwa et al. (2015)
	96	Drought tolerance	10,913 SNPs	27	Hoyos-Villegas et al. (2017)

(continued)

Table 2.3 (continued)

Crop species	Population size of core set	Trait	Type and number of marker used	Number of QTLs/marker trait associations	References
Pea	175	*Aphanomyces euteiches resistance*	13,204 SNPs	52	Desgroux et al. (2016)
	384	Agronomic and quality traits	256 SNPs	72	Cheng et al. (2015)
	179	*Aphanomyces euteiches resistance*	5107 697 SNPs	49	Bonhomme et al. (2014)
	672	Forst tolerance	267 SSRs	7	Liu et al. (2017a); Liu et al. (2017b)
Faba bean	189	Frost tolerance	189 SNPs	5	Sallam et al. (2016a); Sallam et al. (2016b)
	189	Frost tolerance	156 SNPs	67	Sallam and Martsch (2015)
	189	Winter hardiness and yield traits	156 SNPs	25	Sallam et al. (2016a); Sallam et al. (2016b)
	189	Drought and freezing tolerance	175 SNPs, 1147 AFPLs	21	Ali et al. (2016)

(Fig. 2.2). For traits controlled by major genes/QTLs (e.g., disease resistance), the MABC is the appropriate approach (Ribaut and Hoisington 1998). However, the majority of the agronomically important traits like drought tolerance or durable disease resistance are governed by many small effect QTLs, and hence pyramiding of these QTLs through MABC approach is a challenging task. Therefore, under such conditions, MARS has been considered as the better approach (Ribaut and Ragot 2006). Furthermore, for highly complex traits which are controlled by many minor effect genes as well as having low heritability and high G × E interaction, the genomic selection/genome-wide selection has emerged as a powerful approach for identifying desirable progenies for making the crosses. This approach became possible due to the availability of genome-wide high-density marker data through the use of high-throughput genotyping or NGS approaches (Jannink et al. 2010; Bhat et al. 2016). These approaches are being used in the pulse crops for improving a range of traits. Some of the examples have been listed below.

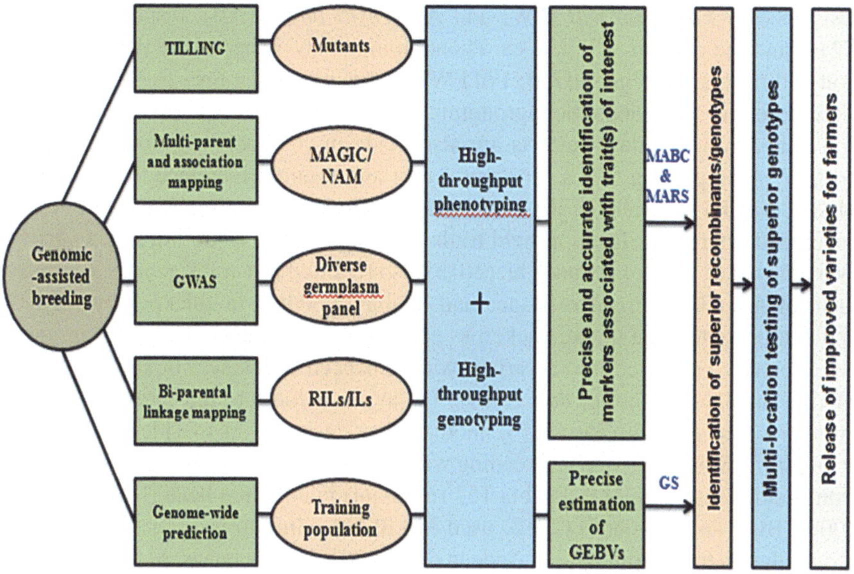

Fig. 2.2 Importance of high-throughput phenotyping (HTP) in genomics-assisted breeding for crop improvement. HTP occupies critical position in a genomics-assisted breeding pipeline; it helps to increase the precision and accuracy in trait mapping to identify genes and QTLs that are targets of MAS as well as increase the precision of genomic selection (GS) to calculate GEBVs that predict the breeding value of individuals in a breeding population

2.4.1 Marker-Assisted Backcrossing (MABC)

MABC involves introgression of the specific trait(s) into the genetic background of an elite recurrent genotype from a donor parent possessing the trait of interest (Hospital 2005). The MABC results in a cultivar, which possesses only the major gene/QTL from the donor parent while retaining the whole genome of the recurrent parent (Gupta et al. 2010). In this approach, limited number of QTLs/genes/loci including transgenes can be transferred into the recipient parent from the other genetic background (donor genotype). This approach has an extra advantage of being used for development of near-isogenic lines (NILs) or chromosome segment substitution lines (CSSLs), which are efficient genetic resources for genetic analysis of genes/QTLs and alien gene introgressions (Varshney et al. 2010a, b). Furthermore, MABC has an important application of gene pyramiding that involves the transfer of few genes either belonging to the same trait (e.g., resistance to different races of a pathogen) or different traits into the single genetic background using molecular markers.

Moreover, the availability of molecular markers associated with the trait of interest has provided an opportunity to initiate MABC for several traits in the pulse crops. For example, Varshney et al. (2014) used two parallel MABC programs by targeting *foc 1* locus and two QTL regions (ABQTL-I and ABQTL-II) to introgress resistance to *Fusarium wilt* (FW) and *Ascochyta blight* (AB), respectively, into C214, an elite cultivar of chickpea. Phenotyping analysis has led to the identification of three resistant lines to race 1 of FW and seven resistant lines to AB, and these were tested for yield and other agronomic traits under multi-location trials for general cultivation. Similarly, efforts are also made in chickpea to introgress a *QTL hotspot* encompassing QTLs for several root and drought tolerance traits through MABC into JG 11, from the ICC 4958 (donor parent). The introgression lines (ILs) developed after multi-location field trails may be released as an improved variety (Varshney et al. 2013c). Taran et al. (2013) also reported fast-track genetic improvement of *Ascochyta blight* resistance and doubled podding in chickpea by MABC and demonstrated that in two backcross generations an elite/adapted cultivar could be efficiently converted into a variety with enhanced resistance. In another study, MABC involved the use of *desi* cultivar (Vijay) as a donor to introgress FW resistance for race *Foc2*into Pusa 256, another elite *desi* cultivar of chickpea (Pratap et al. 2017). Similarly, MABC breeding was used to transfer root rot (RR) resistance from Eagle (susceptible) to Puebla 152 (resistant) in common bean (Navarro et al. 2009). Blair and Izquierdo (2012) used MABC breeding to transfer seed mineral accumulation traits from wild to Andean cultivated common beans. MABC has also been used for pyramiding of three rust resistance genes in common bean (Souza et al. 2014), introgression of bacterial blight resistance (Dinesh et al. 2016), and drought tolerance (Batieno et al. 2016) in cowpea. Moreover, many MABC programs in different pulses are underway aiming the introgression of resistance to various biotic and abiotic stresses as well as improved yield and quality.

2.4.2 Marker-Assisted Recurrent Selection (MARS)

MARS allows identification and selection of several genomic regions (up to 20 or even more)for complex traits within a single population. It involves genotyping of the F_2/F_3 population which provides estimates of marker effects and phenotyping of F_4/F_5 progenies (F_2 derived), followed by two or three recombination cycles based on the presence of marker alleles for small effect QTLs (Eathington and Swenson 2007). The breeding population derived from good × good crosses is first subjected to de novo QTL identification, and subsequently, the lines carrying superior alleles for maximum QTLs are crossed to pyramid superior alleles in one genetic background. Recombined lines (RILs) are subjected to a final phenotypic screening to select the best lines for multi-location field-testing to release them as varieties. The MARS is particularly useful for capturing the several genomic regions especially to target more number of minor as well as major QTLs. Hence, MARS offer's greater scope for achieving higher genetic gain compared to MABC program (Bohra 2013).

The recurrent selection breeding method was routinely used for the improvement of cross-pollinated crops such as maize, but recently it has been improved as well as facilitated by molecular markers which are now known as marker-assisted recurrent selection (MARS). In private breeding programs, MARS has proven to be successful as well as an effective approach for enhancing the genetic gain of quantitative traits in maize, sunflower, and soybean (Thudi et al. 2014a). In short, MARS is a modern breeding approach, which allows breeders to increase the frequency of several beneficial alleles with cumulative and additive effects in recurrent crosses (Bernardo and Charcosset 2006). Although some multinational companies were successful in MARS for crops like maize and soybean, the public sector institutes are in the initial phase of MARS in crops like wheat (Charmet et al. 2001), sorghum (Abdallah et al. 2009), and rice (Grenier et al. 2012). In this regard, some efforts have been initiated to use MARS in pulse crops such as chickpea, for assembling favorable alleles for drought tolerance using ICCV 04112 × ICCV 93954 and ICCV 05107 × ICCV 94954 crosses (Thudi et al. 2014a). IARI and IIPR have also initiated MARS in chickpea by using Pusa 372 × JG130 and DCP92–3 × ICCV 10 crosses. In addition, many projects involving MARS are underway for chickpea and pigeon pea at ICRISAT and IARI. These efforts are expected to result in the development of superior lines with enhanced crop performance. However, when high-throughput phenotyping facility becomes cost-effective, successful reports can be expected even in public sector for more pulse crops as well.

2.4.3 Genomic Selection (GS)

Presently, the cost of generating precise phenotypic data in every generation of the breeding population is much expensive. However, the recent advances in NGS/genomic technologies have offered a wide range of low-cost genotyping platforms for crop plants, which led the reduced cost of genotyping compared to phenotyping (Bhat et al. 2016). GS unlike to MABC and MARS is a recent breeding approach that targets identification of superior lines with higher breeding value based on genome-wide marker profile data. In contrast to MAS, which is based on few markers, the GS uses a large set of marker information distributed across the whole genome to estimate the genetic worth of the individuals. The GS uses the genotypic and phenotypic data of training population (TP) to develop the prediction model. Then the prediction model used to determine genomic estimated breeding values (GEBVs) for all the individuals of the breeding population using their genomic profile (Meuwissen et al. 2001). The GEBVs provide information about the individuals that will be used as a parent for the breeding program (Bhat et al. 2016).

In summary, GS is an emerging breeding approach that minimizes time and cost by reducing the frequency of extensive phenotyping and bypasses the need for QTL mapping. Nevertheless, GS can be effectively implemented along with the QTL mapping or GWAS programs (Deshmukh et al. 2014). GS can also reduce the selection cycle length of a breeding program that could take several seasons to develop reliable phenotypes. However, use of the appropriate statistical model is very critical for estimating the GEBVs with higher precision. In this context, Bayesian or BLUP methods are among the best models of GS for predicting the accurate GEBVs according to simulation studies (Zhang et al. 2010a, b). Moreover, Bayes B is another method in addition to a Bayesian approach, which is also referred to as wBSR (weighted Bayesian shrinkage regression), which minimizes computational burden on MCMC-based Bayesian methods and is considered as a method of choice for GS (Hayashi and Iwata 2010). Although GS has already been proven successful in many animal programs (Schefers and Weigel 2012; Eggen 2012), recently, good results have also been observed in crop plants like maize, wheat, rice, canola, soybean, etc. (see Bhat et al. 2016). In this context, efforts are also made in pulses, for example, Roorkiwal et al. (2016) attempted GS in chickpea for yield and yield-related traits using DArTseq genotyping, and genome-wide prediction accuracies were investigated to vary from 0.138 (seed yield) to 0.912 (100 seed weight) and suggested immediate implementation of GS in chickpea breeding programs. Additionally, the efforts to deploy genomic selection in chickpea and other pulse crops for various other agriculturally important traits are underway (Pandey et al. 2016).

2.5 Phenomics: Present Challenge in Pulse Breeding

The unprecedented development and advances seen in NGS technologies during the last decade have made it easy to unravel and understand crop genomes and offered exciting opportunities for crop genetics and breeding (Edwards et al. 2012). These developments in the technologies have led to the sequencing of crop and plant genomes a routine work (Jackson et al. 2011), and draft genome sequences of many pulse crops, viz., pigeon pea, chickpea, lupin, mung bean, common bean, etc. are now available in public databases (Legume Information System, https://legumeinfo.org). Furthermore, resequencing of crop genomes for identification of allelic variation has also been reported in pulse crop species such as chickpea (Thudi et al. 2016), pigeon pea (Varshney et al. 2017; Singh et al. 2012), and mung bean (Chen et al. 2015a). Besides, the cost for genotyping the pulse genomes also decreased due to evolution and emergence of high-throughput markers and genotypic platforms (Pandey et al. 2016). But, to reap full benefits from the wealth of genomic information, it is necessary to link and integrate this information with the phenotype in a real-world environment (Furbank and Tester 2011; Fig. 2.2). However, fewer efforts have been made to advance high-throughput phenomics compared to high-throughput genotyping approaches. This has become the bottleneck in capitalizing on a huge amount of available genomic data affecting breeding in crop plants (Edwards et al. 2012; Stamp and Visser 2012). This creates a gap between genotype and phenotype (GP-Gap). Hence, there is an immediate need for phenomics revolution to harness the full advantage of genomic data especially in changing the climate and increasing world population. In this regard, efforts are being made to alleviate the phenotyping bottleneck by developing plant phenomics facilities to scan and record data for thousands of plants in a day in a sophisticated manner (Cobb et al. 2013). Considering the importance, an International Plant Phenomics Initiative was launched recently (http://www.plantphenomics.org/). These high-throughput phenomics facilities make use of sophisticated noninvasive imaging, spectroscopy, image analysis, robotics, and high-performance computing facilities, thus saving time, labor, and cost. The present era of phenomics will help us to collect high-quality, accurate phenotyping data, which is necessary and useful for meaningful genetic dissection and genomics-assisted breeding, including (i) QTL interval mapping, (ii) candidate gene-based association mapping, (iii) genome-wide association studies (GWAS), (iv) QTL cloning, (v) QTL meta-analysis, (vi) marker-assisted selection (MAS), (vii) marker-assisted recurrent selection (MARS), (viii) TILLING (Targeting-Induced Local Lesions in Genomes), and (ix) genomic selection (GS) or genome-wide selection (GWS) (Welcker et al. 2011; Tuberosa 2012; Cobb et al. 2013; Fig. 2.2). In the future for promoting the application of high-throughput plant phenotyping, the less expensive, less laborious, and well-sophisticated data analysis infrastructure such as HTPheno (Hartmann et al. 2011) and IAP (Klukas et al. 2012) incorporating the open-source software image J needs to be developed and popularized.

2.6 Transcriptomics: Utility in Marker-Assisted Crop Breeding

The transcriptomics offers immense utility for accelerating marker-assisted breeding and genetic enhancement of crop species. It aids crop breeding in three ways: (a) development of functional/genic markers [EST-SSRs, SNPs, intron-spanning regions (ISRs), etc.], (b) identification of candidate genes associated with different traits of interest, and (c) validation of identified candidate genes for direct use in crop improvement. The transcriptomic-derived genic SSRs offer unique advantages over genomic SSRs as they represent variation in the expressed portion of the genome and will provide "perfect" marker-trait associations in gene tagging experiment, and unlike genomic SSRs, these markers once developed may be used across a number of related species. (Kalia et al. 2011; Jain et al. 2015). Furthermore, genic SSRs are a valuable resource for genetic studies and assessing functional diversity, as they regulate gene expression and function (Zhou et al. 2016). Based on alignment of related genome sequences, ISR (flanking intron junctions) markers were identified in several studies (Hiremath et al. 2011; Kudapa et al. 2012) and were used in generating highly informative dense genetic maps with well-distributed markers (Bordat et al. 2011). Moreover, by means of high-throughput array-based NGS and marker genotyping technologies, many successful efforts were made toward large-scale validation and high-throughput genotyping of known/candidate gene-based markers in diverse natural germplasm collections (core and mini-core accessions) and mapping/mutant populations of crop plants (Morris et al. 2013; Spindel et al. 2013). This, in turn, has propelled the large-scale mining and evaluation of novel functional allelic variation/molecular diversity in diverse crop genotypes for marker-assisted breeding applications. In this regard, several efforts were made recently in both model and non-model pulse crops to develop genic SSR and genic SNP markers through transcriptome sequencing using various NGS technologies. For example, Jhanwar et al. (2012) reported transcriptome sequencing of the wild chickpea, *C. reticulatum* (PI489777), by using GS-FLX 454 technology, and identified 561 polymorphic SSRs and 36,446 single-nucleotide polymorphisms (SNPs). Similarly, Sanger and second-generation sequencing platforms have been used for transcriptome sequencing in pigeonpea, and the alignment of the CcTA v2 transcriptome assembly with the soybean genome predicted 10,009 intron-spanning regions (ISRs) (Kudapa et al. 2012). Vatanparast et al. (2016) also made efforts to strengthen genetic resources for winged bean (an orphan pulse crop) by using two Sri Lankan accessions, viz., CPP34 and CPP37 subjected to de novo transcriptome assembly and annotation, and identified many SSR and SNP markers between these geographically separated genotypes. In addition, various other researchers have used transcriptome sequencing for developing genic markers of many other pulse crops such as cowpea (Chen et al. 2017), mungbean (Liu et al. 2016), common bean (Blair et al. 2011, 2013), lupin (Parra-González et al. 2012), rice bean (Chen et al. 2016a, b), etc.

The gene expression profiling data from the developed transcriptome assemblies provides a valuable source for the candidate genes identification particularly associated with different traits of interest including biotic and abiotic stress-responsive genes (Pandey et al. 2016; Sonah et al. 2016). In common bean, from a set of 2678 transcription factors (TFs) belonging to 59 TF families, a total of 441 salt-responsive TFs were identified (Hiz et al. 2014). Furthermore, integration of high-resolution QTL mapping with differential expression profiling emerged as a powerful technique to identify the corresponding candidate genes underlying the QTL governing the phenotypic trait (Song et al. 2017). For example, Kujur et al. (2016) combined high-resolution QTL mapping with genetic association analysis and differential expression profiling and delineated natural allelic variants in five candidate genes (encoding cytochrome-biosynthesis protein, *malic oxidoreductase*, NADH *dehydrogenase* iron-sulfur protein, expressed protein, and *bZIP* transcription factor) regulating plant height in chickpea. Similarly, Zhang et al. (2016) combined QTL mapping and transcriptome profiling to identify candidate genes underlying QTLs associated with low-phosphorus tolerance and suggested that integration of QTL mapping and transcriptome profiling is an useful tool for rapidly identifying candidate genes underlying complex traits and also reported that *GmACP2* (*phosphatase*-encoding gene) plays an essential role in low-P stress tolerance in soybean. Furthermore, the putative candidate genes identified in various genomic studies will be subjected to RT-qPCR expression analysis to elaborate their actual involvement in the phenotypic trait expression. Hence, transcriptomic resources and candidate gene information provide insights into the molecular mechanisms involved in stress tolerance that can ultimately help to develop an improved stress tolerant pulse varieties.

2.7 Proteomics, Metabolomics, and Ionomics: Efforts in Pulse Crops

Proteomics, metabolomics, and ionomics are three important "omics" techniques in the post-genomic era (Fernandez-Garcia et al. 2011). In breeding programs, we although use QTLs/candidate genes/alleles, but it has been demonstrated that variations (structural/expression) detected at the genetic level are not always translated into the "predicted" phenotype. In addition, the involvement of metabolites or ions, multigenes, and posttranslational modifications (PTM) complicates the mechanism involved in stress tolerance, which cannot be investigated by genomics or transcriptomics approaches (Mazzucotelli et al. 2008; Weckwerth 2011). In this context, proteomics, metabolomics, and ionomics are emerging approaches that improve our understanding of functional molecules instead of analyzing the genetic code (DNA) or transcript (RNA) abundance, which may not correlate with their corresponding proteins (Hossain et al. 2011). Proteomics is a large-scale study of proteins encoded by the genome of an organism (Zhang et al. 2013). It is a powerful technique for delineating whole proteomes at tissue/organ/cell/organelle levels. In addition, proteomics is also used to compare the differential protein expression of two samples under various conditions of biotic and abiotic stresses (Witzel et al. 2009; Fernandez-Garcia et al. 2011). Metabolomics deals with the study of global profile of the metabolites with low molecular weight (1000 Da), which are the end products of metabolisms in biofluids, tissues, and even whole organism (Brosché et al. 2005). It provides us information regarding the metabolic status of an organism as well as assesses the biological responses stimulated by external factors (Liu et al. 2011). The perturbations induced by external factors in stress-responsive proteins, metabolic pathways, and corresponding enzymes are frequently characterized by proteomics and metabolomics (Zhang et al. 2012). Ionomics deals with complete ionome study, involving the simultaneous and quantitative measurement of organism/tissue elemental composition and variations in this composition in response to physiological processes (Salt et al. 2008). Obviously, a combination of "omics" techniques like proteomics, metabolomics, and ionomics would complement and potentially validate one another while studying the effects of biotic/abiotic stresses in organisms. In this regard, several such studies have been performed in the various pulse crops.

For example, root proteome was analyzed using 2-DGE and MALDI-TOF (matrix-assisted laser desorption ionization time-of-flight) mass spectrometry in chickpea inoculated with *Fusarium oxysporum* f. sp. *ciceri* race 1(Foc1) (Chatterjee et al. 2014). They identified 100 differentially or uniquely expressed proteins between susceptible (JG62) and resistant (WR315) chickpea genotypes infected with Foc1 and indicated their involvement in early defense signaling in the host. Parankusam et al. (2017) compared the expression pattern of proteins of ICC16374 (heat-sensitive) and JG14 (heat-tolerant) genotypes of chickpea (*Cicer arietinum* L.) subjected to heat stress at an thesis stage and identified a set of 482 heat-responsive proteins in the tolerant genotype using comparative gel-free proteomics.

Similarly, Krishnan et al. (2017) analyzed the proteome of pigeon pea seed by employing different extraction media and high-resolution two-dimensional (2-D) electrophoresis followed by MALDI-TOF-MS/MS analysis. The analysis resulted in the identification of 373 pigeon pea seed proteins, and large numbers of stress-related proteins identified were presumably due to its adaption to drought-prone environments. In addition, proteomic studies have also been carried out in other pulse crops for various stress-related traits in common bean (Parreira et al. 2016; Zadražnik et al. 2013), mungbean (Lin et al. 2016), lupin (Cabello-Hurtado et al. 2016), cowpea (Varela et al. 2017), pea (Castillejo 2016), etc.

Kumar et al. (2015a, b) used nontargeted metabolic profiling of resistant and susceptible chickpea cultivars at different time intervals using high-resolution liquid chromatography-mass spectrometry to understand the mechanistic basis of wilt resistance or susceptibility. The study revealed discriminating metabolites in chickpea root tissue after Foc inoculation. The metabolites dominantly include flavonoids, isoflavonoids, alkaloids, amino acids, and sugars. Interestingly, more flavonoids and isoflavonoids along with their malonyl conjugates observed in the Foc-inoculated-resistant plants. Similarly, metabolic profiling was also reported by many other authors in different pulse crops such as chickpea (Dias et al. 2015), common bean (Tawaraya et al. 2014), mungbean (Hashiguchi et al. 2017), lupin (Muller et al. 2015), pea (Hradilova et al. 2017), etc. Furthermore, the recent advances in ionomics approach provide new ways to comprehensively examine the mineral networks and elemental composition of an organism in a cost-effective and rapid manner. These technologies will efficiently guide the research community to increase the bioavailable minerals in the edible parts of grain legumes. It is believed that the use of these new-generation tools will provide crop-based solutions to hidden hunger problem existing worldwide and help to achieve global nutritional security.

2.8 Integration of "Omics" Approaches: Relevance in Crop Breeding

The most crucial science arise in the present *era* is "omics" science, which is defined as a suite of approaches like genomics, transcriptomics, proteomics, metabolomics, ionomics, and phenomics that plays vital role in crop improvement by facilitating the identification of genes, proteins, and metabolites associated with crop traits and also by characterizing their functions (Singh et al. 2015; Fig. 2.3). The recent advances in "omics" approaches produce new data sets for crop plant in a cost-effective manner, and these improvements will assure greater integration of "omics" information with crop breeding heading to evolving from GAB to omics-assisted breeding (OAB) in coming years. Forward genomic/genetic studies will provide information regarding the genetic determinants governing the trait expression, whereas functional genomics relying on expression profiles and assigns the function

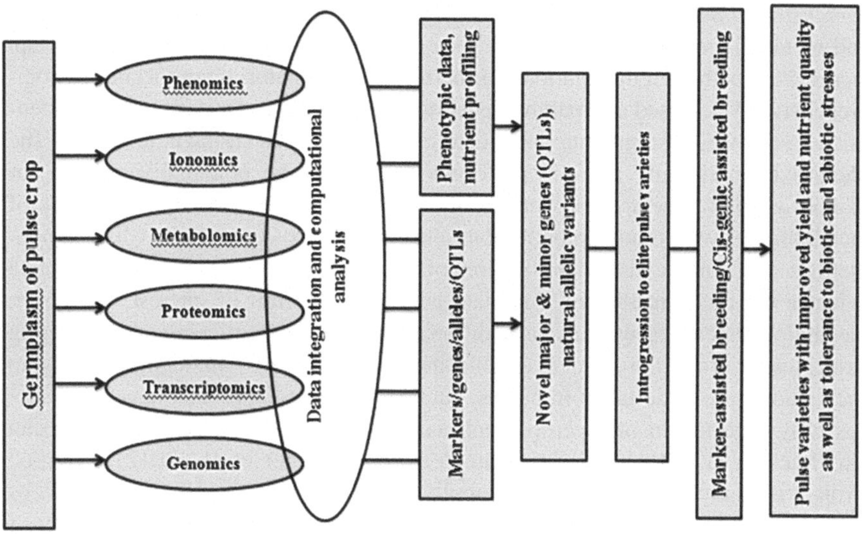

Fig. 2.3 Flow chart demonstrating the role of integrated "omics"-based approaches in unraveling genetic architecture of crop traits as well as their efficiency in the development of improved crop varieties

to possible candidate genes/QTLs. Toward this end, proteomics and metabolomics pinpoint the ultimate amino acid/protein product of the given gene(s) and the various metabolites produced under different key metabolic pathways in response to specific stress (Arbona et al. 2013; Rodziewicz et al. 2014). Ultimately, the outcome of these complex networks enables in characterizing the phenotype of the plant under stress. In parallel, efforts should be made to link all these pieces of high-throughput information obtained from various studies to better comprehend the complex gene network underlying the stress tolerance (Zivy et al. 2015). Hence, combined data analysis approach is important to fully understand and investigate the biochemical, physiological, and molecular interactions. For example, identification of genomic regions associated with phenotypic traits either through linkage mapping or GWAS is a handy tool, but a large number of genes are underlying this region, making it difficult to identify promising candidate genes. Integration of QTL mapping with transcriptomic profiling helps to prioritize the candidate to perform map-based cloning or further gene effect validation experiments. Several studies have been demonstrated the effectiveness of combined omics approaches (Deshmukh et al. 2014). Because of plenty availability of QTL as well as transcriptomic information, both the approaches are equally beneficial to each other (Vuong et al. 2015).

In pulse crops, a combination of these approaches has led to several successful discoveries with their direct effect on crop improvement, for instance, Kumar et al. (2016) combined gene expression, proteomic data, and metabolite data to elucidate the genetic mechanism of *Fusarium oxysporum tolerance in chickpea and demonstrated that* expression analysis of candidate genes supported the proteomic and metabolic variations in the chickpea roots upon *Fusarium oxysporum* f. sp. *ciceri* (Foc) pathogen inoculation. Furthermore, combinatorial approaches of comprehensive QTL-based comparative genome mapping and transcript profiling identified a seed weight-regulating candidate gene in chickpea (Bajaj et al. 2015). The identified functionally relevant molecular tags could be potentially utilized for marker-assisted genetic improvement of chickpea. Furthermore, a combination of metabolomics with transcriptomics datasets obtained from genetically diverse populations will enable the identification of novel metabolic QTLs (mQTLs) and enhance the identification of candidate genes for the trait of interest. Moreover, metabolomics can be used as an additional tool with genomics-assisted selection strategies for crop improvement (Fernie and Schauer 2009). Although, currently no such studies are undertaken in pulses, but this integrated approach have been successfully reported in other crops such as maize (Wen et al. 2017) and soybean (Kovinich et al. 2011) etc. For example, Riedelsheimer et al. (2012) identified strong associations of 26 distinct metabolites with SNPs that explain up to 32.0% of the observed genetic variation in a GWAS experiment, in which they used a set of 289 diverse maize inbred lines, assayed for 118 biochemical compounds in the young leaves and agronomic traits of mature plants in field trials as well as genotyped with 56,110 SNPs. They suggested that metabolites represent promising connecting links for narrowing the genotype-phenotype gap of complex agronomic traits. Hence, the above studies illustrated that an integrated "omics" approach needs to be applied for a better understanding of trait variation in pulse crops.

References

Abdallah AA, Ali AM, Geiger HH, Parzies HK (2009) Marker-assisted recurrent selection for increased out crossing in Caudatum–race Sorghum. International Conference on Applied Biotechnology (ICAB), 28–30th September; Sudan

Akibode S, Maredia M (2011) Global and regional trends in production, trade and consumption of food legume crops. Report Submitted to CGIAR Special Panel on Impact Assessment, 27 March 2011, 83

Ali M, Welna GC, Sallam A, Martsch R, Balko C, Gebser B, Sass O, Link W (2016) Association analyses to genetically improve drought and freezing tolerance of Faba Bean (L.). Crop Sci 56:1036–1048

Andeden EE, Derya M, Baloch FS, Kilian B, Ozkan H (2013) Development of SSR markers in lentil. In: Proceedings of plant and animal genome conference 21:P0351

André LD, Sonah H, Dias WP, Marcelino-Guimarães FC, Belzile F, Abdelnoor RV (2017) Genome-wide association study for resistance to the southern root-knot nematode (Meloidogyne incognita) in soybean. Mol Breed 37(12):148

Arbaoui M, Balko C, Link W (2008a) Study of faba bean (*Vicia faba* L.) winter-hardiness and development of screening methods. Field Crop Res 106:60–67

Arbaoui M, Link W, Satovic Z, Torres AM (2008b) Quantitative trait loci of frost tolerance and physiologically related trait in faba bean (*Vicia faba* L.). Euphytica 164:93–104

Arbona V, Manzi M, Ollas CD, Gómez-Cadenas A (2013) Metabolomics as a tool to investigate abiotic stress tolerance in plants. Inter J Mol Sci 14:4885–4911

Ariani A, Teran JCBM, Gepts P (2016) Genome-wide identification of SNPs and copy number variation in common bean (*Phaseolus vulgaris* L.) using genotyping-by-sequencing (GBS). Mol Breed 36:1–11

Atienza SG, Palomino C, Gutiérrez N, Alfaro CM, Rubiales D, Torres AM, Ávila CM (2016) QTLs for Ascochyta blight resistance in faba bean (*Vicia faba* L.): validation in field and controlled conditions. Crop Pasture Sci 67:216–224

Avila CM, Atienza SG, Moreno MT, Torres AM (2007) Development of a new diagnostic marker for growth habit selection in faba bean (*Vicia faba* L.) breeding. Theor Appl Genet 115:1075

Bajaj D, Upadhyaya HD, Khan Y, Das S, Badoni S, Shree T, Kumar V, Tripathi S, Gowda CLL, Singh S, Sharma S (2015) A combinatorial approach of comprehensive QTL-based comparative genome mapping and transcript profiling identified a seed weight-regulating candidate gene in chickpea. Sci Rep 5:9264

Barkley NA, Wang ML, Gillaspie AG, Dean RE, Pederson GA, Jenkins TM (2008) Discovering and verifying DNA polymorphisms in a mung bean [*V. radiata* (L.) R. Wilczek] collection by EcoTILLING and sequencing. BMC Res Not 1:28

Bassi D, Briñez B, Rosa JS, Oblessuc PR, Almeida CPD, Nucci SM, Silva LCDD, Chiorato AF, Vianello RP, Camargo LEA, Blair MW (2017) Linkage and mapping of quantitative trait loci associated with angular leaf spot and powdery mildew resistance in common beans. Genet Mol Biol 40:109–122

Batieno BJ et al (2016) Application of marker-assisted backcrossing to improve cowpea (*Vigna unguiculata* L. Walp) for drought tolerance. J Plant Breed Crop Sci 8:273–286

Beebe SE, Rojas-Pierce M, Yan X, Blair MW, Pedraza F, Munoz F, Tohme J, Lynch JP (2006) Quantitative trait loci for root architecture traits correlated with phosphorus acquisition in common bean. Crop Sci 46:413–423

Bernardo R, Charcosset A (2006) Usefulness of gene information in marker-assisted recurrent selection: a simulation appraisal. Crop Sci 46:614–621

Bhat JA, Ali S, Salgotra RK, Mir ZA, Dutta S, Jadon V, Tyagi A, Mushtaq M, Jain N, Singh PK, Singh GP (2016) Genomic selection in the era of next generation sequencing for complex traits in plant breeding. Front Genet 7:221

Blair MW, Izquierdo P (2012) Use of the advanced backcross-QTL method to transfer seed mineral accumulation nutrition traits from wild to Andean cultivated common beans. Theor Appl Genet 125:1015–1031

Blair MW, Giraldo MC, Buendia HF, Tovar E, Duque MC, Beebe SE (2006) Microsatellite marker diversity in common bean (*Phaseolus vulgaris L.*). Theor Appl Genet 113:100–109

Blair MW, Torres MM, Giraldo MC, Pedraza F (2009a) Development and diversity of Andean-derived, gene-based microsatellites for common bean (*Phaseolus vulgaris* L.). BMC Plant Biol 9:100

Blair MW, Astudillo C, Grusak MA, Graham R, Beebe SE (2009b) Inheritance of seed iron and zinc concentrations in common bean (*Phaseolus vulgaris L.*). Mol Breed 23:197–207

Blair MW, Hurtado N, Chavarro CM, Muñoz-Torres MC, Giraldo MC, Pedraza F, Tomkins J, Wing R (2011) Gene-based SSR markers for common bean (*Phaseolus vulgaris* L.) derived from root and leaf tissue ESTs: an integration of the BMc series. BMC Plant Biol 11:50

Blair MW, Hurtado N, Sharma P (2012) New gene-derived simple sequence repeat markers for common bean (*Phaseolus vulgaris* L.). Mol Ecol Resour 12:661–668

Blair MW, Cortés AJ, Penmetsa RV, Farmer A, Carrasquilla-Garcia N, Cook DR (2013) A high-throughput SNP marker system for parental polymorphism screening, and diversity analysis in common bean (*Phaseolus vulgaris L.*). Theor Appl Genet 126:535–548

Bohra A (2013) Emerging paradigms in genomics-based crop improvement. Sci World J585467:1–17

Bohra A, Dubey A, Saxena RK, Penmetsa RV, Poornima KN, Kumar N, Farmer AD, Srivani G, Upadhyaya HD, Gothalwal R, Ramesh S (2011) Analysis of BAC-end sequences (BESs) and development of BES-SSR markers for genetic mapping and hybrid purity assessment in pigeonpea (*Cajanus* spp.). BMC Plant Biol 11:56

Bohra A, Saxena RK, Gnanesh BN, Saxena K, Byregowda M, Rathore A, KaviKishor PB, Cook DR, Varshney RK (2012) An intra-specific consensus genetic map of pigeonpea [*Cajanus cajan* (L.) Millspaugh] derived from six mapping populations. Theor Appl Genet 125:1325–1338

Bohra A, Pandey MK, Jha UC, Singh B, Singh IP, Datta D, Chaturvedi SK, Nadarajan N, Varshney RK (2014) Genomics-assisted breeding in four major pulse crops of developing countries: present status and prospects. Theor Appl Genet 127:1263–1291

Bonhomme M, André O, Badis Y, Ronfort J, Burgarella C, Chantret N, Prosperi JM, Briskine R, Mudge J, Debéllé F, Navier H (2014) High-density genome-wide association mapping implicates an F-box encoding gene in *Medicago truncatula* resistance to *Aphanomyces euteiches*. New Phytol 201:1328–1342

Bordat A, Savois V, Nicolas M, Salse J, Chauveau A, Bourgeois M, Potier J, Houtin H, Rond C, Murat F, Marget P, Aubert G, Burstin J (2011) Translational genomics in legumes allowed placing in silico 5460 unigenes on the pea functional map and identified candidate genes in *Pisum sativum* L. G3 (Bethesda) 2:93–103

Boukar O, Kong L, Singh BB, Murdock L, Ohm HW (2004) AFLP and AFLP-derived SCAR markers associated with resistance in cowpea. Crop Sci 44:1259–1264

Braich S, Sudheesh S, Forster JW, Kaur S (2017) Characterisation of Faba bean (*Vicia faba* L.) transcriptome using RNA-Seq: sequencing, de novo assembly, annotation, and expression analysis. Agronomy 7:53

Brinez B, Blair MW, Kilian A, Carbonell SAM, Chiorato AF, Rubiano LB (2012) A whole genome DArT assay to assess germplasm collection diversity in common beans. Mol Breed 30:181–193

Brosché M, Vinocur B, Alatalo ER, Lamminmäki A, Teichmann T, Ottow EA, Djilianov D, Afif D, Bogeat-Triboulot MB, Altman A, Polle A (2005) Gene expression and metabolite profiling of *Populus euphratica* growing in the Negev desert. Genome Biol 6:R101

Broughton WJ, Hernandez G, Blair M, Beebe S, Gepts P, Vanderleyden J (2003) Beans (Phaseolus spp.)–model food legumes. Plant Soil 252:55–128

Buso GSC, Amaral ZPS, Brondani RPV, Ferreira ME (2006) Microsatellite markers for the common bean *Phaseolus vulgaris*. Mol Ecol Resour 6:252–254

Cabello-Hurtado F, Keller J, Ley J, Sanchez-Lucas R, Jorrín-Novo JV, Aïnouche A (2016) Proteomics for exploiting diversity of lupin seed storage proteins and their use as nutraceuticals for health and welfare. J Proteome 143:57–68

Castillejo MA (2016) Label-free quantitative proteomic analysis of tolerance to drought in Pisum sativum. Proteomics 17:2776–2787

Catt SC, Braich S, Kaur S, Paull JG (2017) QTL detection for flowering time in faba bean and the responses to ambient temperature and photoperiod. Euphytica 213:125

Chankaew S, Somta P, Sorajjapinun W, Srinives P (2011) Quantitative trait loci mapping of Cercospora leaf spot resistance in mungbean, *Vigna radiata* (L.) Wilczek. Mol Breed 28:255–264

Charmet G, Robert N, Perretant MR, Gay G, Sourdille P, Groos C, Bernard S, Bernard M (2001) Marker assisted recurrent selection for cumulating QTLs for bread-making related traits. In: Wheat in a global environment. Springer Netherlands, Dordrecht, pp 211–217

Chatterjee M, Gupta S, Bhar A, Chakraborti D, Basu D, Das S (2014) Analysis of root proteome unravels differential molecular responses during compatible and incompatible interaction between chickpea (*Cicer arietinum L.*) and *Fusarium oxysporum* f. sp. ciceri Race1 (Foc1). BMC Genomics 15:949

Chaudhary J, Patil GB, Sonah H, Deshmukh RK, Vuong TD, Valliyodan B, Nguyen HT (2015) Expanding omics resources for improvement of soybean seed composition traits. Front Plant Sci 6:1021

Chen H, Wang L, Wang S, Liu C, Blair MW, Cheng X (2015a) Transcriptome sequencing of mung bean (*Vigna radiate* L.) genes and the identification of EST-SSR markers. PLoS One 10:e0120273

Chen H, Qiao L, Wang L, Wang S, Blair MW, Cheng X (2015b) Assessment of genetic diversity and population structure of mung bean (*Vigna radiata*) germplasm using EST-based and genomic SSR markers. Gene 566:175–183

Chen H, Chen X, Tian J, Yang Y, Liu Z, Hao X, Wang L, Wang S, Liang J, Zhang L, Yin F (2016a) Development of gene-based SSR markers in rice bean (*Vigna umbellata L.*) based on transcriptome data. PLoS One 11:e0151040

Chen J, Somta P, Chen X, Cui X, Yuan X, Srinives P (2016b) Gene mapping of a mutant mungbean (Vigna radiata L.) using new molecular markers suggests a gene encoding a YUC4-like protein regulates the Chasmogamous flower trait. Front Plant Sci 7:830

Chen H, Wang L, Liu X, Hu L, Wang S, Cheng X (2017) De novo transcriptomic analysis of cowpea (*Vigna unguiculata L. Walp.*) for genic SSR marker development. BMC Genet 18:65

Cheng P, Holdsworth W, Ma Y, Coyne CJ, Mazourek M, Grusak MA, Fuchs S, McGee RJ (2015) Association mapping of agronomic and quality traits in USDA pea single-plant collection. Mol Breed 35:75

Cho S, Kumar J, Shultz JL, Anupama K, Tefera F, Muehlbauer FJ (2002) Mapping genes for double podding and other morphological traits in chickpea. Euphytica 128:285–292

Cho S, Chen W, Muehlbauer FJ (2004) Pathotype-specific genetic factors in chickpea (*Cicer arietinum* L.) for quantitative resistance to ascochyta blight. Theor Appl Genet 109:733–739

Chowdhury MA, Andrahennadi CP, Slinkard AE, Vandenberg A (2001) RAPD and SCAR markers for resistance to Ascochyta blight in lentil. Euphytica 118:331–337

Cichy KA, Wiesinger JA, Mendoza FA (2015) Genetic diversity and genome-wide association analysis of cooking time in dry bean (*Phaseolus vulgaris* L.). Theor Appl Genet 128:1555–1567

Cobb JN, DeClerck G, Greenberg A, Clark R, McCouch S (2013) Next-generation phenotyping: requirements and strategies for enhancing our understanding of genotype–phenotype relationships and its relevance to crop improvement. Theor Appl Genet 126:867–887

Cobos MJ, Fernández MJ, Rubio J, Kharrat M, Moreno MT, Gil J, Millán T (2005) A linkage map of chickpea (*Cicer arietinum* L.) based on populations from Kabuli× desi crosses: location of genes for resistance to fusarium wilt race 0. Theor Appl Genet 110:1347–1353

Coles GD, Wratten SD, Porter JR (2016) Food and nutritional security requires adequate protein as well as energy, delivered from whole-year crop production. Peer J 4:2100

Córdoba JM, Chavarro C, Schlueter JA, Jackson SA, Blair MW (2010) Integration of physical and genetic maps of common bean through BAC-derived microsatellite markers. BMC Genomics 11:436

Cottage A, Gostkiewicz K, Thomas JE, Borrows R, Torres AM, O'Sullivan DM (2012) Heterozygosity and diversity analysis using mapped single nucelotide polymorphisms in a faba bean inbreeding programme. Mol Breed 30:1799–1809

Cowley R, Luckett DJ, Ash GJ, Harper JD, Vipin CA, Raman H, Ellwood S (2014) Identification of QTLs associated with resistance to Phomopsis pod blight (*Diaporthe toxica*) in Lupinus albus. Breed Sci 64:83–89

Daba K, Deokar A, Banniza S, Warkentin TD, Tar'an B (2016) QTL mapping of early flowering and resistance to Ascochyta blight in chickpea. Genome 59:413–425

Dalmais M, Schmidt J, Le Signor C, Moussy F, Burstin J, Savois V, Aubert G, Brunaud V, de Oliveira Y, Guichard C, Thompson R (2008) UTILLdb, a *Pisum sativum* in silico forward and reverse genetics tool. Genome Biol 9:43

Das G, Patra JK, Baek KH (2017) Insight into MAS: a molecular tool for development of stress resistant and quality of rice through gene stacking. Front Plant Sci 8:985

de Faria Müller BS, Sakamoto T, de Menezes IPP, Prado GS, Martins WS, Brondani C, de Barros EG, Vianello RP (2014) Analysis of BAC-end sequences in common bean (*Phaseolus vulgaris* L.) towards the development and characterization of long motifs SSRs. Plant Mol Biol 86:455–470

Desgroux AL, Anthoëne V, Roux-Duparque M, Rivière JP, Aubert G, Tayeh N, Moussart A, Mangin P, Vetel P, Piriou C, McGee RJ (2016) Genome-wide association mapping of partial resistance to *Aphanomyces euteiches* in pea. BMC Genomics 17:124

Deshmukh R, Sonah H, Patil G, Chen W, Prince S, Mutava R, Vuong T, Valliyodan B, Nguyen HT (2014) Integrating omic approaches for abiotic stress tolerance in soybean. Front Plant Sci 5:244

Diapari M, Sindhu A, Bett K, Deokar A, Warkentin TD, Tar'an B (2014) Genetic diversity and association mapping of iron and zinc concentrations in chickpea (*Cicer arietinum* L.). Genome 57:459–468
Dias DA, Hill CB, Jayasinghe NS, Atieno J, Sutton T, Roessner U (2015) Quantitative profiling of polar primary metabolites of two chickpea cultivars with contrasting responses to salinity. J Chromato B 1000:1–13
Diaz LM, Ricaurte J, Cajiao C, Galeano CH, Rao I, Beebe S, Raatz B (2017) Phenotypic evaluation and QTL analysis of yield and symbiotic nitrogen fixation in a common bean population grown with two levels of phosphorus supply. Mol Breed 37:76
Díaz-Ruiz R, Torres A, Gutierrez MV, Rubiales D, Cubero JI, Kharrat M, Satovic Z, Román B (2009) Mapping of quantitative trait loci controlling *Orobanche foetida Poir.*resistance in faba bean (*Vicia faba* L.). Afr J Biotechnol 8:2718–2724
Dinesh HB, Lohithaswa HC, Viswanatha KP, Singh P, Rao AM (2016) Identification and marker-assisted introgression of QTL conferring resistance to bacterial leaf blight in cowpea (Vigna unguiculata (L.) Walp.). Plant Breed 135:506–512
Dubey A, Farmer A, Schlueter J, Cannon SB, Abernathy B, Tuteja R, Woodward J, Shah T, Mulasmanovic B, Kudapa H, Raju NL (2011) Defining the transcriptome assembly and its use for genome dynamics and transcriptome profiling studies in pigeonpea (*Cajanus cajan L.*). DNA Res 18:153–164
Durán Y, Fratini R, Garcia P, De la Vega MP (2004) An intersubspecific genetic map of Lens. Theor Appl Genet 108:1265–1273
Dutta S, Kumawat G, Singh BP, Gupta DK, Singh S, Dogra V, Gaikwad K, Sharma TR, Raje RS, Bandhopadhya TK, Datta S (2011) Development of genic-SSR markers by deep transcriptome sequencing in pigeonpea [*Cajanus cajan* (L.) Millspaugh]. BMC Plant Biol 11:17
Eathington L, Swenson DA (2007) Dude, Where's my corn? Constraints on the location of ethanol production in the Corn Belt. Department of Economics, Iowa State University. 46th Annual Meeting of the Southern Regional Science Association, Charleston, SC, March 29–31, 2007
Edwards D, Henry RJ, Edwards KJ (2012) Preface: advances in DNA sequencing accelerating plant biotechnology. Plant Biotechnol J 10:621–622
Eggen A (2012) The development and application of genomic selection as a new breeding paradigm. Animal Front 2:10–15
FAO (2012) The State of Food Insecurity in the World 2012. Economic growth is necessary but not sufficient to accelerate reduction of hunger and malnutrition. Rome, FAO
FAOSTAT (2012) FAO Statistics Division. http://faostat.fao.org/beta/en/#home
Fedoruk MJ, Vandenberg A, Bett KE (2013) Quantitative trait loci analysis of seed quality characteristics in lentil using single nucleotide polymorphism markers. Plant Genome 6:1–10
Fernandez-Garcia n, Hernandez M, Casado-Vela J, Bru R, Elortza F, Hedden P, Olmos E (2011) Changes to the proteome and targeted metabolites of xylem sap in *Brassica oleracea* in response to salt stress. Plant Cell Environ 34:821–836
Fernie AR, Schauer N (2009) Metabolomics-assisted breeding: a viable option for crop improvement. Trends in Genet 25:39–48
Franssen SU, Shrestha RP, Bräutigam A, Bornberg-Bauer E, Weber AP (2011) Comprehensive transcriptome analysis of the highly complex Pisum sativum genome using next generation sequencing. BMC Genomics 12:227
Fratini R, Durán Y, García P, De La Vega MP (2007) Identification of quantitative trait loci (QTL) for plant structure, growth habit and yield in lentil. Spanish J Agric Res 5:348–356
Furbank RT, Tester M (2011) Phenomics–technologies to relieve the phenotyping bottleneck. Trends Plant Sci 16:635–644
Galeano CH, Fernandez AC, Franco-Herrera N, Cichy KA, McClean PE, Vanderleyden J, Blair MW (2011) Saturation of an intra-gene pool linkage map: towards a unified consensus linkage map for fine mapping and synteny analysis in common bean. PLoS One 6:e28135
Gao LL, Hane JK, Kamphuis LG, Foley R, Shi BJ, Atkins CA, Singh KB (2011) Development of genomic resources for the narrow-leafed lupin (Lupinus angustifolius): construction of a bacterial artificial chromosome (BAC) library and BAC-end sequencing. BMC Genomics 12:521

Garg R, Patel RK, Tyagi AK, Jain M (2011a) De novo assembly of chickpea transcriptome using short reads for gene discovery and marker identification. DNA Res 18:53–63

Garg R, Patel RK, Jhanwar S, Priya P, Bhattacharjee A, Yadav G, Bhatia S, Chattopadhyay D, Tyagi AK, Jain M (2011b) Gene discovery and tissue-specific transcriptome analysis in chickpea with massively parallel pyrosequencing and web resource development. Plant Physiol 156:1661–1678

Gaur PM, Thudi M, Samineni S, Varshney RK (2014) Advances in chickpea genomics. In: Legumes in the omic era. Springer, New York, pp 73–94

Gnanesh BN, Bohra A, Sharma M, Byregowda M, Pande S, Wesley V, Saxena RK, Saxena KB, Kishor PK, Varshney RK (2011) Genetic mapping and quantitative trait locus analysis of resistance to sterility mosaic disease in pigeonpea [*Cajanus cajan* (L.) Millsp.]. Field Crop Res 123:53–61

Godfray HCJ, Beddington JR, Crute IR, Haddad L, Lawrence D, Muir JF, Pretty J, Robinson S, Thomas SM, Toulmin C (2010) Food security: the challenge of feeding 9 billion people. Science 327:812–818

González AM, Yuste-Lisbona FJ, Rodiño AP, De Ron AM, Capel C, García-Alcázar M, Lozano R, Santalla M (2015) Uncovering the genetic architecture of *Colletotrichum lindemuthianum* resistance through QTL mapping and epistatic interaction analysis in common bean. Front Plant Sci 6:141

Grenier C, Châtel M, Ospina Y, Cao TV, Guimaraes EP, Martinez C, Tohmé J, Courtois B, Ahmadi N (2012) Population improvement through recurrent selection in rice. Prospect for maker assisted recurrent selection and genome-wide selection. W011

Gupta PK, Langridge P, Mir RR (2010) Marker-assisted wheat breeding: present status and future possibilities. Mol Breed 26:145–161

Gupta SK, Bansal R, Gopalakrishna T (2014) Development and characterization of genic SSR markers for mungbean (*Vigna radiata* (L.) Wilczek). Euphytica 195:245–258

Gutiérrez N, Palomino C, Satovic Z, Ruiz-Rodríguez MD, Vitale S, Gutiérrez MV, Rubiales D, Kharrat M, Amri M, Emeran AA, Cubero JI (2013) QTLs for Orobanche spp. resistance in faba bean: identification and validation across different environments. Mol Breed 32:909–922

Gwag JG, Chung JW, Chung HK, Lee JH, Ma KH, Dixit A, Park YJ, Cho EG, Kim TS, Lee SH (2006) Characterization of new microsatellite markers in mung bean, *Vigna radiata* (L.). Mol Ecol Resour 6:1132–1134

Hamwieh A, Udupa SM, Choumane W, Sarker A, Dreyer F, Jung C, Baum M (2005) A genetic linkage map of Lens sp. based on microsatellite and AFLP markers and the localization of fusarium vascular wilt resistance. Theor Appl Genet 110:669–677

Hamwieh A, Udupa SM, Sarker A, Jung C, Baum M (2009) Development of new microsatellite markers and their application in the analysis of genetic diversity in lentils. Breed Sci 59:77–86

Hane JK, Ming Y, Kamphuis LG, Nelson MN, Garg G, Atkins CA, Bayer PE, Bravo A, Bringans S, Cannon S, Edwards D (2017) A comprehensive draft genome sequence for lupin (*Lupinus angustifolius*), an emerging health food: insights into plant–microbe interactions and legume evolution. Plant Biotechnol J 15:318–330

Hartmann A, Czauderna T, Hoffmann R, Stein N, Schreiber F (2011) HTPheno: an image analysis pipeline for high-throughput plant phenotyping. BMC Bioinform 12:148

Hashiguchi A, Hitachi K, Zhu W, Tian J, Tsuchida K, Komatsu S (2017) Mung bean (*Vigna radiata* (L.)) coat extract modulates macrophage functions to enhance antigen presentation: a proteomic study. J Proteome 161:26–37

Hayashi T, Iwata H (2010) EM algorithm for Bayesian estimation of genomic breeding values. BMC Genet 11:3

Hiremath PJ, Farmer A, Cannon SB, Woodward J, Kudapa H, Tuteja R, Kumar A, Bhanu Prakash A, Mulaosmanovic B, Gujaria N, Krishnamurthy L (2011) Large-scale transcriptome analysis in chickpea (*Cicer arietinum L.*), an orphan legume crop of the semi-arid tropics of Asia and Africa. Plant Biotechnol J 9:922–931

Hiz MC, Canher B, Niron H, Turet M (2014) Transcriptome analysis of salt tolerant common bean (*Phaseolus vulgaris L.*) under saline conditions. PLoS One 9:e92598

Hospital F (2005) Selection in backcross programmes. Philos Trans Roy Soc Lond B Biol Sci 360:1503–1511

Hossain S, Ford R, McNeil D, Pittock C, Panozzo JF (2010) Inheritance of seed size in chickpea (*'Cicer arietinum'*L.) and identification of QTL based on 100-seed weight and seed size index. Aus J Crop Sci 4:126

Hossain Z, Nouri MZ, Komatsu S (2011) Plant cell organelle proteomics in response to abiotic stress. J Proteome Res 11:37–48

Hoyos-Villegas V, Song Q, Kelly JD (2017) Genome-wide association analysis for drought tolerance and associated traits in common bean. Plant Genome 10:1–12

Hradilová I, Trněný O, Válková M, Cechova M, Janská A, Prokešová L, Aamir K, Krezdorn N, Rotter B, Winter P, Varshney RK (2017) A combined comparative transcriptomic, metabolomic, and anatomical analyses of two key domestication traits: pod dehiscence and seed dormancy in pea (*Pisum* sp.). Front Plant Sci 8:542

Humphry ME, Magner T, McIntyre CL, Aitken EA, Liu CJ (2003) Identification of a major locus conferring resistance to powdery mildew (*Erysiphe polygoni* DC) in mungbean (*Vigna radiata* L. Wilczek) by QTL analysis. Genome 46:738–744

Humphry ME, Lambrides CJ, Chapman A, Imrie BC, Lawn RJ, Mcintyre CL, Lili CJ (2005) Relationships between hard-seededness and seed weight in mungbean (*Vigna radiata*) assessed by QTL analysis. Plant Breed 124:292–298

Hyten DL, Song Q, Fickus EW, Quigley CV, Lim JS, Choi IY, Hwang EY, Pastor-Corrales M, Cregan PB (2010) High-throughput SNP discovery and assay development in common bean. BMC Genomics 11:475

Iglesias-García R, Prats E, Fondevilla S, Satovic Z, Rubiales D (2015) Quantitative trait loci associated to drought adaptation in pea (*Pisum sativum L.*). Plant Mol Biol Rep 33:1768–1778

Isemura T, Kaga A, Tabata S, Somta P, Srinives P, Shimizu T, Jo U, Vaughan DA, Tomooka N (2012) Construction of a genetic linkage map and genetic analysis of domestication related traits in mungbean (*Vigna radiata*). PLoS One 7:e41304

Jackson SA, Iwata A, Lee SH, Schmutz J, Shoemaker R (2011) Sequencing crop genomes: approaches and applications. New Phytol 191:915–925

Jadhav AA, Rayate SJ, Mhase LB, Thudi M, Chitikineni A, Harer PN, Jadhav AS, Varshney RK, Kulwal PL (2015) Marker-trait association study for protein content in chickpea (*Cicer arietinum*). J Genet 94:279–286

Jaganathan D, Thudi M, Kale S, Azam S, Roorkiwal M, Gaur PM, Kishor PK, Nguyen H, Sutton T, Varshney RK (2015) Genotyping-by-sequencing based intra-specific genetic map refines a "*QTL-hotspot*" region for drought tolerance in chickpea. Mol Gen Genomics 290:559–571

Jain M, Pole AK, Singh VK, Ravikumar RL, Garg R (2015) Discovery of molecular markers for fusarium wilt via transcriptome sequencing of chickpea cultivars. Mol Breed 35(10):198

Jannink JL, Lorenz AJ, Iwata H (2010) Genomic selection in plant breeding: from theory to practice. Brief Funct Genomics 9:166–177

Javid M, Noy D, Sudheesh S, Forster JW, Kaur S (2017) Identification of QTLs associated with metribuzin tolerance in field pea (*Pisum sativum L.*). Euphytica 213:91

Jha R, Bohra A, Jha UC, Rana M, Chahota RK, Kumar S, Sharma TR (2017) Analysis of an intraspecific RIL population uncovers genomic segments harbouring multiple QTL for seed relevant traits in lentil (*Lens culinaris L.*). Physiol Mol Biol Plants 23:675–684

Jhanwar S, Priya P, Garg R, Parida SK, Tyagi AK, Jain M (2012) Transcriptome sequencing of wild chickpea as a rich resource for marker development. Plant Biotechnol J 10:690–702

Kahraman A, Tanyolac B, Ozatay S, Muehlbauer F (2010) Linkage mapping of lentil (*Lens culinaris L.*) genome using recombinant inbred lines revealed by AFLP, ISSR, RAPD and some morphologic markers. J Agric Biotechnol Sus Dev 2:1

Kalavacharla V, Liu Z, Meyers BC, Thimmapuram J, Melmaiee K (2011) Identification and analysis of common bean (*Phaseolus vulgaris* L.) transcriptomes by massively parallel pyrosequencing. BMC Plant Biol 11:135

Kalia RK, Rai MK, Kalia S, Singh R, Dhawan AK (2011) Microsatellite markers: an overview of the recent progress in plants. Euphytica 177:309–334

Kamfwa K, Cichy KA, Kelly JD (2015) Genome-wide association study of agronomic traits in common bean. Plant Genome 8:1–10

Kamphuis LG, Hane JK, Nelson MN, Gao L, Atkins CA, Singh KB (2015) Transcriptome sequencing of different narrow-leafed lupin tissue types provides a comprehensive uni-gene assembly and extensive gene-based molecular markers. Plant Biotechnol J 13:14–25

Kang YJ, Kim SK, Kim MY, Lestari P, Kim KH, Ha BK, Jun TH, Hwang WJ, Lee T, Lee J, Shim S (2014) Genome sequence of mungbean and insights into evolution within *Vigna* species. Nat Commun 5:5443

Kasettranan W, Somta P, Srinives P (2010) Mapping of quantitative trait loci controlling powdery mildew resistance in mungbean (*Vigna radiata* (L.) Wilczek). J Crop Sci Biotechnol 13:155–161

Kaur S, Cogan NO, Pembleton LW, Shinozuka M, Savin KW, Materne M, Forster JW (2011) Transcriptome sequencing of lentil based on second-generation technology permits large-scale unigene assembly and SSR marker discovery. BMC Genomics 12:265

Kaur S, Pembleton LW, Cogan NO, Savin KW, Leonforte T, Paull J, Materne M, Forster JW (2012) Transcriptome sequencing of field pea and faba bean for discovery and validation of SSR genetic markers. BMC Genomics 13:104

Kaur S, Cogan NO, Stephens A, Noy D, Butsch M, Forster JW, Materne M (2014a) EST-SNP discovery and dense genetic mapping in lentil (*Lens culinaris Medik*) enable candidate gene selection for boron tolerance. Theor Appl Genet 127:703–713

Kaur S, Kimber RB, Cogan NO, Materne M, Forster JW, Paull JG (2014b) SNP discovery and high-density genetic mapping in faba bean (*Vicia faba* L.) permits identification of QTLs for Ascochyta blight resistance. Plant Sci 217:47–55

Keller B, Manzanares C, Jara C, Lobaton JD, Studer B, Raatz B (2015) Fine-mapping of a major QTL controlling angular leaf spot resistance in common bean (*Phaseolus vulgaris* L.). Theor Appl Genet 128:813–826

Klukas C, Pape JM, Entzian A (2012) Analysis of high-throughput plant image data with the information system IAP. J Integr Bioinform 9:191

Kovinich N, Saleem A, Arnason JT, Miki B (2011) Combined analysis of transcriptome and metabolite data reveals extensive differences between black and brown nearly-isogenic soybean (Glycine max) seed coats enabling the identification of pigment isogenes. BMC Genomics 12: 381

Krishnan HB, Natarajan SS, Oehrle NW, Garrett WM, Darwish O (2017) Proteomic analysis of Pigeonpea (*Cajanus cajan*) seeds reveals the accumulation of numerous stress-related proteins. J Agric Food Chem 65:4572–4581

Kudapa H, Bharti AK, Cannon SB, Farmer AD, Mulaosmanovic B, Kramer R, Bohra A, Weeks NT, Crow JA, Tuteja R, Shah T, Dutta S, Gupta DK, Singh A, Gaikwad K, Sharma TR, May GD, Singh NK, Varshney RK (2012) A comprehensive transcriptome assembly of pigeonpea (*Cajanus cajan L.*) using sanger and second-generation sequencing platforms. Mol Plant 5:1020–1028

Kudapa H, Azam S, Sharpe AG, Taran B, Li R, Deonovic B, Cameron C, Farmer AD, Cannon SB, Varshney RK (2014) Comprehensive transcriptome assembly of chickpea (*Cicer arietinum* L.) using sanger and next generation sequencing platforms: development and applications. PLoS One 9:e86039

Kujur A, Upadhyaya HD, Bajaj D, Gowda CLL, Sharma S, Tyagi AK, Parida SK (2016) Identification of candidate genes and natural allelic variants for QTLs governing plant height in chickpea. Sci Repo 6:27968

Kulwal PL, Thudi M, Mhase LB, Varshney RK (2016) Association analysis identifies important QTL for seed weight in chickpea. IORE. J Genet 1:1–11

Kumar S, Rajendran K, Kumar J, Hamwieh A, Baum M (2015a) Current knowledge in lentil genomics and its application for crop improvement. Front Plant Sci 6:78

Kumar Y, Dholakia BB, Panigrahi P, Kadoo NY, Giri AP, Gupta VS (2015b) Metabolic profiling of chickpea-fusarium interaction identifies differential modulation of disease resistance pathways. Phytochemistry 116:120–129

Kumar Y, Zhang L, Panigrahi P, Dholakia BB, Dewangan V, Chavan SG, Kunjir SM, Wu X, Li N, Rajmohanan PR, Kadoo NY (2016) *Fusarium oxysporum* mediates systems metabolic reprogramming of chickpea roots as revealed by a combination of proteomics and metabolomics. Plant Biotechnol J 14:1589–1603

Kumawat G, Raje RS, Bhutani S, Pal JK, Mithra AS, Gaikwad K, Sharma TR, Singh NK (2012) Molecular mapping of QTLs for plant type and earliness traits in pigeonpea (*Cajanus cajan* L. Millsp.). BMC Genet 13:84

Leonforte A, Sudheesh S, Cogan NO, Salisbury PA, Nicolas ME, Materne M, Forster JW, Kaur S (2013) SNP marker discovery, linkage map construction and identification of QTLs for enhanced salinity tolerance in field pea (*Pisum sativum L.*). BMC Plant Biol 13:161

Li Y, Ruperao P, Batley J, Edwards D, Davidson J, Hobson K, Sutton T (2017) Genome analysis identified novel candidate genes for Ascochyta blight resistance in chickpea using whole genome re-sequencing data. Front Plant Sci 8:359

Lin WJ, Ko CY, Liu MS, Kuo CY, Wu DC, Chen CY, Schafleitner R, Chen LFO, Lo HF (2016) Transcriptomic and proteomic research to explore Bruchid-resistant genes in Mungbean isogenic lines. J Agric Food Chem 64:6648–6658

Liu X, Yang C, Zhang L, Li L, Liu S, Yu J, You L, Zhou D, Xia C, Zhao J, Wu H (2011) Metabolic profiling of cadmium-induced effects in one pioneer intertidal halophyte Suaeda salsa by NMR-based metabolomics. Ecotoxicology 20:1422–1431

Liu MS, Kuo TCY, Ko CY, Wu DC, Li KY, Lin WJ, Lin CP, Wang YW, Schafleitner R, Lo HF, Chen CY (2016) Genomic and transcriptomic comparison of nucleotide variations for insights into bruchid resistance of mungbean (*Vigna radiata* [L.] R. Wilczek). BMC Plant Biol 16:46

Liu R, Fang L, Yang T, Zhang X, Hu J, Zhang H, Han W, Hua Z, Hao J, Zong X (2017a) Marker-trait association analysis of frost tolerance of 672 worldwide pea (*Pisum sativum* L.) collections. Sci Rep 7:5919

Liu C, Wu J, Wang L, Fan B, Cao Z, Su Q, Zhang Z, Wang Y, Tian J, Wang S (2017b) Quantitative trait locus mapping under irrigated and drought treatments based on a novel genetic linkage map in mungbean (*Vigna radiata* L.). Theor ApplGenet 130:2375–2393

Ma Y, Bao SY, Yang T, Hu JG, Guan JP, He YH, Wang XJ, Wan YL, Sun XL, Jiang JY, Gong CX (2013) Genetic linkage map of Chinese native variety faba bean (*Vicia faba* L.) based on simple sequence repeat markers. Plant Breed 132:397–400

Ma Y, Coyne CJ, Grusak MA, Mazourek M, Cheng P, Main D, McGee RJ (2017) Genome-wide SNP identification, linkage map construction and QTL mapping for seed mineral concentrations and contents in pea (*Pisum sativum* L.). BMC Plant Biol 17:43

Macas J, Neumann P, Navrátilová A (2007) Repetitive DNA in the pea (*Pisum sativum* L.) genome: comprehensive characterization using 454 sequencing and comparison to soybean and *Medicago truncatula*. BMC Genomics 8:427

Mackay I, Powell W (2007) Methods for linkage disequilibrium mapping in crops. Trends Plant Sci 12:57–63

Mallikarjuna BP, Samineni S, Thudi M, Sajja SB, Khan AW, Patil A, Viswanatha KP, Varshney RK, Gaur PM (2017) Molecular mapping of flowering time major genes and QTLs in chickpea (*Cicer arietinum* L.). Front Plant Sci 8:1140

Masari A, Kaewwongwal A, Somta P, Srinives P (2017) Inheritance and a major quantitative trait locus of seed starch content in mungbean (*Vigna radiata* (L.) Wilczek). Euphytica 213:166

Mazzucotelli E, Mastrangelo AM, Crosatti C, Guerra D, Stanca AM, Cattivelli L (2008) Abiotic stress response in plants: when post-transcriptional and post-translational regulations control transcription. Plant Sci 174:420–431

Mc Phee KE, Inglis DA, Gundersen B, Coyne CJ (2012) Mapping QTL for fusarium wilt race 2 partial resistance in pea (*Pisum sativum*). Plant Breed 131:300–306

McCallum CM, Comai L, Greene EA, Henikoff S (2000) Targeting induced locallesions in genomes (TILLING) for plant functional genomics. Plant Physiol 123:439–442

McPhee KE (2007) Pea. In Pulses, Sugar and Tuber Crops. Springer, Berlin, Heidelberg, pp. 33-47
Meuwissen THE, Hayes BJ, Goddard ME (2001) Prediction of total genetic value using genome-wide dense marker maps. Genetics 157:1819–1829
Mir RR, Saxena RK, Saxena KB, Upadhyaya HD, Kilian A, Cook DR, Varshney RK (2013) Whole-genome scanning for mapping determinacy in Pigeonpea (*Cajanus* spp.). Plant Breed 132:472–478
Mishra RR, Sahu AR, Rath SC, Panigrahi J (2016) Genetic linkage mapping of loci conferring resistance to blue butterfly (*Lampides boeticus* L.) and plume moth (*Exelastis atomosa* Wals.) on chromosome 2 (CcLG02) in Pigeonpea. Bot Lett 163:217–230
Mitchell-Olds T (2010) Complex-trait analysis in plants. Genome Biol 11:113
Miyagi M, Humphry M, Ma ZY, Lambrides CJ, Bateson M, Liu CJ (2004) Construction of bacterial artificial chromosome libraries and their application in developing PCR-based markers closely linked to a major locus conditioning bruchid resistance in mungbean (*Vigna radiata* L. Wilczek). Theor Appl Genet 110:151–156
Morris GP, Ramu P, Deshpande SP, Hash CT, Shah T, Upadhyaya HD, Riera-Lizarazu O, Brown PJ, Acharya CB, Mitchell SE, Harriman J, Glaubitz JC, Buckler ES, Kresovich S (2013) Population genomic and genome-wide association studies of agroclimatic traits in sorghum. Proc Natl Acad Sci U S A 110:453–458
Müller J, Gödde V, Niehaus K, Zörb C (2015) Metabolic adaptations of white lupin roots and shoots under phosphorus deficiency. Front Plant Sci 6:1014
Murray LJ, Dincă M, Long JR (2009) Hydrogen storage in metal–organic frameworks. Chem Soc Rev 38:1294–1314
Navarro FM, Sass ME, Nienhuis J (2009) Marker-facilitated selection for a major QTL associated with root rot resistance in snap bean (*Phaseolus vulgaris* L). Crop Sci 49:850–856
Pandey MK, Roorkiwal M, Singh VK, Ramalingam A, Kudapa H, Thudi M, Chitikineni A, Rathore A, Varshney RK (2016) Emerging genomic tools for legume breeding: current status and future prospects. Front Plant Sci 7:455
Parankusam S, Bhatnagar-Mathur P, Sharma KK (2017) Heat responsive proteome changes reveal molecular mechanisms underlying heat tolerance in chickpea. Environ Exp Bot 141:132–144
Park SO, Coyne DP, Steadman JR, Skroch PW (2001) Mapping of QTL for resistance to white mold disease in common bean. Crop Sci 41:1253–1262
Parra-González LB, Aravena-Abarzúa GA, Navarro-Navarro CS, Udall J, Maughan J, Peterson LM, Salvo-Garrido HE, Maureira-Butler IJ (2012) Yellow lupin (*Lupinus luteusL.*) transcriptome sequencing: molecular marker development and comparative studies. BMC Genomics 13:425
Parreira JR, Bouraada J, Fitzpatrick MA, Silvestre S, Da Silva AB, Da Silva JM, Almeida AM, Fevereiro P, Altelaar AFM, Araújo SS (2016) Differential proteomics reveals the hallmarks of seed development in common bean (*Phaseolus vulgaris L.*). J Proteome 143:188–198
Patil BL, Dangwal M, Mishra R (2017a) Variability of Emaravirus species associated with sterility mosaic disease of pigeonpea in India provides evidence of segment Reassortment. Viruses 9:183
Patil PG, Dubey J, Bohra A, Mishra RK, Saabale PR, Das A, Rathore M, Singh NP (2017b) Association mapping to discover significant marker-trait associations for resistance against fusarium wilt variant 2 in pigeonpea [*Cajanus cajan* (L.) Millspaugh] using SSR markers. J Appl Genet 58:307–319
Pereira G, Marques C, Ribeiro R, Formiga S, Dâmaso M, Sousa MT, Farinhó M, Leitão JM (2010) Identification of DNA markers linked to an induced mutated gene conferring resistance to powdery mildew in pea (*Pisum sativum* L.). Euphytica 171:327
Perez-de-Castro AM, Vilanova S, Cañizares J, Pascual LM, Blanca JJ, Diez M, Prohens J, Picó B (2012) Application of genomic tools in plant breeding. Curr Genomics 13:179–195
Pérez-Vega E, Pañeda A, Rodríguez-Suárez C, Campa A, Giraldez R, Ferreira JJ (2010) Mapping of QTLs for morpho-agronomic and seed quality traits in a RIL population of common bean (*Phaseolus vulgaris* L.). Theor Appl Genet 120:1367–1380

Perseguini JMKC, Oblessuc PR, Rosa JRBF, Gomes KA, Chiorato AF, Carbonell SAM, Garcia AAF, Vianello RP, Benchimol-Reis LL (2016) Genome-wide association studies of anthracnose and angular leaf spot resistance in common bean (*Phaseolus vulgaris* L.). PLoS One 11:e0150506

Phan HT, Ellwood SR, Adhikari K, Nelson MN, Oliver RP (2007) The first genetic and comparative map of white lupin (*Lupinus albus* L.): identification of QTLs for anthracnose resistance and flowering time, and a locus for alkaloid content. DNA Res 14:59–70

Porch TG, Blair MW, Lariguet P, Galeano C, Pankhurst CE, Broughton WJ (2009) Generation of a mutant population for TILLING common bean genotype BAT 93. J Amer Soc Hort Sci 134:348–355

Pratap A et al (2017) Marker-assisted introgression of resistance to fusarium wilt race 2 in Pusa 256, an elite cultivar of desi chickpea. Mol Genet Genomics. https://doi.org/10.1007/s00438-017-1343-z

Prioul S, Frankewitz A, Deniot G, Morin G, Baranger A (2004) Mapping of quantitative trait loci for partial resistance to Mycosphaerella pinodes in pea (*Pisum sativum L.*), at the seedling and adult plant stages. Theor Appl Genet 108:1322–1334

Raju NL, Gnanesh BN, Lekha P, Jayashree B, Pande S, Hiremath PJ, Byregowda M, Singh NK, Varshney RK (2010) The first set of EST resource for gene discovery and marker development in pigeonpea (*Cajanus cajan L.*). BMC Plant Biol 10:45

Reynolds MP, Hays D, Chapman S (2010) Breeding for adaptation to heat and drought stress. In: Reynolds MP (ed) Climate change and crop production. CABI, London, pp 447–454

Ribaut JM, Hoisington D (1998) Marker-assisted selection: new tools and strategies. Trends Plant Sci 3:236–239

Ribaut JM, Ragot M (2006) Marker-assisted selection to improve drought adaptation in maize: the backcross approach, perspectives, limitations, and alternatives. J Exp Bot 58:351–360

Riedelsheimer C, Lisec J, Czedik-Eysenberg A, Sulpice R, Flis A, Grieder C, Altmann T, Stitt M, Willmitzer L, Melchinger AE (2012) Genome-wide association mapping of leaf metabolic profiles for dissecting complex traits in maize. Proc Natl Acad Sci U S A 109:8872–8877

Rodziewicz P, Swarcewicz B, Chmielewska K, Wojakowska A, Stobiecki M (2014) Influence of abiotic stresses on plant proteome and metabolome changes. Acta Physiol Plant 36:1–19

Román B, Torres AM, Rubiales D, Cubero JI, Satovic Z (2002) Mapping of quantitative trait loci controlling broomrape (*Orobanche crenata* Forsk.) resistance in faba bean (*Vicia faba* L.). Genome 45:1057–1063

Roorkiwal M, Rathore A, Das RR, Singh MK, Jain A, Srinivasan S, Gaur PM, Chellapilla B, Tripathi S, Li Y, Hickey JM (2016) Genome-enabled prediction models for yield related traits in chickpea. Front Plant Sci 7:1666

Rubeena Taylor PWJ, Ades PK, Ford R (2006) QTL mapping of resistance in lentil (*Lens culinaris ssp. culinaris*) to Ascochyta blight (*Ascochyta lentis*). Plant Breed 125:506–512

Rubiales D, Ambrose MJ, Domoney C, Burstin J (2011) Pea (*Pisum sativum* L.). In: de la Vega MP, Torres AM, Cubero JI, Kole C (eds) Genetics, genomics and breeding of cool season grain legumes (genetics, genomics and breeding in crop plants). Science Pub, New Hampshire, pp 1–49

Sabbavarapu MM, Sharma M, Chamarthi SK, Swapna N, Rathore A, Thudi M, Gaur PM, Pande S, Singh S, Kaur L, Varshney RK (2013) Molecular mapping of QTLs for resistance to fusarium wilt (race 1) and Ascochyta blight in chickpea (*Cicer arietinum* L.). Euphytica 193:121–133

Saha GC, Sarker A, Chen W, Vandemark GJ, Muehlbauer FJ (2010) Inheritance and linkage map positions of genes conferring resistance to Stemphylium blight in lentil. Crop Sci 50:1831–1839

Sahu AR, Mishra RR, Rath SC, Panigrahi J (2015) Construction of interspecific genetic linkage map of pigeonpea using SCoT, RAPD, ISSR markers and simple inherited trait loci. Nucleus 58:23–31

Sallam A, Martsch R (2015) Association mapping for frost tolerance using multi-parent advanced generation inter-cross (MAGIC) population in faba bean (*Vicia faba*). Genetica 143:501–514

Sallam A, Arbaoui M, El-Esawi M, Abshire N, Martsch R (2016a) Identification and verification of QTL associated with frost tolerance using linkage mapping and GWAS in winter faba bean. Front Plant Sci 7:1098

Sallam A, Dhanapal AP, Liu S (2016b) Association mapping of winter hardiness and yield traits in faba bean (*Vicia faba* L.). Crop Pasture Sci 67:55–68

Salt DE, Baxter I, Lahner B (2008) Ionomics and the study of the plant ionome. Annu Rev Plant Biol 59:709–733

Santra DK, Tekeoglu M, Ratnaparkhe M, Kaiser WJ, Muehlbauer FJ (2000) Identification and map ping of QTLs conferring resistance to Ascochyta blight in chickpea. Crop Sci 40:1606–1612

Saxena RK, Obala J, Sinjushin A, Kumar CS, Saxena KB, Varshney RK (2017a) Characterization and mapping of *Dt1* locus which co-segregates with *CcTFL1* for growth habit in pigeonpea. Theor Appl Genet 130:1773–1784

Saxena RK, Singh VK, Kale SM, Tathineni R, Parupalli S, Kumar V, Garg V, Das RR, Sharma M, Yamini KN, Muniswamy S (2017b) Construction of genotyping-by-sequencing based high-density genetic maps and QTL mapping for *Fusarium* wilt resistance in pigeonpea. Sci Rep 7:1911

Schafleitner R, Huang SM, Chu SH, Yen JY, Lin CY, Yan MR, Krishnan B, Liu MS, Lo HF, Chen CY, Long-fang OC (2016) Identification of single nucleotide polymorphism markers associated with resistance to bruchids (*Callosobruchus* spp.) in wild mungbean (*Vigna radiata var. sublobata*) and cultivated *V. radiata* through genotyping by sequencing and quantitative trait locus analysis. BMC Plant Biol 16:159

Schefers JM, Weigel KA (2012) Genomic selection in dairy cattle: integration of DNA testing into breeding programs. Animal Front 2:4–9

Schmutz J, McClean PE, Mamidi S, Wu GA, Cannon SB, Grimwood J, Jenkins J, Shu S, Song Q, Chavarro C, Torres-Torres M (2014) A reference genome for common bean and genome-wide analysis of dual domestications. Nat Genet 46:707–713

Schmutz J, McClean PE, Sujan Mamidi S, G Albert Wu GA, Steven B Cannon SB, Jane Grimwood J, Jerry Jenkins J, Shengqiang Shu S, Qijian Song Q, Carolina Chavarro C, Torres-Torres M, Geffroy V, Moghaddam SM, Gao D, Abernathy B, Barry K, Blair M, Brick MA, Chovatia M, Gepts P, Goodstein DM, Gonzales M, Hellsten U, Hyten DL, Jia G, Kelly JD, Kudrna D, Lee R, Richard MMS, Miklas PN, Osorno JM, Rodrigues J, Thareau V, Urrea CA, Wang M, Yu Y, Zhang M, Wing RA, Cregan PB, Rokhsar DS, Jackson SA (2014) A reference genome for common bean and genome-wide analysis of dual domestications. Nature Gene 46:707–713

Sharpe AG, Ramsay L, Sanderson LA, Fedoruk MJ, Clarke WE, Li R, Kagal S, Vijayan P, Vandenberg A, Bett KE (2013) Ancient orphan crop joins modern era: gene-based SNP discovery and mapping in lentil. BMC Genomics 14:192

Singh NK, Gupta DK, Jayaswal PK, Mahato AK, Dutta S, Singh S, Bhutani S, Dogra V, Singh BP, Kumawat G, Pal JK (2012) The first draft of the pigeonpea genome sequence. J Plant Biochem Biotechnol 21:98–112

Singh S, Parihar P, Singh R, Singh VP, Prasad SM (2015) Heavy metal tolerance in plants: role of transcriptomics, proteomics, metabolomics, and ionomics. Front Plant Sci 6:1143

Singh VK, Khan AW, Saxena RK, Kumar V, Kale SM, Sinha P, Chitikineni A, Pazhamala LT, Garg V, Sharma M, Kumar S (2016) Next-generation sequencing for identification of candidate genes for *Fusarium* wilt and sterility mosaic disease in pigeonpea (*Cajanus cajan*). Plant Biotechnol J 14:1183–1194

Smýkal P, Coyne CJ, Ambrose MJ, Maxted N, Schaefer H, Blair MW, Berger J, Greene SL, Nelson MN, Besharat N, Vymyslický T (2015) Legume crops phylogeny and genetic diversity for science and breeding. Crit Rev Plant Sci 34:43–104

Sompong U, Somta P, Raboy V, Srinives P (2012) Mapping of quantitative trait loci for phytic acid and phosphorus contents in seed and seedling of mungbean (*Vigna radiata* (L.) Wilczek). Breed Sci 62:87–92

Sonah H, Zhang X, Deshmukh RK, Borhan MH, Fernando WD, Bélanger RR (2016) Comparative transcriptomic analysis of virulence factors in Leptosphaeria maculans during compatible

and incompatible interactions with canola. Front Plant Sci 7:1784. https://doi.org/10.3389/fpls.2016.01784
Song Q, Yan L, Quigley C, Jordan BD, Fickus E, Schroeder S, Song BH, Charles An YQ, Hyten D, Nelson R, Rainey K (2017) Genetic characterization of the soybean nested association mapping population. Plant Genome 10:1–14
Souza TLP, Ragagnin VA, Dessaune SN, Sanglard DA, Carneiro JES, Moreira MA, Barros EG (2014) DNA marker-assisted selection to pyramid rust resistance genes in "carioca" seeded common bean lines. Euphytica 199:303–316
Spindel J, Wright M, Chen C, Cobb J, Gage J, Harrington S et al (2013) Bridging the genotyping gap: using genotyping by sequencing (GBS) to add high-density SNP markers and new value to traditional bi-parental mapping and breeding populations. Theor Appl Genet 126:2699–2716
Srivastava R, Singh M, Bajaj D, Parida SK (2016) A high-resolution InDel (insertion–deletion) markers-anchored consensus genetic map identifies major QTLs governing pod number and seed yield in chickpea. Front Plant Sci 7:1362
Stamp P, Visser R (2012) The twenty-first century, the century of plant breeding. Euphytica 186:585–591
Tar'an B, Warkentin T, Somers DJ, Miranda D, Vandenberg A, Blade S, Woods S, Bing D, Xue A, DeKoeyer D, Penner G (2003) Quantitative trait loci for lodging resistance, plant height and partial resistance to mycosphaerella blight in field pea (*Pisum sativum L.*). Theor Appl Genet 107:1482–1491
Tar'an B, Warkentin T, Somers DJ, Miranda D, Vanenberg A, Blade S, Bing D (2004) Identification of quantitative trait loci for grain yield, seed protein concentration and maturity in field pea (*Pisum sativum L.*). Euphytica 136:297–306
Taran B, Warkentin TD, Vandenberg A (2013) Fast track genetic improvement of Ascochyta blight resistance and double podding in chickpea by marker-assisted backcrossing. Theor Appl Genet 126:1639–1647
Tardivel A, Sonah H, Belzile F, O'Donoughue LS (2014) Rapid identification of alleles at the soybean maturity gene E3 using genotyping by sequencing and a haplotype-based approach. Plant Genome 7(2). https://doi.org/10.3835/plantgenome2013.10.0034
Tawaraya K, Horie R, Saito S, Wagatsuma T, Saito K, Oikawa A (2014) Metabolite profiling of root exudates of common bean under phosphorus deficiency. Meta 4:599–611
Teixeira FF, dos Santos JB, Ramalho MA, Abreu ÂD, Guimarães CT, de Oliveira AC (2005) QTL mapping for angular leaf spot in common bean using microsatellite markers. Crop Breed Appl Biotechnol 5:272–278
Temel HY, Gol D, Kahriman A, Tanyolac MB (2014) Single nucleotide polymorphism discovery through Illumina- based transcriptome sequencing and mapping in lentil. Turk J Agric For 38:1–19
Tester M, Langridge P (2010) Breeding technologies to increase crop production in a changing world. Science 327:818–822
Thudi M, Bohra A, Nayak SN, Varghese N, Shah TM, Penmetsa RV, Thirunavukkarasu N, Gudipati S, Gaur PM, Kulwal PL, Upadhyaya HD (2011) Novel SSR markers from BAC-end sequences, DArT arrays and a comprehensive genetic map with 1,291 marker loci for chickpea (*Cicer arietinum* L.). PLoS One 6:e27275
Thudi M, Gaur PM, Krishnamurthy L, Mir RR, Kudapa H, Fikre A, Kimurto P, Tripathi S, Soren KR, Mulwa R, Bharadwaj C (2014a) Genomics-assisted breeding for drought tolerance in chickpea. Funct Plant Biol 41:1178–1190
Thudi M, Upadhyaya HD, Rathore A, Gaur PM, Krishnamurthy L, Roorkiwal M, Nayak SN, Chaturvedi SK, Basu PS, Gangarao NVPR, Fikre A (2014b) Genetic dissection of drought and heat tolerance in chickpea through genome-wide and candidate gene-based association mapping approaches. PLoS One 9:e96758
Thudi M, Khan AW, Kumar V, Gaur PM, Katta K, Garg V, Roorkiwal M, Samineni S, Varshney RK (2016) Whole genome re-sequencing reveals genome-wide variations among parental lines of 16 mapping populations in chickpea (*Cicer arietinum* L.). BMC Plant Biol 16:10

Timmerman-Vaughan GM, Frew TJ, Russell AC, Khan T, Butler R, Gilpin M, Murray S, Falloon K (2002) QTL mapping of partial resistance to field epidemics of Ascochyta blight of pea. Crop Sci 42:2100–2111

Timmerman-Vaughan GM, Frew TJ, Butler R, Murray S, Gilpin M, Falloon K, Johnston P, Lakeman MB, Russell A, Khan T (2004) Validation of quantitative trait loci for Ascochyta blight resistance in pea (*Pisum sativum* L.), using populations from two crosses. Theore Appl Genet 109:1620–1631

Tock AJ, Fourie D, Walley PG, Holub EB, Soler A, Cichy KA, Pastor-Corrales MA, Song Q, Porch TG, Hart JP, Vasconcellos RC (2017) Genome-wide linkage and association mapping of halo blight resistance in common bean to race 6 of the globally important bacterial pathogen. Front Plant Sci 8:1170

Torres AM, Avila C, Gutierrez N, Palomino C, Moreno MT, Cubero J (2010) Marker-assisted selection in faba bean (Vicia faba L.). Field Crops Res 115(3):243–252

Torres AM, Román B, Avila CM, Satovic Z, Rubiales D, Sillero JC, Cubero JI, Moreno MT (2006) Faba bean breeding for resistance against biotic stresses: towards application of marker technology. Euphytica 147:67–80

Triques K, Sturbois B, Gallais S, Dalmais M, Chauvin S, Clepet C, Aubourg S, Rameau C, Caboche M, Bendahmane A (2007) Characterization of Arabidopsis thaliana mismatch specific endonucleases: application to mutation discovery by TILLING in pea. Plant J 51:1116–1125

Tuberosa R (2012) Phenotyping for drought tolerance of crops in the genomics era. Front Physiol 3:347

Upadhyaya HD, Thudi M, Dronavalli N, Gujaria N, Singh S, Sharma S, Varshney RK (2011) Genomic tools and germplasm diversity for chickpea improvement. Plant Genet Resour 9:45–58

Upadhyaya HD, Bajaj D, Das S, Kumar V, Gowda CLL, Sharma S, Tyagi AK, Parida SK (2016a) Genetic dissection of seed-iron and zinc concentrations in chickpea. Sci Rep 6:24050

Upadhyaya HD, Bajaj D, Narnoliya L, Das S, Kumar V, Gowda CLL, Sharma S, Tyagi AK, Parida SK (2016b) Genome-wide scans for delineation of candidate genes regulating seed-protein content in chickpea. Front Plant Sci 7:302

Van K, Kang YJ, Han KS, Lee YH, Gwag JG, Moon JK, Lee SH (2013) Genome-wide SNP discovery in mungbean by Illumina HiSeq. Theor Appl Genet 126:2017–2027

Varela ALN, Komatsu S, Wang X, Silva RG, Pedro Filho NS, Lobo AKM, Vasconcelos IM, Silveira JA, Oliveira JT (2017) Gel-free/label-free proteomic, photosynthetic, and biochemical analysis of cowpea (*Vigna unguiculata* [*L.*] *Walp.*) resistance against cowpea severe mosaic virus (CPSMV). J Proteome 163:76–91

Varshney RK, Glaszmann JC, Leung H, Ribaut JM (2010a) More genomic resources for less-studied crops. Trends Biotechnol 28:452–460

Varshney RK, Penmetsa RV, Dutta S, Kulwal PL, Saxena RK, Datta S, Sharma TR, Rosen B, Carrasquilla-Garcia N, Farmer AD, Dubey A (2010b) Pigeonpea genomics initiative (PGI): an international effort to improve crop productivity of pigeonpea (*Cajanus cajan L.*). Mol Breed 26:393–408

Varshney RK, Chen W, Li Y, Bharti AK, Saxena RK, Schlueter JA, Donoghue MT, Azam S, Fan G, Whaley AM, Farmer AD (2012) Draft genome sequence of pigeonpea (*Cajanus cajan*), an orphan legume crop of resource-poor farmers. Nat Biotechnol 30:83–89

Varshney RK, Mohan SM, Gaur PM, Gangarao NVPR, Pandey MK, Bohra A, Sawargaonkar SL, Chitikineni A, Kimurto PK, Janila P, Saxena KB (2013a) Achievements and prospects of genomics-assisted breeding in three legume crops of the semi-arid tropics. Biotechnol Adv 31:1120–1134

Varshney RK, Song C, Saxena RK, Azam S, Yu S, Sharpe AG, Cannon S, Baek J, Rosen BD, Tar'an B, Millan T (2013b) Draft genome sequence of chickpea (*Cicer arietinum*) provides a resource for trait improvement. Nat Biotechnol 31:240–246

Varshney RK, Gaur PM, Chamarthi SK, Krishnamurthy L, Tripathi S, Kashiwagi J, Samineni S, Singh VK, Thudi M, Jaganathan D (2013c) Fast-track introgression of "QTL-hotspot" for root traits and other drought tolerance traits in JG 11, an elite and leading variety of chickpea. Plant Genome 6:1–13

Varshney RK, Mohan SM, Gaur PM, Chamarthi SK, Singh VK, Srinivasan S, Swapna N, Sharma M, Pande S, Singh S, Kaur L (2014) Marker-assisted backcrossing to introgress resistance to fusarium wilt race 1 and Ascochyta blight in C 214, an elite cultivar of chickpea. Plant Genome 7:1–10

Varshney RK, Saxena RK, Upadhyaya HD, Khan AW, Yu Y, Kim C, Rathore A, Kim D, Kim J, An S, Kumar V (2017) Whole-genome resequencing of 292 pigeonpea accessions identifies genomic regions associated with domestication and agronomic traits. Nature Genet 49:1082. https://doi.org/10.1038/ng.3872

Vatanparast M, Shetty P, Chopra R, Doyle JJ, Sathyanarayana N, Egan AN (2016) Transcriptome sequencing and marker development in winged bean (*Psophocarpus tetragonolobus; Leguminosae*). Sci Rep 6:29070

Verma P, Shah N, Bhatia S (2013) Development of an expressed gene catalogue and molecular markers from the de novo assembly of short sequence reads of the lentil (*Lens culinaris* Medik.) transcriptome. Plant Biotechnol J 11:894–905

Verma P, Goyal R, Chahota RK, Sharma TR, Abdin MZ, Bhatia S (2015a) Construction of a genetic linkage map and identification of QTLs for seed weight and seed size traits in lentil (*Lens culinaris Medik.*). PLoS One 10:e0139666

Verma S, Gupta S, Bandhiwal N, Kumar T, Bharadwaj C, Bhatia S (2015b) High-density linkage map construction and mapping of seed trait QTLs in chickpea (*Cicer arietinum* L.) using genotyping-by-sequencing (GBS). Sci Rep 5:17512

Vipin CA, Luckett DJ, Harper JD, Ash GJ, Kilian A, Ellwood SR, Phan HT, Raman H (2013) Construction of integrated linkage map of a recombinant inbred line population of white lupin (*Lupinus albus* L.). Breed Sci 63:292–300

Vuong TD, Sonah H, Meinhardt CG, Deshmukh R, Kadam S, Nelson RL, Shannon JG, Nguyen HT (2015) Genetic architecture of cyst nematode resistance revealed by genome-wide association study in soybean. BMC Genomics 16(1):593

Weckwerth W (2011) Green systems biology—from single genomes, proteomes and metabolomes to ecosystems research and biotechnology. J Proteome 75:284–305

Welcker C, Sadok W, Dignat G, Renault M, Salvi S, Charcosset A, Tardieu F (2011) A common genetic determinism for sensitivities to soil water deficit and evaporative demand: meta- analysis of quantitative trait loci and introgression lines of maize. Plant Physiol 157:718–729

Wen W, Li D, Li X, Gao Y, Li W, Li H, Liu J, Liu H, Chen W, Luo J, Yan J (2017) Metabolome-based genome-wide association study of maize kernel leads to novel biochemical insights. Nature Commun 5:3438

Witzel K, Weidner A, Surabhi GK, Börner A, Mock HP (2009) Salt stress-induced alterations in the root proteome of barley genotypes with contrasting response towards salinity. J Exp Bot 60:3545–3557

Wu J, Wang L, Li L, Wang S (2014) De novo assembly of the common bean transcriptome using short reads for the discovery of drought-responsive genes. PLoS One 9:e109262

Yan X, Liao H, Beebe SE, Blair MW, Lynch JP (2004) QTL mapping of root hair and acid exudation traits and their relationship to phosphorus uptake in common bean. Plant Soil 265:17–29

Yang H, Tao Y, Zheng Z, Zhang Q, Zhou G, Sweetingham MW, Howieson JG, Li C (2013) Draft genome sequence, and a sequence-defined genetic linkage map of the legume crop species *Lupinus angustifolius* L. PLoS One 8:e64799

Yang H, Jian J, Li X, Renshaw D, Clements J, Sweetingham MW, Tan C, Li C (2015) Application of whole genome re-sequencing data in the development of diagnostic DNA markers tightly linked to a disease-resistance locus for marker-assisted selection in lupin (*Lupinus angustifolius*). BMC Genomics 16:660

Yu K (2011) BioMed Res Inter 2012: Bacterial artificial chromosome libraries of pulse crops: characteristics and applications. BioMed Res Int ID493186: 1–8

Yu J, Buckler ES (2006) Genetic association mapping and genome organization of maize. Curr Opin Biotechnol 17:155–160

Zadražnik T, Hollung K, Egge-Jacobsen W, Meglič V, Šuštar-Vozlič J (2013) Differential proteomic analysis of drought stress response in leaves of common bean (*Phaseolus vulgaris L.*). J Proteome 78:254–272

Zargar SM, Raatz B, Sonah H, Bhat JA, Dar ZA, Agrawal GK, Rakwal R (2015) Recent advances in molecular marker techniques: insight into QTL mapping, GWAS and genomic selection in plants. J Crop Sci Biotech 18(5):293–308

Zeid M, Mitchell S, Link W, Carter M, Nawar A, Fulton T, Kresovich S (2009) Simple sequence repeats (SSRs) in faba bean: new loci from Orobanche-resistant cultivar 'Giza 402'. Plant Breed 128:149–155

Zhang D, Li H, Wang J, Zhang H, Hu Z, Chu S, Lv H, Yu D (2016) High-density genetic mapping identifies new major loci for tolerance to low-phosphorus stress in soybean. Front Plant Sci 7:372

Zhang X, Scheuring CF, Zhang M, Dong JJ, Zhang Y, Huang JJ, Lee MK, Abbo S, Sherman A, Shtienberg D, Chen W (2010a) A BAC/BIBAC-based physical map of chickpea, *Cicer arietinum* L. BMC Genomics 11:501

Zhang ZW, Ersoz E, Lai CQ, Todhunter RJ, Tiwari HK, Gore MA, Bradbury PJ, Yu JM, Arnett DK, Ordovas JM, Buckler ES (2010b) Mixed linear model approach adapted for genome-wide association studies. Nat Genet 42:355–360

Zhang S, Gao M, Zaitlin D (2012) Molecular linkage mapping and marker-trait associations with NlRPT, a downy mildew resistance gene in Nicotiana langsdorffii. Front Plant Sci 3:185

Zhang Y, Lin H, Yang Z, Wang J, Li Y, Xu B (2013) Protein complex prediction in large ontology attributed protein-protein interaction networks. IEEE/ACM Trans Comput Biol Bioinform 10:729–741

Zhou Q, Luo D, Ma L, Xie W, Wang Y, Wang Y, Liu Z (2016) Development and cross-species transferability of EST-SSR markers in Siberian wildrye (*Elymus sibiricus* L.) using Illumina sequencing. Sci Rep 6:20549

Zhu H, Choi HK, Cook DR, Shoemaker RC (2005) Bridging model and crop legumes through comparative genomics. Plant Physiol 137:1189–1196

Zivy M, Wienkoop S, Renaut J, Pinheiro C, Goulas E, Carpentier S (2015) The quest for tolerant varieties: the importance of integrating "omics" techniques to phenotyping. Front Plant Sci 6:448

Chapter 3
Molecular and Genomic Approaches to Peanut Improvement

Jeffrey N. Wilson and Ratan Chopra

3.1 Introduction

Peanut ranks second to soybean in the world of legume oilseeds both in terms of area grown and tonnage produced. Peanut is grown in over 100 countries (Nwokolo 1996) with a total production of 41.18 million tons from 24.70 million ha in 2012 (FAO 2014), with a mean productivity of 1.66 tons/ha. The five largest producers in the world in 2012, based on pod tonnage, were China, India, Nigeria, the USA, and Myanmar. The crop is a rich source of oil (36–54%), proteins (16–36%), and carbohydrates (10–20%) (Knauft and Ozias-Akins 1995). Peanuts are utilized for seed, which supplies essential minerals such as zinc, iron, phosphorus, and calcium and vitamins – riboflavin, thiamine, niacin, and vitamin E. Peanut seed cotyledons are also a major source of vegetable oil. As with numerous other members of the Fabaceae (Leguminosae) family, peanuts can convert atmospheric nitrogen into ammonia by symbiotic nitrogen fixation. Therefore, peanuts can increase soil fertility (Pimratch et al. 2004).

In 1753, Linnaeus described the domesticated peanut as *Arachis hypogaea*, depicting peanut as a weed with underground fruits, unlike most angiosperms. The *Arachis* genus is placed within the Leguminosae family. Major grain legumes are in the Papilionoideae subfamily, which is further subdivided into several clades including *phaseoloids* (milletoids or warm-season legumes), *galegoids* (cool-season legumes), and *genistoids*. Within *phaseoloids* are multiple agronomically important genera including *Glycine*, *Phaseolus*, and *Vigna*. Important *galegoids* include

J. N. Wilson (✉)
Ready Roast Nut Company, Madera, CA, USA
e-mail: jeffwils@aggienetwork.com

R. Chopra
University of Minnesota, Department of Agronomy and Plant Genetics, St. Paul, MN, USA

S. H. Wani, M. Jain (eds.), *Pulse Improvement*,
https://doi.org/10.1007/978-3-030-01743-9_3

Pisum, *Medicago*, *Lens*, and *Vicia*, while *genistoids* include the genus *Lupinus* (Doyle and Luckow 2003). *Arachis* belongs to the *dalbergioids* clade, which also includes members of the genus *Stylosanthes*.

Wild species of *Arachis* have been collected in Brazil, Bolivia, Paraguay, Argentina, and Uruguay (Krapovickas and Gregory 1994; Singh and Simpson 1994). Based on morphological and cross-compatibility data and geographic distribution, the *Arachis* genus has been divided into nine taxonomic sections (Krapovickas and Gregory 1994). The nine sections include Trierectoides, Erectoides, Extranervosae, Triseminatae, Heteranthae, Caulorrhizae, Procumbentes, Rhizomatosae, and Arachis. Krapovickas and Gregory (1994) described a total of 69 *Arachis* species, and 11 additional species were later described by Valls and Simpson (2005).

The Arachis section consists of two tetraploids, *A. hypogaea* and *A. monticola*. The remaining species of Arachis section are diploids. Twenty A-genome diploid species have been described (Krapovickas and Gregory 1994), including perennials (*A. cardenasii*, *A. diogoi*, *A. helodes*, *A. villosa*, and *A. correntina*) and annuals (*A. duranensis* and *A. stenosperma*). Based on cytological evidence and cross-hybridization data, *A. cardenasii* was originally considered the most probable A-genome ancestor of *A. hypogaea* (Smartt and Stalker 1982). However, more recent studies have proposed that *A. duranensis* is the A-genome donor to *A. hypogaea* (Robledo et al. 2009).

B-genome species *A. batizocoi* was identified by Krapovickas et al. (1974) and was later proposed as the B-genome donor to the cultigen (Smartt and Stalker 1982). Studies on cross-compatibility, molecular genetics, and cytogenetics provided evidence for up to ten B-genome species (Burow et al. 2009; Kochert et al. 1996; Krapovickas and Gregory 1994; Milla et al. 2005; Tallury et al. 2005; Valls and Simpson 2005). Based on FISH, GISH, and geographic origin data, the B-genome classification was deemed inadequate and subsequently split into three genome types (Robledo and Seijo 2008). *Arachis ipaensis*, *A. magna*, *A. gregoryi*, *A. vallsii*, and *A. williamsii* remained in the B-genome, while *A. batizocoi*, *A. cruziana*, and *A. krapovickasii* were reclassified as K-genome types. *A. benensis* and *A. trinitensis* were classified as F-genome types.

The D-genome consists of one species, *A. glandulifera*. This species is characterized by extensive genome rearrangements relative to other section Arachis species, as observed cytologically (Stalker 1991). In addition, there are 3 diploid species that possess 18 rather than 20 chromosomes. These species have been identified as *A. decora*, *A. palustris*, and *A. praecox* (Lavia 1996; Krapovickas and Gregory 1994).

A. hypogaea is considered to be an AABB tetraploid ($2n = 4x = 40$), arising from hybridization between A- and B-diploid species (Smartt and Stalker 1982). Krapovickas and Gregory (1994) classified *A. hypogaea* into two subspecies and six botanical varieties. *A. hypogaea* subsp. *hypogaea* and the botanical varieties *hypogaea* (Runner and Virginia) and *hirsute* are characterized by a spreading growth habit and lack of flowers on the main stem. *A. hypogaea* subsp. *fastigiata* has an erect growth habit, flowers on the main stem, and includes the botanical varieties *fastigiata* (Valencia), *vulgaris* (Spanish), *peruviana*, and *aequatoriana*.

3.2 Utilization of Wild Relatives

Recent speciation, coupled with self-pollination biology, limited capacity for seed dispersal, and selection, has limited genetic variation in tetraploid peanuts. This lack of genetic variation and the complex genome structure of *A. hypogaea* have slowed the rate of progress in peanut genetic and genomic research compared to many domesticated crop species. Wild species chromatin has been utilized to introgress novel genes into *A. hypogaea* and broaden its genetic base for QTL mapping. Wild species have evolved to thrive in a wide variety of soil and climatic conditions (Brasileiro et al. 2015; Stalker 2017) Therefore, wild *Arachis* genotypes harbor novel resistance/tolerances to a wide array of biotic and abiotic stresses not available in the *A. hypogaea* genome (Stalker 2017). However, efficiently integrating genetic diversity from wild species is an ongoing challenge to crop researchers. Barriers to the use of diploid species include genomic incompatibilities, progeny sterility, and low cross success rates with *A. hypogaea.*

A unique amphidiploid introgression pathway, proposed by Simpson (1991) at Texas A&M University, led to the introduction of peanut cultivars with resistance to root-knot nematodes (*Meloidogyne arenaria*) (Simpson and Starr 2001; Simpson et al. 2003). Strong yield drag associated with these cultivars compared with other commercial runner-type cultivars (Holbrook et al. 2008a, b) indicates nontarget chromatin from wild species introduced unfavorable alleles (Nagy et al. 2010). However, two newer cultivars containing the *Rma* nematode marker from TxAG-6, 'Tifguard' (Holbrook et al. 2008a) and 'Webb' (Simpson et al. 2013), offer yields comparable to commercial peanut cultivars combined with other desirable agronomic traits.

Tanksley and Nelson (1996) proposed an advanced backcross QTL (AB-QTL) approach to integrate valuable agronomic traits from exotic donor genotypes into elite germplasm. This approach has been utilized to locate and transfer specific QTLs in a variety of crops (Bernacchi et al. 1998; Liu et al. 2006; Tian et al. 2006). AB-QTL and chromosome segment substitution line (CCSL) populations have been successfully utilized for the introgression of amphidiploid chromatin in *A. hypogaea* mapping populations (Fonceka et al. 2009, 2012a, b; Sharma et al. 2013; Wilson et al. 2017). These studies demonstrated that genetic diversity in *A. hypogaea* can be broadened through wild species introgression while minimizing the introduction of nontarget chromatin.

3.3 Genome Resources

Peanut genome resources have been well supported by bioinformatic tools that allow researchers to look up the gene orientation, QTL regions, molecular markers, and synteny between the two sub-genomes. Diploid genome sequences have been generated for both A-genome and B-genome species (Bertioli et al. 2016; Chen et al. 2016; Lu et al. 2018). Both A- and B-genome sequences are available through https://peanutbase.org.

These genome sequences will provide additional insights into *Arachis* evolution and polyploidization. The biological interpretation of these genome sequences enhances understanding of fruit development (Chen et al. 2016), oil biosynthesis (Lu et al. 2018), and evolution of retrotransposons (Bertioli et al. 2016). These sequences will also serve as a basis for developing molecular markers (SSR or SNP) for investigating genetic diversity and aid breeding programs for peanut improvement.

3.4 Transcriptome Resources

Transcriptome sequencing is an important method of determining gene expression in various stages of plant development. Developing tissue-specific gene expression is crucial for understanding the gene function and its effect on specific traits. There are reports demonstrating the differences in expression of genes in peanut. For example, express sequence tags (ESTs) have been generated to identify the genes regulating biotic stress resistance in peanut (Guo et al. 2008, 2009;Bi et al. 2010; Liu et al. 2013). The development of new sequencing technologies has led to differential expression analyses using RNA-Seq for both biotic and abiotic stresses. Differential expression studies in peanut have been performed for water stress (Guimaraes et al. 2012), drought stress (Ding et al. 2014; Brasileiro et al. 2015), and bacterial wilt (Chen et al. 2014). Numerous transcription factors and enzymes have been identified in these studies which can serve as a marker for selecting elite breeding lines. In addition to understanding the differences in expression under environmental stress, Peng et al. (2017) and Clevenger et al. (2017b) have highlighted the tissue-specific differences in expression. These tissue-specific expression atlases will provide crucial information required for understanding the role of each gene in plant development. These resources will be beneficial to peanut biologists and breeders in developing gene expression markers and gene-specific mutation markers.

3.5 Molecular Markers and Genetic Maps

The goal of genetic mapping and marker development in crops is to identify genetic markers that are associated with traits of interest. Once tightly linked/perfect/functional markers are developed using these resources, these markers can be deployed in marker-assisted peanut improvement. The narrow genetic base of the cultivated peanut has provided a substantial obstacle to genetic mapping using only cultivated germplasm. Thus, initial *Arachis* genetic maps were generated using crosses involving wild species. The greater DNA polymorphisms found in wild relatives allow for higher-resolution mapping, and diploid genetics simplify genetic analyses. Molecular markers have been developed using transcriptome sequencing (Chopra et al. 2016) and whole-genome sequencing (Clevenger et al. 2017a) in both diploid and tetraploid peanuts.

Genetic mapping in peanut has been performed using marker systems such as RFLP (Kochert et al. 1991,1996; Halward et al. 1993), RAPD (Garcia et al. 1996), SSR (Moretzsohn et al. 2005; Leal-Bertioli et al. 2009), and SNP (Leal-Bertioli et al. 2009; Nagy et al. 2010; Chopra et al. 2018) in AA-genome crosses. Use of SNP markers in combination with SSRs had resulted in the development of a dense F_2 population genetic map derived from the cross (*A. duranenis* × *A. duranensis*) (Nagy et al. 2012) . The map contains 2319 markers (971 SSRs, 221 single-stranded DNA conformation polymorphism (SSCP) markers, and 1127 SNPs) mapped on 10 linkage groups. This map is currently the densest genetic map made from a single cross among peanut diploid and tetraploid *Arachis* maps.

Only two maps have been reported for the BB-genome. Moretzsohn et al. (2009) developed a genetic map with 149 SSR loci on 11 linkage groups covering 1294 cM genome which was developed based on an F_2 population (93 lines) derived from the cross between *A. ipaensis* (KG30076) and *A. magna* (KG30097). A genetic map was constructed with 449 SSR loci using an F_2 population derived from the cross *A. batizocoi* (PI298639) × *A. batizocoi* (PI468327) (Guo et al. 2012).

The first effort to construct a tetraploid genetic map for peanut utilized RFLP markers (Burow et al. 2001). Several partial genetic maps in tetraploids have since been developed using SSRs (Varshney et al. 2009; Hong et al. 2010; Khedikar et al. 2010; Sarvamangala et al. 2011; Wang et al. 2012; Wilson et al. 2017). Attempts to develop maps with higher densities have required screening of several thousand SSR markers. Recently, Qin et al. (2012) screened 4576 markers and constructed an integrated map with 324 marker loci covering a genome distance of 1352 cM. The highest density map of cultivated peanut to date contains more than 1000 markers utilized in silico analysis of DNA sequence data from the different parent lines (Shirasawa et al. 2012).

A reference consensus genetic map was constructed by Gautami et al. (2012a), which contains 897 marker loci. Interestingly, this reference consensus genetic map was divided into 20,320-cM-long bins, which carry 1 to 20 loci with an average of 4 marker loci per bin. This consensus genetic map has recently been improved by international research partners. The addition of mapping information from 5 new genetic maps (total of 16 individual genetic maps) increased markers from 897 to 3693, spanning 2651 cM of the genome and 20 linkage groups (Shirasawa et al. 2013). These consensus maps have the potential to be useful in aligning new genetic and physical maps and QTL analysis and determining genetic background effect on QTL expression.

3.5.1 Mapping of Important Agronomic Traits

QTL and association mapping are powerful tools to assess genetic diversity and identify QTLs for agronomic traits based on linkage equilibrium and disequilibrium, respectively, in crop germplasm. Utilizing the markers and genetic maps developed in both cultivated and wild-derived population, both approaches have been used to dissect the genomic regions for important agronomic traits in peanut. Few of the important traits in peanut production are explained below.

3.5.1.1 Drought Tolerance

Peanuts are commonly produced in seasonally rain-fed regions without supplemental irrigation. Escape and avoidance (also referred to as tolerance) (Zhang et al. 2001) are primary drought adaptations in peanut. Early maturing varieties designed to escape late-season drought have been released in Africa (Nigam and Aruna 2008). As noted by Brasileiro et al. (2015), the *Arachis* genus is native to areas of South America with pronounced wet and dry seasons, particularly *A. duranensis* (Guimaraes et al. 2012). Although many desirable drought resistance traits in wild species have likely been lost due to tetraploidization (Leal-Bertioli et al. 2012), variability in drought responses does exist within *A. hypogaea*. The US mini-core peanut germplasm collection has been screened at Texas A&M University, and five accessions were identified with drought resistance traits (Selvaraj et al. 2010). These accessions generated SSR markers which were linked to drought resistance via association mapping (Belamkar et al. 2010). The ICRISAT mini-core collection has also been screened for several drought tolerance measurements, and improved accessions have been identified (Upadhyaya 2005).

Measuring drought tolerance in crops is challenging at best. Researchers have long struggled with reproducibility of results and the relationship between yield and various drought parameters. Components of yield under moisture stress are water-use efficiency (WUE), water use (WU), and harvest index (HI) (Passioura 1996). Many studies in row crops are aimed at improving WUE, which is a function of transpiration efficiency (TE) of the leaf (Blum 2009). TE is commonly measured indirectly through carbon 13 discrimination ratio, specific leaf area (SPLA), and SPAD chlorophyll content. Several studies in peanut were aimed at discovering QTLs for drought parameters and related yield traits. Ravi et al. (2011) reported 105 main-effect QTLs explaining 3.5–33.4% PVE for multiple drought traits. Multiple QTLs were identified for transpiration, TE, SLA, canopy conductance, dry matter, pod yield, and seed weight. Gautami et al. (2012b) reported 153 main-effect QTLs and 25 epistatic QTLs explaining from 4.8 to 22.4 percent of phenotypic variation for drought traits in 2 peanut populations. None of the drought QTLs identified by Gautami et al. (2012b) were considered "major" QTLs, since they explained a relatively small percentage of observed phenotypic variation.

3.5.1.2 Disease and Pest Resistance

A scarcity of genetic polymorphisms has caused susceptibility to a diverse range of diseases in cultivated peanut, including those caused by bacterial, viral, and fungal pathogens. Host plant resistance to pathogens may be classified as qualitative, which tends to be highly specific to pathogen strains, or quantitative, i.e., multi-gene resistance. Disease resistance (*R*) proteins, which are generally qualitative, typically contain a nucleotide-binding site (NBS) domain and C-terminal leucine-rich repeats (LRRs). *R* proteins recognize a corresponding *avr* gene in the pathogen, inducing a cascade of defense signaling in the host, inducing the hypersensitive response. This is often referred to as the gene-for-gene relationship. NBS-LRR gene families are among the largest found in higher plant genomes. *R*-gene candidates, including numerous NBS-LRR gene analogs, have been identified in *Arachis* (Bertioli et al. 2003; Yuksel et al. 2005; Ratnaparkhe et al. 2011; Liu 2013). Sequencing of the A- and B-genome progenitor species has provided insight into the genomic distribution of NBS-LRR disease resistance-like genes in the *Arachis* genome. Bertioli et al. (2016) found 397 NBS-LRR-encoding genes in *A. duranensis* (A-genome) versus 345 in *A. ipaensis* (B-genome).

The unique geocarpic nature of peanut reproduction that protects pods also leaves fruit vulnerable to soilborne pests. Nematode (*Meloidogyne* spp.) resistance in close *Arachis* relatives is often, but not always, mediated by the hypersensitive response (Milligan et al. 1998; Cook et al. 2012; Liu et al. 2012). Resistance to *M. arenaria* in *A. hypogaea* cultivar COAN (Simpson and Starr 2001) is primarily derived from a single dominant gene, designated *Rma* (Nagy et al. 2010) from *A. cardenasii* (Choi et al. 1999; Burow et al. 2001). A series of markers linked to *Rma* have been developed, including a tightly linked SCAR markers (Chu et al. 2007). Subsequent research has identified additional homologous QTLs associated with root-knot nematode resistance in 'COAN' (Burow et al. 2014). Guimaraes et al. (2015) utilized qRT-PCR to identify candidate root-knot nematode resistance genes including NBS-LRR R-genes and induced resistance genes involved in the jasmonic acid signaling pathway. The expressed gene sequence contributing to *Rma* has yet to be confirmed. However, RNA-Seq technology has identified a NBS-LRR R candidate gene introgressed from *A. cardenasii* that conveys nematode resistance in cultivar 'Tifguard' (Clevenger et al. 2017b).

QTL data is very limited for several important soilborne fungal diseases of peanut. One major QTL has been reported for *Sclerotinia* blight (caused by *Sclerotinia minor*) (Chenault et al. 2009) and stem rot disease (caused by *Sclerotium rolfsii*) (Bera et al. 2016). There are no published reports on QTLs for resistance to *Cylindrocladium* black rot (caused by *Cylindrocladium parasiticum*), although partial resistance exists in *A. hypogaea* (Wynne et al. 1975) and wild relatives (Tallury et al. 2014). Marker-assisted selection for quantitatively inherited disease resistance traits has the potential to improve the speed of introgression, particularly when diverse or wild species chromatin is utilized. Additional research is needed to identify genomic regions associated with soilborne fungal diseases in peanut.

Peanuts are susceptible to foliar diseases such as early and late leaf spot (caused by *Cercospora arachidicola* and *Cercospora personatum*, respectively); tomato spotted wilt virus (TSWV), which is vectored by tobacco thrips (*Frankliniella* spp.); and rust (caused by *Puccinia arachidis*). Consistently identifying QTLs for foliar peanut diseases is difficult due to annual variations in natural epidemics (Pandey et al. 2017a). However, recent studies have identified major-effect QTLs for TSWV and leaf spots with PVE values over 10% in RIL populations (Khera et al. 2016;

Pandey et al. 2017b). The vast majority of QTLs for TSWV and leaf spots identified by Pandey et al. (2017b) mapped to the A sub-genome. Inheritance of rust resistance in *A. hypogaea* is not well understood but is thought to be more complex than the traditional rust resistance gene-for-gene model (Subrahmanyam et al. 1993). Genomic regions associated with differing components of rust resistance have been identified in a population derived from *A. magna*, a B-genome wild peanut relative (Leal-Bertioli et al. 2015).

3.5.1.3 Aflatoxin

Aflatoxin, a poisonous carcinogen produced by the fungal mold *Aspergillus flavus*, has emerged as a major obstacle to peanut production in the developing world. Resistance to *A. flavus* in *A. hypogaea* seeds is very limited, highly influenced by environment, and little is known regarding mechanisms of resistance (Nigam et al. 2009). Aflatoxin is highly opportunistic; therefore insect and nematode damage can encourage contamination (Lynch and Wilson 1991; Timper et al. 2004). Preharvest resistance to aflatoxin contamination has been linked to drought resistance (Holbrook et al. 2000). Korani et al. (2017) identified peanut cultivar effects on post-harvest aflatoxin production despite susceptibility in all cultivars screened to *Aspergillus* infection. Expression of $PnAG_3$, a NBS-LRR resistance analog, increased in response to *A. flavus* challenge in seed coat, kernel, and pericarp tissue in a resistant peanut cultivar. This increase in expression was two to three times greater than in a susceptible cultivar. Variation in resistance to aflatoxin contamination exists among wild species accessions (Thakur et al. 2000). Several interspecific tetraploids derived from *A. cardenasii* exhibited partial resistance to aflatoxin accumulation after challenge with *A. flavus* in a study by Xue et al. (2004).

3.5.1.4 Oil Chemistry

Peanut oil, derived from seed cotyledons, is an important worldwide commodity. Vegetable oil chemistry has been manipulated for human consumption and industrial purposes by altering fatty acid ratios. The ratio of oleic (18:1) to linoleic (18:2) acid in peanut oil has been of particular interest to researchers, processors, and consumers. A high oleic to linoleic acid ratio in peanut improves self-life by reducing the rate of oxidization and provides numerous human health benefits. In cultivated peanut, the high-oleic trait is controlled by two recessive alleles that code for the fatty acid enzyme desaturase (FAD2), *ahFAD2A* and *ahFAD2B* (Moore and Knauft 1989; Jung et al. 2000). SNPs for the *ahFAD2A* and *ahFAD2B* alleles based on transcriptome sequences have been developed for *A. hypogaea* (Chopra et al. 2014), and molecular markers for *ahFAD2* alleles have been utilized in the development of new high-oleic cultivars (Janila et al. 2016).

Besides oleic and linoleic acid, at least six other fatty acids are present in measureable quantities in peanut oil. Variation in saturation of these fatty acids also plays an important role in the nutritional and industrial properties of peanut oil. Five studies have identified QTLs for minor fatty acids in peanut, including a study utilizing transposable element markers that detected comparatively high numbers of minor-effect QTLs (Hake et al. 2017). Wang et al. (2015) identified 34 major QTLs (PVE >10%) for 6 fatty acids in 2 RIL populations, Huang et al. (2015) found 11 QTLs for 6 fatty acids explaining between 3.01 and 81.51% of observed variation, and Shasidhar et al. (2017) identified 20 QTLs for 7 fatty acids. An advanced backcross peanut population derived from TxAG-6 was utilized to identify 15 unique QTLs for 8 different fatty acids across multiple environments (Wilson et al. 2017). In order to determine gene function, additional work is needed to refine map and sequence QTLs associated with minor fatty acids in peanut.

3.5.1.5 Other Traits

Pandey et al. (2014a) revealed 524 significant marker-trait associations for 36 agronomic traits in a diverse mapping panel consisting of over 300 individuals using SSR markers. Association mapping using Chinese germplasm (Jiang et al. 2014; Zhao et al. 2017) and QTL mapping (Fonceka et al. 2012a; Chopra et al. 2018) using biparental populations were utilized to identify marker-trait association for multiple seed-related traits. More recently, Lu et al. (2018) and Chopra et al. (2018) identified QTLs for plant height (Table 3.1).

3.5.2 Challenges and Opportunities in Peanut Improvement

Recent speciation, coupled with self-pollination biology, limited capacity for seed dispersal, and selection, has limited genetic variation in tetraploid peanuts. This lack of genetic variation and the complex genome structure of *A. hypogaea* have slowed the rate of progress in peanut genetic and genomic research compared to many domesticated crop species. However, the development of genomic and transcriptome-based approaches has accelerated the discovery of new genes and gene pathways in *A. hypogaea* and wild relatives.

In the near future, positive agronomic alleles from *Arachis* will be fine-mapped, sequenced, and introgressed into *A. hypogaea*, aided by new technologies such as transcriptome sequencing and SNP genotyping arrays. The next step for peanut breeders is developing new peanut cultivars utilizing a marker-assisted backcross (MABC) approach that pyramids wild species-derived positive alleles. The MABC approach has been successfully utilized at ICRISAT to introgress a large-effect rust QTL into elite peanut germplasm (Varshney et al. 2014). For quantitative traits

Table 3.1 List of QTL mapping studies in *Arachis* conducted from 2014 to 2017

Trait	Reference	QTL number	Phenotypic variance explained (%)	Population
Disease/insect resistance				
Bruchid pod damage	Mondal et al. (2014a)	3	32–37	VG 9514 × TAG24
Bruchid development	Mondal et al. (2014a)	4	36–67	VG 9514 × TAG24
Pod weight loss due to bruchid	Mondal et al. (2014a)	1	33	VG 9514 × TAG24
Rust resistance	Mondal et al. (2014b)	1	65–75	VG 9514 × TAG24
Rust resistance	Leal-Bertioli et al. (2015)	13	5.8–59.3	*I. ipaensis* K 30076 × *A. magna* GKSSc 30097)
Rust resistance	Pandey et al. (2017a)	8	42.7–83.6	TAG 24 × GPBD 4
Leaf spot resistance	Liang et al. (2017)	8	8–20	Tamrun OL07 × Tx9464117
Early leaf spot resistance	Pandey et al. (2017b)	6	6.26–13.20	Tifrunner × GT-C20
Late leaf spot resistance	Pandey et al. (2017a)	3	9.0–14.9	TAG 24 × GPBD 4
Late leaf spot resistance	Pandey et al. (2017b)	22	6.40–15.55	Tifrunner × GT-C20
TSWV resistance	Pandey et al. (2017b)	11	6.74–14.41	Tifrunner × GT-C20
Plant morphology				
Plan height	Faye et al. (2015)	5	4.04–8.16	TAG 24 × ICGV 86031
Plant height (main stem)	Huang et al. (2015)	3	6.12–8.9	Zhonghua 10 × ICG 12625
Plant height (main stem)	Li et al. (2017)	11	6.00–22.53	79266 × D893
Plant height (main stem)	Chen et al. (2017)	6[b]	6.72–47.1	Fuchuan Dahuasheng × ICG6375
Number of branches	Huang et al. (2015)	2	6.11–7.46	Zhonghua 10 × ICG 12625
Number of branches	Chen et al. (2017)	3[b]	5.09–6.04	Fuchuan Dahuasheng × ICG6375
Primary branch number	Faye et al. (2015)	5	0.04–8.58	TAG 24 × ICGV 86031
Primary branch number	Hake et al. (2017)	1[a]	1.0	TMV 2 × TMV 2-NLM
Secondary branch number	Hake et al. (2017)	1[a]	10.8	TMV 2 × TMV 2-NLM

(continued)

Table 3.1 (continued)

Trait	Reference	QTL number	Phenotypic variance explained (%)	Population
First lateral branch length	Li et al. (2017)	16	2.83–21.63	79266 × D893
Pod traits				
100 pod weight	Huang et al. (2015)	2	2.11–18.7	Zhonghua 10 × ICG 12625
100 pod weight	Chen et al. (2017)	8[b]	5.29–8.45	Fuchuan Dahuasheng × ICG6375
Pod width	Huang et al. (2015)	2	2.11–18.7	Zhonghua 10 × ICG 12625
Pod width	Chen et al. (2016)	4	4.48–8.78	Xuhua 13 × Zhonghua 6
Pod width	Chen et al. (2016)	8	5.16–16.14	Fuchuan Dahuasheng × ICG 6375
Pod width	Luo et al. (2017)	11	5.26–14.12	Yuanza 9102 × Xuzhou 68–4
Pod width	Hake et al. (2017)	3[a]	9.3	TMV 2 × TMV 2-NLM
Pod width	Chen et al. (2017)	7[b]	3.94–8.54	Fuchuan Dahuasheng × ICG6375
Pod length	Huang et al. (2015)	1	11.23	Zhonghua 10 × ICG 12625
Pod length	Chen et al. (2016)	4	1.25–7.79	Xuhua 13 × Zhonghua 6
Pod length	Chen et al. (2016)	6	5.7–24.29	Fuchuan Dahuasheng × ICG 6375
Pod length	Luo et al. (2017)	15	3.68–27.84	Yuanza 9102 × Xuzhou 68–4
Pod length	Chen et al. (2017)	12[b]	5.72–13.21	Fuchuan Dahuasheng × ICG6375
Seed traits				
100 seed weight	Faye et al. (2015)	2	3.3–7.41	TAG 24 × ICGV 86031
100 seed weight	Huang et al. (2015)	3	1.69–17.88	Zhonghua 10 × ICG 12625
100 seed weight	Luo et al. (2017)	16	4.12–26.82	Yuanza 9102 × Xuzhou 68–4
100 seed weight	Chen et al. (2017)	9[b]	5.24–10.24	Fuchuan Dahuasheng × ICG6375
Seed length	Huang et al. (2015)	3	9.86–10.48	Zhonghua 10 × ICG 12625
Seed length	Chen et al. (2016)	2	3.03–4.87	Xuhua 13 × Zhonghua 6

(continued)

Table 3.1 (continued)

Trait	Reference	QTL number	Phenotypic variance explained (%)	Population
Seed length	Chen et al. (2016)	8	5.66–20.8	Fuchuan Dahuasheng × ICG 6375
Seed length	Chen et al. (2017)	12[b]	4.04–13.21	Fuchuan Dahuasheng × ICG6375
Seed width	Huang et al. (2015)	4	6.39–12.2	Zhonghua 10 × ICG 12625
Seed width	Chen et al. (2016)	3	3.77–9.76	Xuhua 13 × Zhonghua 6
Seed width	Chen et al. (2016)	4	7.42–14.43	Fuchuan Dahuasheng × ICG 6375
Seed width	Chen et al. (2017)	10[b]	6.07–10.42	Fuchuan Dahuasheng × ICG6375
Sound mature kernels	Faye et al. (2015)	6	3.3–7.41	TAG 24 × ICGV 86031
Test weight	Hake et al. (2017)	1[a]	8.9	TMV 2 × TMV 2-NLM
Drought traits				
SPAD chlorophyll	Faye et al. (2015)	12	2.96–10.4	TAG 24 × ICGV 86031
Oil and protein				
Protein content	Hake et al. (2017)	1[a]	25.4	TMV 2 × TMV 2-NLM
Oil content	Pandey et al. (2014b)	9	3.93–14.07	Tifrunner × GT-C20
Oil content	Pandey et al. (2014b)	6	3.07–10.23	SunOleic 97R × NC94022
Oil content	Huang et al. (2015)	1	14.36	Zhonghua 10 × ICG 12625
Oil content	Wilson et al. (2017)	3	16–31	Florunner^2 // (TxAG-6 / Florunner BC_3)
Oil content	Hake et al. (2017)	1[a]	10.4	TMV 2 × TMV 2-NLM
Oil content	Shasidhar et al. (2017)	8	5.6–22.1	ICGV 07369 × ICGV 06420
Palmitic acid	Huang et al. (2015)	1	17.02	Zhonghua 10 × ICG 12625
Palmitic acid	Wang et al. (2015)	19	3.06–37.37	Tifrunner × GT-C20
Palmitic acid	Wang et al. (2015)	11	1.7–22.04	SunOleic 97R × NC94022
Palmitic acid	Hake et al. (2017)	1[a]	14.8	TMV 2 × TMV 2-NLM

(continued)

Table 3.1 (continued)

Trait	Reference	QTL number	Phenotypic variance explained (%)	Population
Palmitic acid	Shasidhar et al. (2017)	5	10.0–20.1	ICGV 06420 × SunOleic 95R
Stearic acid	Huang et al. (2015)	2	2.52–18.31	Zhonghua 10 × ICG 12625
Stearic acid	Wang et al. (2015)	15	2.63–40.57	Tifrunner × GT-C20
Stearic acid	Wang et al. (2015)	6	3.26–5.9	SunOleic 97R × NC94022
Stearic acid	Wilson et al. (2017)	3	12–28	Florunner[2] // (TxAG-6 / Florunner BC_3)
Stearic acid	Hake et al. (2017)	1[a]	12.2	TMV 2 × TMV 2-NLM
Stearic acid	Shasidhar et al. (2017)	1	78.6	ICGV 06420 × SunOleic 95R
Oleic acid	Huang et al. (2015)	1	1.72	Zhonghua 10 × ICG 12625
Oleic acid	Pandey et al. (2014b)	9	3.63–28.98	Tifrunner × GT-C20
Oleic acid	Pandey et al. (2014b)	8	1.59–27.54	SunOleic 97R × NC94022
Oleic acid	Wilson et al. (2017)	3	13–24	Florunner[2] // (TxAG-6 / Florunner BC_3)
Oleic acid	Hake et al. (2017)	1[a]	12.4	TMV 2 × TMV 2-NLM
Oleic acid	Shasidhar et al. (2017)	3	17.4–34.2	ICGV 06420 × SunOleic 95R
Linoleic acid	Pandey et al. (2014b)	9	3.91–25.49	Tifrunner × GT-C20
Linoleic acid	Pandey et al. (2014b)	7	1.46–28.22	SunOleic 97R × NC94022
Linoleic acid	Wilson et al. (2017)	2	17–26	Florunner[2] // (TxAG-6 / Florunner BC_3)
Linoleic acid	Hake et al. (2017)	1[a]	15.5	TMV 2 × TMV 2-NLM
Linoleic acid	Shasidhar et al. (2017)	3	12.1–41.0	ICGV 06420 × SunOleic 95R
Arachidic acid	Huang et al. (2015)	2	8.1–20.2	Zhonghua 10 × ICG 12625
Arachidic acid	Wang et al. (2015)	14	3.05–36.93	Tifrunner × GT-C20
Arachidic acid	Wang et al. (2015)	6	3.6–6.4	SunOleic 97R × NC94022
Arachidic acid	Wilson et al. (2017)	3	13–32	Florunner[2] // (TxAG-6 / Florunner BC_3)

(continued)

Table 3.1 (continued)

Trait	Reference	QTL number	Phenotypic variance explained (%)	Population
Arachidic acid	Shasidhar et al. (2017)	2	12.4–23.4	ICGV 06420 × SunOleic 95R
Eicosenoic acid	Huang et al. (2015)	3	3.8–7.51	Zhonghua 10 × ICG 12625
Eicosenoic acid	Wang et al. (2015)	20	2.98–15.11	Tifrunner × GT-C20
Eicosenoic acid	Wang et al. (2015)	7	2.55–8.77	SunOleic 97R × NC94022
Eicosenoic acid	Wilson et al. (2017)	1	15	Florunner^2 // (TxAG-6 / Florunner BC_3)
Behenic acid	Huang et al. (2015)	2	15.76–18.85	Zhonghua 10 × ICG 12625
Behenic acid	Wang et al. (2015)	16	4.74–13.56	Tifrunner × GT-C20
Behenic acid	Wang et al. (2015)	5	2.88–6.95	SunOleic 97R × NC94022
Behenic acid	Wilson et al. (2017)	1	12	Florunner^2 // (TxAG-6 / Florunner BC_3)
Behenic acid	Hake et al. (2017)	1[a]	4.5	TMV 2 × TMV 2-NLM
Behenic acid	Shasidhar et al. (2017)	2	8.4–12.0	ICGV 06420 × SunOleic 95R
Lignoceric acid	Wang et al. (2015)	13	3.85–12.61	Tifrunner × GT-C20
Lignoceric acid	Wang et al. (2015)	5	2.89–6.58	SunOleic 97R × NC94022
Lignoceric acid	Shasidhar et al. (2017)	5	10.3–13.4	ICGV 06420 × SunOleic 95R
O/L ratio	Pandey et al. (2014b)	5	5.7–14.9	Tifrunner × GT-C20
O/L ratio	Pandey et al. (2014b)	6	1.04–42.33	SunOleic 97R × NC94022
O/L ratio	Wilson et al. (2017)	3	12–35	Florunner^2 // (TxAG-6 / Florunner BC_3)
O/L ratio	Hake et al. (2017)	1[a]	10.8	TMV 2 × TMV 2-NLM

[a]Major-effect QTLs only
[b]Repeated QTLs only

governed by small-effect QTLs such as pod yield and drought tolerance, approaches such as marker-assisted recurrent selection and genome-wide selection based on transcript data are appropriate (Varshney and Dubey 2009). These breeding strategies will likely become commonplace in peanut improvement programs in the future and will be accelerated by the anticipated release of the *A. hypogaea* genome sequence.

Literature Cited

Belamkar V, Selvaraj MG, Ayers JL et al (2010) Identification of SSR markers for drought tolerance-related traits in the U.S. peanut minicore collection using association mapping. Int Cong Leg Genet Genomics

Bera SK, Kamdar JH, Kasundra SV et al (2016) A novel QTL governing resistance to stem rot disease caused by Sclerotium rolfsii in peanut. Australas Plant Pathol 45:637–644

Bernacchi D, Beck-Bunn T, Lopez EJ et al (1998) Advanced backcross QTL analysis in tomato I. Identification of QTLs for traits of agronomic importance from *Lycopersicon hirsutum*. Theor Appl Genet 97:381–397

Bertioli DJ, Cannon SB, Froenicke L et al (2016) The genome sequences of *Arachis duranensis* and *Arachis ipaensis*, the diploid ancestors of cultivated peanut. Nat Genet 48:438–446

Bertioli DJ, Leal-Bertioli SCM, Lion MB et al (2003) A large scale analysis of resistance gene homologues in *Arachis*. Mol Gen Genomics 270:34

Bi YP, Liu W, Xia H et al (2010) EST sequencing and gene expression profiling of cultivated peanut (*Arachis hypogaea* L.). Genome 53:832–839

Brasileiro AC, Morgante CV, Araujo ACG et al (2015) Transcriptome profiling of wild Arachis from water-limited environments uncovers drought tolerance candidate genes. Plant Mol Biol Rep 33:1876–1892

Blum A (2009) Effective use of water (EUW) and not water-use efficiency (WUE) is the target of crop yield improvement under drought stress. Field Crops Res 112:119–112

Burow MD, Simpson CE, Faries MW et al (2009) Molecular biogeography study of recently described B- and A-genome *Arachis* species, also providing new insights into the origins of cultivated peanut. Genome 52:107–119

Burow MD, Simpson CE, Starr JL et al (2001) Transmission genetics of chromatin from a synthetic amphidiploid to cultivated peanut (*Arachis hypogaea* L.): broadening the gene pool of a monophyletic polyploid species. Genetics 159:823–837

Burow MD, Starr JL, Park C-H et al (2014) Introgression of homologous quantitative trait loci (QTLs) for resistance to the root-knot nematode [*Meloidogyne arenaria* (Neal) Chitwood] in advanced backcross-QTL population of peanut (*Arachis hypogaea* L.). Mol Breeding 34(2):393–406

Chen W, Jiao Y, Cheng L et al (2016) Quantitative trait locus analysis for pod and kernel-related traits in cultivated peanut (*Arachis hypogaea* L.). BMC Genet 17:25

Chen Y, Ren X, Zheng Y et al (2017) Genetic mapping of yield traits using RIL population derived from Fuchuan Dahuasheng and ICG6375 of peanut. (*Arachis hypogaea* L) Mol Breed 37(2):17

Chenault KD, Maas AL, Damicone JP et al (2009) Discovery and characterization of a molecular marker for *Sclerotinia minor* (Jagger) resistance in peanut. Euphytica 166:357–365

Choi K, Burow MD, Church G et al (1999) Genetics and mechanism of resistance to *Meloidogyne arenaria* in peanut germplasm. J Nematol 31:283–290

Chopra R, Burow GB, Farmer A, Mudge JM, Simpson CE, Burow MD (2014) Comparisons of de novo transcriptome assemblers in diploid and polyploid species using peanut (*Arachis* spp.) RNA-Seq data. PLoS One 9(12):e115055

Chopra R, Burow G, Simpson CE, Chagoya J, Mudge J, Burow MD (2016) Transcriptome sequencing of diverse Peanut (Arachis) wild species and the cultivated species reveals a wealth of untapped genetic variability. G3-Genes Genomes Genetics 6(12):3825–3836

Chopra R, Simpson CE, Hillhouse A et al (2018) SNP genotyping reveals major QTLs for plant architectural traits between A-genome peanut wild species. Mol Genet Genomics. https://doi.org/10.1007/s00438-018-1472-z [Epub ahead of print]

Chu Y, Holbrook CC, Timper P, Ozias-Akins P, (2007) Development of a PCR-Based Molecular Marker to Select for Nematode Resistance in Peanut. Crop Science 47(2):841

Clevenger J, Chu Y, Chavarro C et al (2017a) Genome-wide SNP genotyping resolves signatures of selection and tetrasomic recombination in peanut. Mol Plant 10:309–322

Clevenger J, Chu Y, Guimaraes LA et al (2017b) Gene expression profiling describes the genetic regulation of *Meloidogyne arenaria* resistance in *Arachis hypogaea* and reveals a candidate gene for resistance. Sci Rep 7:1317

Cook DE, Lee TG, Guo X (2012) Copy number variation of multiple genes at *Rgh1* mediates nematode resistance in soybean. Science 338:1206–1209

Da Chen, Xin Liu, Chunyu Li, Wei Tian, Qirong Shen, Biao Shen (2014) Isolation of Bacillus amyloliquefaciens S20 and its application in control of eggplant bacterial wilt. Journal of Environmental Management 137:120–127

Ding H, Zhang ZM, Qin FF et al (2014) Isolation and characterization of drought-responsive genes from peanut roots by suppression subtractive hybridization. Electron J Biotechnol 17:304. https://doi.org/10.1016/j.ejbt.2014.09.004

Doyle JJ, Luckow MA (2003) The rest of the iceberg. Legume diversity and evolution in a phylogenetic context. Plant Physiol 131:900–910

FAO (2014) Food and Agriculture Organization of the United Nations. http://www.fao.org/faostat/en/#data/QC

Faye I, Pandey MK, Hamidou F et al (2015) Identification of quantitative trait loci for yield related traits in groundnut (*Arachis hypogaea* L.) under different water regimes in Niger and Senegal. Euphytica 206:631–647

Fonceka D, Tossim H-A, Rivallan R et al (2009) Genetic mapping of wild introgressions into cultivated peanut: a way toward enlarging the genetic basis of a recent allotetraploid. BMC Plant Biol 9:103

Fonceka D, Tossim H-A, Rivallan R et al (2012a) Fostered and left behind alleles in peanut: interspecific QTL mapping reveals footprints of domestication and useful natural variation for breeding. BMC Plant Biol 12:26

Fonceka D, Tossim H-A, Rivallan R et al (2012b) Construction of chromosome segment substitution lines in peanut (*Arachis hypogaea* L.) using a wild synthetic and QTL mapping for plant morphology. PLoS One 7(11):e48642

Garcia GM, Stalker HT, Shroeder E et al (1996) Identification of RAPD, SCAR, and RFLP markers tightly linked to nematode resistance genes introgressed from *Arachis cardenasii* into *Arachis hypogaea*. Genome 39:836–845

Gautami B, Fonceka D, Pandey MK et al (2012a) An international reference consensus genetic map with 897 marker loci based on 11 mapping populations for tetraploid groundnut (*Arachis hypogaea* L.). PLoS One 7(7):e41213

Gautami B, Pandey MK, Vadez V et al (2012b) Quantitative trait locus analysis and construction of consensus genetic map for drought tolerance traits based on three recombinant inbred line populations in cultivated groundnut (*Arachis hypogaea* L.). Mol Breed 30:757–772

Guimaraes PM, Brasileiro AC, Morgante CV et al (2012) Global transcriptome analysis of two wild relatives of peanut under drought and fungi infection. BMC Genomics 13:387

Guimaraes PM, Guimaraes LA, Morgante CV et al (2015) Root transcriptome analysis of wild peanut reveals candidate genes for nematode resistance. PLoS ONE 10(10):e0140937

Guo B, Chen X, Dang P et al (2008) Peanut gene expression profiling in developing seeds at different reproduction stages during *Aspergillus parasiticus* infection. BMC Dev Biol 8:12

Guo B, Chen X, Hong Y et al (2009) Analysis of gene expression profiles in leaf tissues of cultivated peanuts and development of EST-SSR markers and gene discovery. Int J Plant Genomics 2009:1. https://doi.org/10.1155/2009/715605

Guo Y, Khanal S, Tang S et al (2012) Comparative mapping in intraspecific populations uncovers a high degree of macrosynteny between A- and B-genome diploid species of peanut. BMC Genomics 13:608

Hake AN, Shirasawa K, Yadawad A et al (2017) Mapping of important taxonomic and productivity traits using genic and non-genic transposable element markers in peanut (*Arachis hypogaea* L.). PLoS One 12(10):e0186113

Halward TM, Stalker HT, Kochert G (1993) Development of an RFLP linkage map in diploid peanut species. Theor Appl Genet 87:379–384

Holbrook CC, Kvien CK, Ruker KS et al (2000) Pre-harvest aflatoxin contamination in drought-tolerant and drought-intolerant peanut genotypes. Peanut Sci 27:45–48

Holbrook CC, Timper P, Culbreath AK et al (2008a) Registration of 'Tifguard' peanut. J Plant Reg 2:92–94

Holbrook CC, Timper P, Dong W et al (2008b) Development of near-isogenic peanut lines with and without resistance to the peanut root-knot nematode. Crop Sci 48:194–198

Hong Y, Chen X, Liang X et al (2010) A SSR-based composite genetic linkage map for the cultivated peanut (*Arachis hypogaea* L.) genome. BMC Plant Biol 10:17

Huang L, He H, Chen W et al (2015) Quantitative trait locus analysis of agronomic and quality-related traits in cultivated peanut (*Arachis hypogaea* L.). Theor Appl Genet 128:1103–1115

Janila P, Pandey MK, Shasidhar Y et al (2016) Molecular breeding for introgression of fatty acid desaturase mutant alleles (ahFAD2A and ah FAD2B) enhances oil quality in high and low oil containing peanut genotypes. Plant Sci 242:203–221

Jiang H, Huang L, Ren X (2014) Diversity characterization and association analysis of agronomic traits in a Chinese peanut (*Arachis hypogaea* L.) mini-core collection. J Integr Plant Biol 56:159–169

Jung S, Powell G, Moore K et al (2000) The high oleate trait in the cultivated peanut [*Arachis hypogaea* L.] II. Molecular basis and genetics of the trait. Mol Gen Genet 263:806–811

Khedikar YP, Gowda MVC, Sarvamangala C et al (2010) A QTL study on late leaf spot and rust revealed one major QTL for molecular breeding for rust resistance in groundnut (*Arachis hypogaea* L.). Theor Appl Genet 121:971–984

Khera P, Pandey MK, Wang H et al (2016) Mapping quantitative trait loci resistance to Tomato spotted wilt virus and leaf spots in a recombinant inbred line population of peanut (Arachis hypogaea L.) from SunOleic 97R and NC94022. PLoS ONE 11(7): e0158452

Knauft DA, Ozias-Akins P (1995) Recent methodologies for germplasm enhancement and breeding. In: Pattee HE, Stalker HT (eds) Advances in peanut science. American Peanut Research and Education Society, Stillwater, pp 54–94

Kochert G, Halward T, Branch W et al (1991) RFLP variation in peanuts (*Arachis hypogaea* L.) cultivars and wild species. Theor Appl Genet 81:565–570

Kochert G, Stalker HT, Gimenes M et al (1996) RFLP and cytogenetic evidence on the origin and evolution of allotetraploid domesticated peanut, *Arachis hypogaea* (Leguminosae). Am J Bot 83:1282–1291

Korani WA, Chu Y, Holbrook C et al (2017) Genotypic regulation of aflatoxin accumulation but not Aspergillus fungal growth upon post-harvest infection of peanut (*Arachis hypogaea* L.) seeds. Toxins 9:218

Krapovickas A, Fernandez A, Seeligmann P (1974) Recuperacion de la fertilidaden un hiberido interspecifico esteril de *Arachis* (Leguminosae). Bonplandia Corrientes 3:129–142

Krapovickas A, Gregory WC (1994) Taxonomía del género *Arachis* (Leguminosae). Bonplandia 8:1–186

Lavia GI (1996) Estudios cromosomicos en *Arachis* (Leguminosae). Bonplandia 9:111–120

Leal-Bertioli SCM, Bertioli DJ, Guimaraes PM et al (2012) The effect of tetraploidization of wild *Arachis* on leaf morphology and other drought-related traits. Environ Exp Bot 84:17–24

Leal-Bertioli SCM, Cavalcante U, Gouveia EG et al (2015) Identification of QTLs for rust resistance in the peanut wild species *Arachis magna* and the development of KASP markers for marker assisted selection. G3 (Bethesda) 5(7):1403–1413

Leal-Bertioli SCM, Jose ACVF, Alves-Freitas DMT et al (2009) Identification of candidate genome regions controlling disease resistance in Arachis. BMC Plant Biol 9:112–123

Li Y, Li L, Zhang X et al (2017) QTL mapping and marker analysis of main stem height and first lateral branch length in peanut (Arachis hypogaea L.). Euphytica 213:57

Liang Y, Baring MR, Wang S et al (2017) Mapping QTLs for leafspot resistance in peanut using SNP-based next-generation sequencing markers. Plant Breed Biotech 5(2):115–122

Liu Z, Feng S, Pandey MK et al (2013) Identification of expressed resistance gene analogs from peanut (*Arachis hypogaea* L.) expressed sequence tags. J Integr Plant Biol 55:453–461

Liu S, Kandoth PK, Warren SD et al (2012) A soybean cyst nematode resistance gene points to a new mechanism of plant resistance to pathogens. Nature 492:256–260

Liu S, Zhou R, Dong Y et al (2006) Development, utilization of introgression lines using a synthetic wheat as donor. Theor Appl Genet 112:1360–1373

Lu Q, Li H, Hong Y et al (2018) Genome sequencing and analysis of the peanut B-genome progenitor (*Arachis ipaensis*). Front Plant Sci 9:604. https://doi.org/10.3389/fpls.2018.00604

Luo H, Ren X, Li Z et al (2017) Co-localization of major quantitative trait loci for pod size and weight to a 3.7 cM interval on chromosome A05 in cultivated peanut (*Arachis hypogaea* L.). BMC Genomics 18:58

Lynch RE, Wilson DM (1991) Enhanced infection of peanut, *Arachis hypogaea* L., seeds with *Aspergillus flavus* group fungi due to external scarification of peanut pods by the lesser cornstalk borer, Elasmopalpuslignosellus (Zeller). Peanut Sci 18:110–116

Moretzsohn MC, Barbosa AVG, Alves-Freitas DMT et al (2009) A linkage map for the B genome of *Arachis(Fabaceae)* and its synteny to the A-genome. BMC Plant Biol 9:40

Moretzsohn MC, Leoi L, Proite K et al (2005) Microsatellite-based gene-rich linkage map for the AA genome of *Arachis* (Fabaceae). Theor Appl Genet 111:1060–1071

Milla S, Isleib TG, Stalker HT (2005) Taxonomic relationships among *Arachis* sect. *Arachis* species as revealed by AFLP markers. Genome 48:1–117

Milligan SB, Bodeau J, Yaghoobi J et al (1998) The root knot nematode resistance gene *Mi* from tomato is a member of the leucine-rich repeat family of plant genes. Plant Cell 10:1307–1320

Moore KM, Knauft DA (1989) The inheritance of high oleic acid in peanut. J Hered 80:252–253

Mondal S, Hadapad AB, Hande PA et al (2014a) Identification of quantitative trait loci for bruchid (*Caryedon serratus* Olivier) resistance components in cultivated groundnut (*Arachis hypogaea* L.). Mol Breed 33:961–973

Mondal S, Phadke RR, Bradigannavar AM (2014b) Identification of transposable element markers for a rust (*Puccinia arachidis* Speg.) resistance gene in cultivated peanut. J Phytopathol 162:548–552

Nagy ED, Chu Y, Guo Y et al (2010) Recombination is suppressed in an alien introgression in peanut harboring *Rma*, a dominant root-knot nematode gene. Mol Breed 26:357–370

Nagy ED, Guo Y, Tang S, Bowers JE, Okashah RA, Taylor CA, Zhang D, Khanal S, Heesacker AF, Khalilian N, Farmer AD, Carrasquilla-Garcia N, Penmetsa RV, Cook D, Stalker HT, Nielsen N, Ozias-Akins P, Knapp SJ, (2012) A high-density genetic map of Arachis duranensis, a diploid ancestor of cultivated peanut. BMC Genomics 13(1):469

Nigam SN, Aruna R (2008) Improving breeding efficiency for early maturity in peanut. Plant Breed Rev 30:295–322

Nigam SM, Waliyar F, Aruna R et al (2009) Breeding for resistance to aflatoxin contamination at ICRISAT. Peanut Sci 36:42–49

Nwokolo E (1996) Peanut (*Arachis hypogaea* L.). In: Nwokolo E, Smartt J (eds) Food and feed from legumes and oilseeds. Chapman & Hall, London, pp 49–61

Pandey MK, Khan AW, Singh VK et al (2017a) QTL-seq approach identified genomic regions and diagnostic markers for rust and late leaf spot resistance in groundnut (*Arachis hypogaea* L.). Plant Biotechnol J 15:927–941

Pandey MK, Wang H, Khera P et al (2017b) Genetic dissection of novel QTLs for resistance to leaf spots and tomato spotted wilt virus in peanut (*Arachis hypogaea* L.). Front Plant Sci 8:25

Pandey MK, Upadhyaya HD, Rathore A et al (2014a) Genome wide association studies for 50 agronomic traits in peanut using the 'reference set' comprising 300 genotypes from 48 countries of the semi-arid tropics of the world. PLoS One 9:e105228

Pandey MK, Wang ML, Qiao L et al (2014b) Identification of QTLs associated with oil content and mapping FAD2 genes and their relative contribution to oil quality in peanut (*Arachis hypogaea* L.). BMC Genet 15:133

Peng Z, Liu F, Wang L et al (2017) Transcriptome profiles reveal gene regulation of peanut (*Arachis hypogaea* L.) nodulation. Sci Rep 7:40066

Passioura JB (1996) Drought and drought tolerance. Plant Growth Regul 20:79–83

Pimratch S, Jogloy S, Toomsan B et al (2004) Evaluation of seven peanut genotypes for nitrogen fixation and agronomic traits. Songklanakarin J SciTechnol 2:295–304

Qin H, Feng S, Chen C et al (2012) An integrated genetic linkage map of cultivated peanut (*Arachis hypogaea* L.) constructed from two RIL populations. Theor Appl Genet 124:653–664

Ratnaparkhe MB, Wang X, Li J (2011) Comparative analysis of peanut NBS-LRR gene clusters suggests evolutionary innovation among duplicated domains and erosion of gene microsynteny. New Phytol 192:164–178

Ravi K, Vadez V, Isobe S (2011) Identification of several small main-effect QTLs and a large number of epistatic QTLs for drought tolerance related traits in groundnut (*Arachis hypogaea* L.). Theor Appl Genet 122:1119–1132

Robledo G, Lavia GI, Seijo G (2009) Species relations among wild *Arachis* species with the A genome as revealed by FISH mapping of rDNA loci and heterochromatin detection. Theor Appl Genet 118:1295–1307

Robledo G, Seijo JG (2008) Characterization of *Arachis* D genome by FISH chromosome markers and total genome DNA hybridization. Genet Mol Biol 31:717–724

Sarvamangala C, Gowda MVC, Varshney RK (2011) Identification of quantitative trait loci for protein content, oil content and oil quality for groundnut (Arachis hypogaea L.). Field Crops Research 122(1):49–59

Selvaraj MG, Belamkar V, Ayers JL et al (2010) Variability for drought resistance traits in U.S. minicore collection of peanut. In: Plant breeding for drought adaptation symposium. Colorado State University, CO, June 2010

Sharma S, Upadhyaya HD, Varshney RK et al (2013) Pre-breeding for diversification of primary gene pool and genetic enhancement of grain legumes. Front Plant Sci 20:309

Shasidhar Y, Vishwakarma MK, Pandy MK et al (2017) Molecular mapping of oil content and fatty acids using dense genetic maps in groundnut (*Arachis hypogaea* L.). Front Plant Sci 8:794

Shirasawa K, Bertioli DJ, Varshney RK et al (2013) Integrated consensus map of cultivated peanut and wild relatives reveals structures of the A and B genomes of *Arachis* and divergence of legume genomes. DNA Res 20:173–174

Shirasawa K, Koilkonda P, Aoki K et al (2012) In silico polymorphism analysis for the development of simple sequence repeat and transposon markers and construction of linkage map in cultivated peanut. BMC Plant Biol 12:80

Simpson CE (1991) Pathways for introgression of pest resistance into Arachis hypogaea L. Peanut Sci 18:22–26

Simpson CE, Starr JL (2001) Registration of 'COAN' peanut. Crop Sci 41:918

Simpson CE, Starr JL, Baring MR et al (2013) Registration of 'Webb' Peanut. Journal of Plant Research. 7:265–268

Singh AK, Simpson CE (1994) Biosystematics and genetic resources. In: Smartt J. (eds) The Groundnut Crop. World Crop Series. Springer, Dordrecht

Simpson CE, Starr JL, Church GT (2003) Registration of 'NemaTAM' peanut. Crop Sci 43:1561

Smartt L, Stalker HT (1982) Speciation and cytogenetics in Arachis. In: Pattee HE, Young ET (eds) Peanut science and technology. American Peanut Research and Education Society, Yoakum, pp 21–49

Stalker HT (1991) A new species in section Arachis of peanuts with a D genome. Am J Bot 78:630–637

Stalker HT (2017) Utilizing wild species for peanut improvement. Crop Sci 57:1102–1120

Subrahmanyam P, McDonald D, Reddy U et al (1993) Origin and utilization of rust resistance in groundnut. In: Jacobs T, Parlevliet JE (eds) Durability of disease resistance. Kluwer Academic Publishers, Dordrecht, pp 147–158

Tallury SP, Hilu KW, Milla SR et al (2005) Genomic affinities in *Arachis* section *Arachis* (Fabaceae): molecular and cytogenetic evidence. Theor Appl Genet 111:1229–1237

Tallury SP, Hollowell JE, Isleib TG et al (2014) Greenhouse evaluation of section *Arachis* wild species for Sclerotinia blight and Cylindrocladium black rot resistance. Peanut Sci 41:17–24

Tanksley SD, Nelson JC (1996) Advance backcross QTL analysis: a method for the transfer of valuable QTLs from unadapted germplasm into elite breeding lines. Theor Appl Genet 92:191–203

Thakur RP, Rao VP, Reddy SV et al (2000) Evaluation of wild *Arachis* germplasm accessions for in vitro seed colonization and aflatoxin production by *Aspergillus flavus*. Int Arachis Newsl 2000:44–46

Tian F, Li DJ, Fu Q et al (2006) Construction of introgression lines carrying wild rice (*Oryza rufipogon* Griff.) segments in cultivated rice (*O. sativa* L.) background and characterization of introgressed segments associated with yield-related traits. Theor Appl Genet 112:570–580

Timper P, Wilson DM, Holbrook CC et al (2004) Relationship between *Meloidogyne arenaria* and aflatoxin contamination in peanut. J Nematol 36:167–170

Upadhyaya HD (2005) Variability in drought resistance related traits in the mini core collection of peanut. Crop Sci 45:1432–1440

Valls JFM, Simpson CE (2005) New species of Arachis from Brazil, Paraguay, and Bolivia. Bonplandia 14(1&2):35–64

Varshney RK, Dubey A (2009) Novel genetic tools and modern genetic breeding approaches for crop improvement. J Plant Biochem Biotechnol 182:127–138

Varshney RK, Bertioli DJ, Moretzsohn MC, Vadez V, Krishnamurthy L, Aruna R, Nigam SN, Moss BJ, Seetha K, Ravi K, He G, Knapp SJ, Hoisington DA (2009) The first SSR-based genetic linkage map for cultivated groundnut (Arachis hypogaea L.). Theoretical and Applied Genetics 118 (4):729–739

Varshney RK, Pandey MK, Janila P et al (2014) Marker-assisted introgression of a QTL region to improve rust resistance in three elite and popular varieties of peanut (*Arachis hypogaea* L.). Theor Appl Genet 127:1771–1781

Wang ML, Khera P, Pandey MK et al (2015) Genetic mapping of QTLs controlling fatty acids provided insights into the genetic control of fatty acid synthase pathway in peanut (*Arachis hypogaea* L.). PLoS One 10(4):e0119454

Wang H, Penmetsa RP, Yuan M (2012) Development and characterization of BAC-end sequence derived SSRs, and their incorporation into a new higher density genetic map for cultivated peanut (*Arachis hypogaea* L.). BMC Plant Biol 12:10

Wilson JN, Chopra R, Baring MR et al (2017) Advanced backcross quantitative trait loci (QTL) analysis of oil concentration and oil quality traits in peanut (*Arachis hypogaea* L.). Tropical Plant Biol 10:1. https://doi.org/10.1007/s12042-016-9-180-5

Wynne JC, Rowe RC, Beute MK (1975) Resistance of peanut genotypes to *Cylindrocladium crotalariae*. Peanut Sci 2:54–56

Xue HQ, Isleib TG, Payne GA et al (2004) Evaluation of post-harvest aflatoxin production in peanut germplasm with resistance to seed colonization and pre-harvest aflatoxin contamination. Peanut Sci 31:124–134

Yuksel B, Estill JC, Schulze SR et al (2005) Organization and evolution of resistance gene analogs in peanut. Mol Gen Genomics 274:248–263

Zhang J, Zheng HG, Aarti A, Pantuwan G, Nguyen TT, Tripathy JN, Sarial AK, Robin S, Babu RC, Nguyen BD, Sarkarung S, Blum A, Nguyen HT (2001) Locating genomic regions associated with components of drought resistance in rice: comparative mapping within and across species. TAG Theoretical and Applied Genetics 103(1):19–29

Zhanji Liu, Suping Feng, Manish K. Pandey, Xiaoping Chen, Albert K. Culbreath, Rajeev K. Varshney, Baozhu Guo (2013) Identification of Expressed Resistance Gene Analogs from Peanut (L.) Expressed Sequence Tags. Journal of Integrative Plant Biology 55(5):453–461

Zhao J, Huang L, Ren X et al (2017) Genetic variation and association mapping of seed-related traits in cultivated peanut (*Arachis hypogaea* L.) using single-locus simple sequence repeat markers. Front Plant Sci 8:2105

Chapter 4
Response of Pulses to Drought and Salinity Stress Response: A Physiological Perspective

Titash Dutta, Nageswara Rao Reddy Neelapu, Shabir H. Wani, and Surekha Challa

4.1 Introduction

Cereals, legumes, and tubers (root) are the major classes of crops that give rise to a healthy balanced diet. Legumes are a rich source of dietary proteins and have high nutritional value (Table 4.1). Legumes account for 27% of the global crop production and are ranked third after cereal and oilseed productions (Ashraf et al. 2010; Kudapa et al. 2013) For centuries, legumes have been an integral food crop in developing countries and are often called "poor man's meat." Moreover leguminous crops such as soybean and groundnut account for 35% of the global processed vegetable oil (Sharma et al. 2010; Mantri et al. 2013). The other miscellaneous uses of legumes include fodder for animals and green manure (as they are involved in nitrogen fixation). Legumes are being considered worldwide as a sustainable food source in the near future with the United Nations declaring 2016 as the "International Year of Pulses" (FAO 2016).

Taxonomically, legumes fall under the Fabaceae/Leguminosae family comprising over 18,000 species. The dry seeds obtained after harvesting the legumes are termed as pulses. Legumes are classified into two groups based on their ability to thrive in different climate seasons. The first group includes the legumes that grow in the cool season and are termed as cool season food legumes. They include broad bean (*Vicia faba*), lentil (*Lens culinaris*), lupins (*Lupinus* spp.), dry pea (*Pisum sativum*), chickpea (*Cicer arietinum*), and grass pea (*Lathyrus sativus*) (Toker and Yadav 2010). These are grown evenly in all continents excluding Antarctica.

T. Dutta · N. R. R. Neelapu · S. Challa (✉)
Department of Biochemistry and Bioinformatics, Institute of Science, GITAM (Deemed to be University), Visakhapatnam, AP, India
e-mail: surekha.challa@gitam.edu

S. H. Wani
Mountain Research Centre for Field Crops, Khudwani, Sher-e-Kashmir University of Agricultural Sciences and Technology, Srinagar, Jammu and Kashmir, India

S. H. Wani, M. Jain (eds.), *Pulse Improvement*,
https://doi.org/10.1007/978-3-030-01743-9_4

Table 4.1 Nutritional value of pulse

Constituents	Values (per 100 g dry seeds)
Protein (g)	>20–%
Carbohydrate (g)	55–60%
Fat (g)	>1.0%
Fiber (%)	3.2%
Iron (mg)	7–10 mg
Vitamin C (mg)	10–15 mg
Phosphorous	300–500 mg
Calcium (mg)	69–75 mg
Vitamin A	430–489 IU
Energy (kCal)	345–600

The second group is called the tropical season food legumes. These legumes require hot and humid climatic conditions for their growth. This group includes pigeon pea (*Cajanus cajan*), cowpea (*Vigna unguiculata*), soybean (*Glycine max* L.), mung bean (*Vigna radiata* var. *radiata*), and urd bean (*Vigna mungo*) (FAOSTAT 2009; Andrews and Hodge 2010).

The global legume production between 2011 and 2013 was 72.3 million metric tons of grains produced from 80.3 million hectares of crop area and increased to 79 million metric tons in 201 (Joshi and Rao 2017). Dry beans is the major legume grown, accounting for 32% of the global legume production. It is followed by chickpea (17%), dry peas (14.6%), cowpea (8.9%), lentils (6.5%), pigeon pea (6.2%), and broad beans (5.8%). In India, Madhya Pradesh is the largest producer of legumes (20.3%) followed by Maharashtra (13.8%), Rajasthan (16.4%), Uttar Pradesh (9.5%), Karnataka (9.3%), Andhra Pradesh (7.9%), Chhattisgarh (3.8%), Bihar (2.6%), and Tamil Nadu (2.9%) (Singh et al. 2015). India alone accounts for 19% of the world production behind global leader China (Li et al. 2017).

Despite the exceptional nutritional importance and wide application of legumes for human race, the productivity of legumes is significantly affected by abiotic stress conditions in the semiarid tropic (SAT) regions. The SAT region is the major producer of grain legumes and spreads across 55 developing countries with a population count of 1.4 billion (Bray et al. 2000; Varshney et al. 2013)

Abiotic stresses such as salinity, drought, and extreme temperatures lead to significant loss in global crop production (Farooq et al. 2017). According to FAO 2010, drought and salinity affects 60 and 10.5 million km^2 area, respectively, and are collectively responsible for approximately 70% loss in annual crop yield (Wild 2003). It is also reported that 90% of the global arable land is under the effect of one or more abiotic stresses and crop cultivation on such lands leads to production of crops more vulnerable to biotic stress factors (Rao et al. 2015; Surekha et al. 2013; Dita et al. 2006). These factors affect plant growth and minimizes overall seed quality which subsequently leads to crop yield reduction (Cattivelli et al. 2008; Surekha et al. 2014a; Hussein et al. 2017). In addition, changes in general climate pattern, sporadic rainfall along with loss of agricultural land predominantly due to salinity and drought stress conditions threatens global food security. Significant crop yield reduction has been reported in major cereal crops (barley, rice, wheat, maize, and wheat) as well as in legumes (chick pea, groundnut, mung bean, millet, pigeon pea, common bean, etc.) (Jaleel et al. 2009; Farooq et al. 2017).

Plant response to abiotic stresses such as drought, salinity, extreme temperatures, and oxidative stress are often correlated, inducing cellular damage (Surekha et al. 2015). The effects of salinity and drought are expressed by a series of morphological, physiological, metabolic, and molecular changes in plants (Abdelrahman et al. 2018) (Fig. 4.1). Hence it is necessary to develop superior salt- and drought-tolerant cultivars of these legumes with improved seed quality and yield. Moreover, it is important to elucidate the molecular basis of abiotic stress tolerance based on these morphological and physiological traits. In this review, we discuss the salinity and drought stress response in legumes and identify the various physiological traits such as plant-water relations and transpiration efficiency, photosynthesis and stomatal conductance, ion concentrations, yield parameters, etc. in response to the stress conditions. These physiological traits have been identified across major economically important legumes and can be utilized for quantifying plant growth and survival status when subjected to salinity and drought and can also help in developing cultivars with multi-stress tolerance.

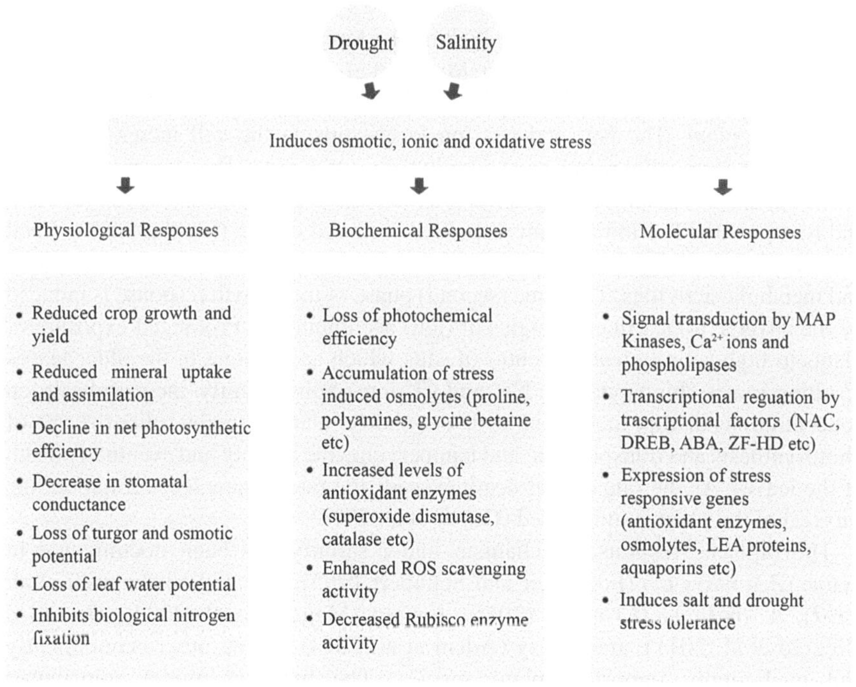

Fig. 4.1 Physiological, biochemical, and molecular responses in plants under salt and drought stress

4.2 Overview of Drought and Salinity

Drought and salinity leads to significant loss in crop production, thereby affecting the world food security index. Geographically 70% of the earth's surface is covered by saline water. The major ions that contribute to salinity include Na^+, Ca^{2+}, Mg^{2+}, K^+, Cl^-, SO_4^{2-}, HCO_3^-, CO_3^{2-}, and NO_3^- (Flowers et al. 2000). Salination is referred to excess accumulation of salts in soil or water. The factors favoring salination include mineral weathering, dust, precipitation, and inward movement of salt toward land surface along with groundwater.

Drought on the other hand is a condition when plants are subjected to inadequate water supply lower than the optimum level essential for growth and survival. Less rainfall, irregular rainfall pattern, and wrong irrigation practices contribute toward drought (Tuberosa and Salvi 2006). When plants are exposed to prolonged drought, it leads to wilting and drop in photosynthetic activities in the water-stressed plants and finally resulting in plant death.

4.2.1 Salt Tolerance: Response Mechanism and Physiological Traits used in Tolerance Quantification

Salt tolerance response is best described by the biphasic response mechanism proposed by Munns and Termaat (1986), Munns et al. (1995), and Munns (2002). The osmotic (first) phase of growth response initiates as high levels of salt accumulates outside the plant. The Na^+ and Cl^- ion levels outside the cell increase rapidly, whereas the intracellular ion concentration does not vary. This lowers the water intake capacity of plants leading to water deficit and is characterized by reduced leaf and root growth. The initial plant responses observed during this phase is stomatal closure. Under drought stress, plants exhibit similar responses in terms of cellular and metabolic activities. The ionic (second) phase of the growth response is induced by the adverse toxic effect of high salt (ion) accumulation. Prolonged exposure of plants to high salinity leads to entry of salts which accumulate in the older leaves resulting in very high cytosolic Na^+ and Cl^- ions. Subsequently, increased salt ion concentrations disrupt membrane structure and cellular organelles, alter the rate of photosynthesis and transpiration, and hampers enzyme activity and eventually death of the leaves. As the rate of leaf death exceeds the rate of new leaf formation, the survival of the plant is minimized (Gilroy et al. 2014).

The biphasic response mechanism under salinity has been documented in maize (*Zea mays* L.) (Fortmeier and Schubert 1995), wheat (Wakabayashi et al. 1997), *A. thaliana* (Li et al. (2015), tomato (Maggio et al. 2007), *O. sativa* (Negrão et al. 2011), and barley (Adem et al. 2014) among other economically and medicinally important plant species. The biphasic model contributed immensely in understanding the possible cross-talks surrounding all abiotic stresses. Moreover it highlighted that gene expression levels also altered in response to abiotic stress conditions (Wani et al. 2017). Adem et al. (2014) compared three salt-stressed barley cultivars with their wild-type counterparts. They observed rapid initial osmotic stress responses that lead to three- to fivefold decrease in its biomass, leading to inhibition in plant growth. The second phase is triggered when the plants were exposed to 4 weeks of saline stress. All barley cultivars exhibited low chlorophyll content, high oxidative stress, and elevated levels of cellular Na^+ accumulation.

Plants evolved three vital strategies that help them overcome biphasic growth inhibition and survive in saline conditions. Plants use ion exclusion, compartmentalization, and osmoprotection to maintain optimum cellular ionic concentrations: Ion exclusion strategy regulates the overall exclusion of toxic ions accumulated via ion exchangers (NHX). Compartmentalization of Na^+ ions into vacuoles improves tissue tolerance and is accomplished by major ion channels (V-PPases, V-ATPases, Na^+/K^+ antiporters). The third strategy involves cellular osmotic adjustment and osmoprotection by upregulating the expression of candidate genes involved in the synthesis of osmolytes, aquaporins, and antioxidant enzymes that contribute to salt tolerance (Dutta et al. 2018a, b; Roychoudhury and Chakraborty 2013).

The physiological traits or parameters that contribute to salinity tolerance and used for quantifying the degree of stress response include the maintenance of plant relative water content (RWC), transpiration (T) and transpiration use efficiency (TUE) (Harris et al. 2010; This et al. 2010; Barbieri et al. 2012), leaf area (Maggio et al. 2007), seed germination (Foolad and Lin 1997), production of antioxidants (Neelapu et al. 2015; Ashraf 2009), early seedling growth (Kingsbury and Epstein 1984), and harvest index (HI) (Gholizadeh et al. 2014). Shoot and root fresh and dry matter, root and shoot length, relative growth rate, net assimilation rate, relative water content, and water use efficiency are the other physiological traits that reflect plant growth and productivity.

4.2.2 Drought Tolerance: Response Mechanism and Physiological Traits Used in Tolerance Quantification

The initial response of plants to drought stress is similar to that observed during the osmotic phase under salinity. Plants experience water deficit conditions that affect the metabolic activities of plants. Drought leads to cellular damage by altering membrane lipid composition and membrane fluidity. The onset of water deficit triggers the signaling cascades responsible for synthesis of gene transcripts associated with protection and chaperone function. Plants generally wilt under water-deficit conditions caused due to various physiological responses such as reduction in turgor pressure, low gaseous exchange, mineral assimilation, and growth inhibition. Reduction in turgor pressure reduces the rate of stomatal CO_2 diffusion as well as from mesophyll cells causing stomatal closure. The rate of photosynthesis is also lowered due to limited CO_2 supply to RUBISCO enzyme, which inhibits the conversion of inorganic carbon dioxide into its organic forms. Prolonged exposure to drought leads to loss of functional activity of photosystem II, thereby decreasing photosynthetic activity and inactivation of RUBISCO enzyme (Nishiyama and Murata 2014; Pandey et al. 2015).

Plants are equipped with drought tolerance mechanisms such as canopy resistance and reduction in leaf area. Both these mechanisms induce tolerance by reducing excessive absorption of indecent light as a result of reducing the surface area exposed to the light rays and radiations. Stomatal closure is the primary plant response under the regulation of chemical (ABA) and hydraulic signals (Tardieu et al. 2010). Moreover it leads to reduction in stomatal exchange, CO_2 diffusion as well as energy consumption rate.

The physiological traits used for evaluating drought stress include root trait characteristics (root length, root density, root biomass, root length density (Serraj et al. 2004), delayed canopy wilting (DCW) and leaf pubescence density (LPD) as studied in soybean (Charlson et al. 2009; Du et al. 2009), delayed leaf senescence (DLS) trait in cowpea (Hall et al. 2002), and recovery ability after wilting (RAW) in chickpea (Toker et al. 2007)). Apart from these measurement of stomatal conductance, chlorophyll concentrations and use of carbon isotope discrimination are also effective screening methods for drought stress tolerance and has been used for some food legumes (Stoddard et al. 2006; Khan et al. 2010a; Swathi et al. 2017).

4.3 Effects of Drought and Salinity on Pulses

Abiotic stress predominantly drought and salinity affects plant growth and survival. They induce significant morphological and molecular changes that are observed during the plant developmental phases. The intensity and duration of these stress factors determine the major physiological changes taking place in the affected plants (Daryanto et al. 2017). Drought and salinity severely affects crop productivity, plant-water relations, photosynthetic capacity, and nitrogen fixation capacity. Therefore valuable insights of the physiological processes occurring during drought and saline stress are essential for mitigating the detrimental effects either by improving existing crop management practices or by alteration of regional cropping patterns. In this section, we highlight the various drought and saline stress effects on the major pulses from the physiological point of view. Tables 4.2, 4.3, 4.4, and 4.5 contain the list of major pulses and their physiological responses in relation to crop production under salt and drought stress.

Table 4.2 Physiological response of pulses to growth and crop yield parameters under drought and saline conditions

Sl no	Pulses	Stress	Selection criteria	Physiological responses	References
1	Chickpea	Salt (3.8 dS m^{-1})	Pod and seed number, pods/ plant, seeds/pod, seed size	72% yield reduction, reduced number of pods/plant, seeds/pod, and seed size	Van Hoorn et al. (2001)
2	Chick pea	25 mM–100 mM NaCl stress	Pods/plant, seeds/pod, seed size	Decrease in the number of pods/plant Decrease in seeds/pod	Flowers et al. (2010)
3	Chick pea	0.5 g NaCl/kg soil	Root and shoot length, root/ shoot ratio, yield rate	10% growth and yield reduction, decrease in leaf size	Hirich et al. (2014)
4	Mung bean	100 mM NaCl	Salinity susceptibility index	Root/shoot ratio decreased and 10–36% reduction in crop dry mass	Ahmed (2009)
5	Mung bean	0, 25 and 50 mM NaCl	Top dry weight, nodule mass,	Reduction of 9% grain yield and nodule weight	Kabir et al. (2004)
6	Mung bean	0, 50, 75 and 100 mM NaCl	Crop dry mass, root/shoot ratio,	Root/shoot ratio decreased and 10–36% reduction in crop dry mass	Alderfasi et al. (2017)
7	Mung bean	NaCl concentrations (0, 4, 7 and 10 dS. m^{-1}	Grain yield, average biomass	47% reduction in grain yield, 43–84% reduction in biomass production	Sehrawat et al. (2015)
8	Mung bean	NaCl concentrations (4, 8, 12 and 16 dS. m^{-1})	Relative growth rate (RGR), plant height, shoot and root dry weight	50% reduction in shoot and root dry weight in both varieties: desi and kabuli Reduced RGR	Saha et al. (2010)
9	Faba bean	NaCl stress (0–80 mM)	Pod and seed number, seed size, flowering time	High salinity delays flowering and maturity, 13–43% reduction in pod number, 23–45% reduction in seed number, no effect on seed size	Pitann et al. (2011)
10	Faba bean	NaCl concentrations (0, 25, 50, 75 and 100 mM)	Root and shoot fresh and dry weight, yield components	50% reduction in fresh and dry weight of root and shoot at 100 mM NaCl. Reduced grain yield.	Abid et al. (2017)

(continued)

Table 4.2 (continued)

Sl no	Pulses	Stress	Selection criteria	Physiological responses	References
11	Common bean	NaCl stress(25, 50, 75, 100 mM)	Pod and seed number, seed size, flowering time	30–50% yield loss, decrease in pod and seed number	Ferri et al. (2000)
12	Soybean	6.3–7 $dS.m^{-1}$ NaCl concentration	Grain yield, average biomass	46–66% loss in yield, less grain weight	Katerji et al. (2000)
13	Pinto bean	8–12 $dS.m^{-1}$ NaCl concentration	Root and shoot fresh and dry weight, yield components	41–86% yield loss, low grain quality, shunted growth	Ghassemi-Golezani et al. (2012)
14	Chick pea	Drought stress imposed by withholding irrigation after 7 and 14 days	Drought susceptibility index (DSI) in 35 genotypes	DSI varied from 0.46 to 1.77, 34.3–78.1% reduction in yield	Talebi et al. (2013)
15	Chick pea	Water withholding after sowing till flowering (108 DAS)	Dry matter partitioning after pod set and yield components	Seed yield was reduced by 50–80%, reduction in seed number and seed size. Pod filling varied from 0 to 60% among genotypes	Toker and Yadav (2010)
16	Pigeon pea, soybean, cow pea, chick pea, and common bean	Water withholding (60–65%)	Root and shoot fresh and dry weight, yield components	Yield reduction highest in common beans (60.8%) and lowest in pigeon pea (21.8%)	Daryanto et al. (2015)

4.3.1 Physiological Effects of Salinity on Pulses

Salinity contributes toward 50% reduction in seed and crop yield, germination rate, and plant growth predominantly in arid and semiarid regions (Toker et al. 2011; Egamberdieva and Lugtenberg 2014). In general pulses are more sensitive to saline stress in comparison to cereals and oilseeds. Kumar et al. (2016a) reported that excess salt accumulation triggers anthocyanin pigmentation in stems and leaves which reduces germination and seedling formation rate in pulses accounting for their salt sensitivity. Saline stress leads to ionic toxicity, nutrient imbalance, and membrane disorganization and reduces cellular division. All these factors collectively reduce plant growth, development, and survival (Rasool et al. 2013; Hameed et al. 2014).

The major salinity induced physiological responses in legumes contributing toward yield reduction. They include decrease in seed germination rate, nutrient regulation, carbon fixation capacity, and photosynthetic efficiency (Ahmed and

Table 4.3 Physiological response of pulses to water relations under drought and saline conditions

Sl no	Pulses	Stress	Selection criteria	Physiological responses	References
1	Chick pea	0.5 g nacl/kg soil	Water use efficiency (WUE)	10% growth reduction and severely hampered water extraction ability	Hirich et al. (2014)
2	Mung bean	NaCl concentrations (0, 4, 7 and 10 dS. m^{-1}	Total water uptake and transpiration efficiency	31–73% reduction in total water uptake, reduced rate of transpiration	Sehrawat et al. (2015)
3	Soybean	100 mM NaCl	Root and shoot length, root/shoot ratio, water use efficiency (WUE)	20–30% decrease in root/shoot length which decreased WUE	Luo et al. (2006)
4	Soybean	6.3–7 dS.m^{-1} NaCl concentration	Total water uptake, average biomass	50% loss in total water content, less grain weight	Katerji et al. (2000)
5	Chick pea	Drought stress imposed by withholding irrigation after 7 and 14 days	Relative water content	7.8–30% reduction in RWC	Talebi et al. (2013)
6	Chick pea	Drought stress imposed by withholding water after sowing	Leaf water potential, pollen viability, soil water content	Reduced leaf water potential and transpiration efficiency	Pang et al. (2017)
7	Pigeon pea	Drought	WUE, crop growth rate (CGR)	40–55% low rate of WUE and growth inhibition	Kaur et al. (2017)
8	Pigeon pea	Drought stress imposed by withholding water after sowing	Leaf water potential, pollen viability, soil water content	50% reduced leaf water potential and transpiration efficiency	Nam et al. (2001)
9	Faba bean	Moisture stress (2–4% watering of field capacity)	Total water uptake, transpiration efficiency, and leaf water potential	Decreased water usage, RWC and osmotic potential Leaf temperature and transpiration efficiency were higher in stressed plants	Khan et al. (2010a)
10	Pea	Drought	Water use efficiency (WUE), relative water content (RWC)	10–20% reduction in RWC and decreased WUE	Alexieva et al. (2001)

Table 4.4 Physiological response of pulses to photosynthetic and gaseous exchange parameters under drought and saline conditions

Sl no	Pulses	Stress	Selection criteria	Physiological responses	References
1	Chick pea	NaCl concentrations (0, 8 and 12 $dS.m^{-1}$	Total leaf chlorophyll content	24% more reduction in chlorophyll content at 8 and 12 $dS.m^{-1}$ NaCl concentration	Arefian and Shafaroudi (2015)
2	Faba bean	5.6 $dS.m^{-1}$ NaCl concentration	Chlorophyll content, primary quantum yield	Shorter leaf life cycle, overall reduction in net photosynthesis, and inhibition of electron flow to P700 (photosystem I)	Van Hoorn et al. (2001)
3	Faba bean	NaCl concentrations (0, 25, 50, 75, and 100 mM)	Chlorophyll a and b content	30% reduction in chlorophyll content and low net photosynthesis	Abid et al. (2017)
4	Common bean	25–100 mM NaCl stress	Total leaf chlorophyll content, salinity susceptibility index	24% more reduction in chlorophyll content at 75 and 100 mM NaCl	Qados (2011)
5	Lentil	2.0–3.1 $dS.m^{-1}$ NaCl concentration	Chlorophyll content	24–88%, decrease in chlorophyll content and net photosynthesis rate	Van Hoorn et al. (2001)
6	Chick pea	Drought stress imposed by withholding irrigation after 7 and 14 days	Chlorophyll and carotenoid concentrations	Reduction in total chlorophyll content	Talebi et al. (2013)
7	Chick pea	Drought stress imposed by withholding water after sowing	Leaf water potential, pollen viability, soil water content, pod and seed number, and stomatal conductance	Decrease in stomatal conductance and net leaf photosynthesis	Pang et al. (2017)
8	Pigeon pea	Drought stress imposed by withholding water after sowing.	Stomatal conductance and chlorophyll content	Decrease in net leaf photosynthesis, stomatal conductance, and chlorophyll a content	Nam et al. (2001)
9	Soybean	Drought	Stomatal conductance and photosynthesis rate	40% reduction in rate of photosynthesis, decreased stomatal conductance	Valentine et al. (2010)
10	Broad bean	NaCl (0.0, 50, 100, 150, 200 mM)	Chlorophyll content and photosynthesis rate	Reduced levels of chlorophyll content and net photosynthesis activity	El Sayed (2014)

(continued)

Table 4.4 (continued)

Sl no	Pulses	Stress	Selection criteria	Physiological responses	References
11	Chick pea	NaCl (0, 50, 75, and 100 mM	Chlorophyll content and photosynthesis rate	Reduced levels of chlorophyll content and net photosynthesis activity at 100 mM	Soussi et al. (1998)
12	Faba bean	Moisture stress (2–4% watering of field capacity)	Stomatal conductance, carbon isotope discrimination	Decreased stomatal conductance and photosynthesis	Khan et al. (2010a)

Table 4.5 Physiological response of pulses to nutrient and mineral assimilation under drought and saline conditions

Sl no	Pulses	Stress	Selection criteria	Physiological responses	References
1.	Mung bean	0, 25 and 50 mM NaCl	Acetylene reduction assay and shoot nitrogen content	Reduction of nitrogen fixation	Kabir et al. (2004)
2.	Mung bean	NaCl concentrations (4, 8, 12 and 16 dS.m^{-1})	Rate of nitrogen fixation and nodule biomass	70% decrease in weight of nodule/plant, low nitrogen fixation rate	Saha et al. (2010)
3.	Chick pea	NaCl (100 mM)	Ion levels	Reduced K^+ /Na^+ levels	Garg and Bhandari 2016
4.	Broad bean	NaCl (0.0, 50, 100, 150, 200 mM)	Ion levels	Increased the accumulation of Na+, P3+, Fe3+, Mn2+ in root, stem, leaf, and legume, uptake of K+ reduced	El Sayed (2014)
5	Mung bean Blackgram	NaCl (0, 50, 75, 100 mM)	Ion levels	Increased uptake of $Ca2^+$ and $Mg2^+$	Raptan et al. (2001)
6.	Common bean	NaCl (0, 50, 75, 100 mM)	Ion levels	Increased uptake of Zn^{2+} and decrease in Mn^{2+} levels	Doering et al. (1984)

Sayed 2011; Latef and Ahmad 2015). In major studies, salt stress led to reduction in shoot growth of soybean, chickpea, pea, faba bean, and mung bean plants (Elsheikh and Wood 1990, 1995; Delgado et al. 1994; Hussain et al. 2010; Saha et al. 2010; Rasool et al. 2013).

The rate of germination is affected due to inhibition of water uptake and ion toxicity (Farooq et al. 2015). Inadequate water supply fails to activate the hydrolytic enzymes necessary for initiation of germination. For example, salt stress inhibited the activities of the hydrolytic enzymes α- and β-amylase in cowpea (*Vigna unguic-*

ulata L.), which substantially reduced seed germination (Enéas Filho et al. 1995). Similarly salinity resulted in reduction of growth and development in common bean (*Phaseolus vulgaris* L.; Ferri et al. 2000), mung bean (*Vigna radiata* L.; Kabir et al. 2004), pigeon pea (*Cajanus cajan* L.; Surekha et al. 2014b), lentil (*Lens culinaris* L.; Bandeoğlu et al. 2004), faba bean (*Vicia faba* L.; Pitann et al. 2011), and soybean (*Glycine max* L.; Luo et al. 2006). Reduction in growth results from decrease in tissue water potential, which indicates shortage of water availability to cells (Sehrawat et al. 2015; Garg and Bhandari 2016). This leads to stomatal closure thereby reducing photosynthesis and eventually growth inhibition (Garg and Manchanda 2009).

The number of pods per plant, grains per pod, and the weight of individual grains are the principal parameters in quantifying grain yield in pulses. Salinity leads to reduction in flower numbers and pollen production (Dhingra and Varghese 1993), which subsequently reduces pod number, grains/pod, and grain weight (Mamo et al. 1996). In chickpea, 50–100 mM salt concentration resulted in reduced pollen tube length, grain numbers, and substantial decline in grain yield (Turner et al. 2013). Similarly all yield-related traits were equally responsible for the salinity-induced yield reduction in soybean (Ghassemi-Golezani et al. 2009). However in mung bean, only fewer grain/pod traits contributed to reduction in grain yield (Ahmed 2009). Salt stress also caused 80–100% yield loss in mung bean, particularly due to salinity-induced desiccation, flower shedding, and pod shattering (Sehrawat et al. 2015). The effect of salinity stress on growth of mung bean was investigated by Saha et al. (2010). They concluded that salinity stress suppressed the early growth of mung bean seedlings by 50%.

Salt stress affects the availability, competitive uptake, and translocation of nutrients to above ground plant parts. Under salt stress, the presence of excessive concentrations of Na^+ and Cl^- ions in the root zone will cause imbalanced nutrition in legumes as these ions interfere with other elements, including boron, zinc, calcium, copper, magnesium, iron, nitrogen, phosphorus, and potassium (Doering et al. 1984; Yadav et al. 1989; El Sayed 2014). For example, the K^+/Na^+ ratio decreased significantly in chickpea (Garg and Bhandari 2016), faba bean (Ullah et al. 1994), and mung bean (Nandwal et al. 2000), due to competition for intercellular Na^+ and K^+ ion flux, resulting in significant yield reductions (Sekeroglu et al. 1999).

Another important physiological response induced by salinity is reduction in net photosynthesis in grain legumes which occurs through either stomatal and/or non-stomatal components (Flexas et al. 2004; Chaves et al. 2009; Khan et al. 2015). Under salt stress, carbon fixation in legumes (C3 photosynthesis) decreases due to a reduction in the availability of CO_2 caused due to stomatal limitations (Flexas et al. 2004). Non-stomatal factors include mesophyll conductance to CO_2 and oxidative damage to the photosynthetic apparatus. Studies conducted in chickpea revealed that decrease in photosynthetic efficiency under salt stress is due to non-stomatal factor: damage to photosystem II (PS II). Similarly in mung bean, reduction in photosynthesis was due to a decline in levels of photosynthetic pigments and damage to electron transport in PS II (Khan et al. 2010b).

Salt stress has diverse effects on the quality and composition of grain legumes (Manchanda and Garg 2008). For instance, salt stress substantially reduced grain protein content in chickpea, mung bean, and faba bean due to imbalance in N_2 metabolism (Ghassemi-Golezani et al. 2010; Qados 2011). In contrast, salt stress reportedly increased protein content in mashbean (Kapoor and Srivastava 2010) and common beans (Qados 2011). In soybean, salt stress severely decreased the protein yields per plant with increasing salt concentrations compared with nonsaline conditions, while the oil percentage/plant increased (Ghassemi-Golezani et al. 2009, 2010).

Salt stress interferes with biological fixation and uptake of nitrogen (Frechilla et al. 2001; Rabie and Almadini 2005), thereby limiting nitrogen supply in grain legumes. Salinity significantly reduced the density and activity of nodule formation in faba bean (Cordovilla et al. 1995; Rabie and Almadini 2005) and pigeon pea (Garg and Manchanda 2008) as a result of premature senescence (Matamoros et al. 1999), thus inhibiting biological N_2 fixation in these grain legumes.

4.3.2 Physiological Effects of Drought on Pulses

Drought or water deficit hampers cell division, expansion, and differentiation. Losses of turgor pressure and xylem water content as a result of water deficit are primarily responsible for reduction in plant growth (Taiz and Zeiger 2006). Drought stress has adverse effects on total biomass, pod number, seed number, seed weight and quality, and seed yield/plant in chickpea (Toker and Yadav 2010), soybean (Valentine et al. 2010), mung bean (Yin et al. 2015), pigeon pea (Deshmukh et al. 2009), and pea (Assefa et al. 2017).

Deshmukh and Mate (2013) evaluated 11 genotypes of pigeon pea grown in rain shelter under drought conditions. They reported that lower drought susceptibility index (DSI), increased dry matter biomass, and harvest index (HI) contributed toward drought tolerance as observed in the drought-tolerant cultivar (JSA-59). In a study on pea, drought stress impaired the rate of germination and early seedling growth of five cultivars tested (Okçu et al. 2005). Water-deficit stress in chickpea, field pea, faba bean, and lentil at soil temperature 35–40 °C resulted in 20–70% yield reductions (Abid et al. 2017; Kumar et al. 2016b),whereas terminal drought stress in chickpea reduced the seed yield by 60%. Similar studies have been carried out among the major agro-economic pulses. A comparative analysis in grain legumes differentiated faba bean and pea as drought sensitive from lentil and chickpea (drought resistant) (Toker and Yadav 2010). In another yield-related analysis, it was observed that lentil (21.7%) and groundnut (28.6%) exhibited the lowest rate yield reduction, while faba bean reported highest yield reduction (40%) when subjected to water reduction (i.e., >65%) (Daryanto et al. 2015). When the same legumes were subjected to moderate water reduction (i.e., 60–65%), the yield reduction was as follows: pigeon pea (21.8%), soybean (28.0%), chickpeas (40.4%), cowpeas (44.3%), and common beans (60.8%).

The physiological response to drought depends on its severity and duration of exposure. Apart from reducing crop growth and grain yield, it reduces biological nitrogen fixation capacity in terms of uptake and assimilation by the grain legumes due to reduction in leghemoglobin in nodules and number of nodule under severe water-deficit conditions. Other important physiological traits include rate of CO_2 assimilation, transpiration efficiency, and stomatal conductance. All these traits contribute toward drought avoidance as well as drought tolerance.

The symbiotic nitrogen fixation (SNF) rate drastically declines in legumes as ureides accumulated heavily in nodules and shoots (Vadez et al. 2000; Charlson et al. 2009) and also decrease in shoot nitrogen demand. The decrease in SNF under drought conditions contributed to reduction of photosynthesis rate in legumes (Ladrera et al. 2007; Valentine et al. 2010). Pulses (common bean and cowpea) under water deficit maintain their leaf water content and avoid tissue dehydration by regulating their stomatal conductance, leaf abscission, and stomatal closure (Polania et al. 2016; Zegaoui et al. 2017). This subsequently leads to decrease in internal CO_2 concentrations, thereby limiting photosynthesis and shoot growth.

Relative water content, leaf water potential, stomatal resistance, rate of transpiration, leaf temperature, and canopy temperature are set of important physiological parameters that helps in quantifying plant-water relations under drought (Choudhary et al. 2018). Exposure to drought stress alters the water status of the crop plants and is characterized by steady decline in stomatal conductance as well as transpiration efficiency (Ribas-Carbo et al. 2005). Leaf water potential and water use efficiency (WUE) are therefore important physiological responses that contribute to drought avoidance strategy. WUE is measured either with respect to the whole plant or the leaf area. In case of the whole plant across the complete growth season, WUE is measured as a ratio of harvested yield to plant transpiration rate (Chaves and Oliveira 2004. In relation to plant leaf, the ratio between net rate of CO_2 assimilation and transpiration is reported as the WUE value. The WUE value reflects on the plant photosynthetic activity. Kashiwagi et al. (2006) evaluated the WUE status of drought-stressed chickpea cultivars using the carbon isotope discrimination (CID) technique and reported that lowered WUE significantly reduced grain formation during reproductive growth phase.

Drought leads to increased ethylene production in roots due to oxidation of 1-amino-cyclopropane 1-caboxylate (ACC) oxidase. Increased ACC oxidase accumulation is responsible for decreased root nodulation and root biomass (Glick et al. 2007). Biofertilizers composed of plant growth-promoting rhizobacteria such as *Bacillus subtilis*, *Pseudomonas stutzeri*, etc. effectively solve this problem. These microbes hydrolyze the accumulated ACC into α-ketobutyrate and ammonia, which can then be used as carbon and nitrogen sources as observed in drought-stressed chick pea (Swarnalakshmi et al. 2016). In pea plants, the fungi arbuscular mycorrhizal improved the WUE by 11–24% under drought (Kumar et al. 2016a).

The plant water status is also dependent on the stomatal conductance. It is calculated using the stomatal size and average size of stomatal opening. A lower stomatal conductance leads to decrease in water loss, thereby improving plant biomass (Lawlor and Tezara 2009) as observed in common bean (Miyashita et al. 2005). Several pulses (chickpea, cowpea, common bean, pigeon pea) maintain cellular water content and turgor by lowering stomatal conductance, while in other beans (pea, faba bean, mung bean), it is achieved by lowering the osmotic potential in response to drought (Amede et al. 2003). The stress responses of the above-discussed physiological traits can be incorporated in developing transgenic stress-tolerant cultivars with improved growth and yield.

4.4 Conclusion

Abiotic stresses are responsible for extensive loss in crop production worldwide. Enhancing salt and drought tolerance is of prime importance for plant breeders in order to strike a balance between crop production and population growth. Salinity and drought are the major obstacles in the path of sustainable crop productivity. Legumes serve as rich sources of protein and these crops have been earmarked to solve the protein requirements of the future generation. Legumes are more susceptible to salinity and drought than cereals and oilseeds. Both these stresses significantly reduce growth, grain yield and quality, turgor, water potential, photosynthesis, carbon fixation, and stomatal conductance in the major pulses. In the last few decades, extensive studies have been conducted on legumes in order to understand the physiological responses of legumes under salinity and drought. These physiological traits when coupled with superior crop, soil fertility, and pest management practices have the potential to enhance stress tolerance and pulse productivity, thereby mitigating the global food crisis. Moreover with the advent of molecular biology and genomics, these physiological traits or parameters can serve as markers for stress-tolerant legumes, whose genotype can be introduced in other plants or legumes, thereby developing transgenic varieties with enhanced stress tolerance.

Therefore, future studies should focus on identifying potential salinity and drought-responsive genes and their expression pattern in legumes in all their developmental stages. Such studies are essential for unraveling their stress tolerance mechanisms and incorporating these insights to improve legume production and grain quality under existing abiotic stresses.

Acknowledgment The authors are grateful to GITAM (Deemed to be University) for providing necessary facilities to carry out the research work and for extending constant support in writing this review. TD is thankful for financial support in the form of DST Inspire Fellowship (IF 160964), Department of Science and Technology, New Delhi.

References

Abdelrahman M, Jogaiah S, Burritt DJ, Tran LSP (2018) Legume genetic resources and transcriptome dynamics under abiotic stress conditions. Plant Cell Environ. https://doi.org/10.1111/pce.13123

Abid G, Hessini K, Aouida M, Aroua I, Baudoin JP, Muhovski Y, Mergeai G, Sassi K, Machraoui M, Souissi F, Jebara M (2017) Agro-physiological and biochemical responses of faba bean (*Vicia faba* L. var.'minor') genotypes to water deficit stress. Biotechnol Agron Soc Environ 21(2):1–13

Adem GD, Roy SJ, Zhou M, Bowman JP, Shabala S (2014) Evaluating contribution of ionic, osmotic and oxidative stress components towards salinity tolerance in barley. BMC Plant Biol 14(1):113

Ahmed HES, Sayed E (2011) Influence of Sodium chloride and Sodium sulphate treatments on growth development of broad bean (*Vicia Faba*, L.) plant. J Life Sci 5(7):152–163

Ahmed S (2009) Effect of soil salinity on the yield and yield components of mungbean. Pak J Bot 41(1):263–268

Alderfasi AA, Alzarqaa AA, AL-Yahya FA, Roushdy SS, Dawabah AA, Alhammad BA (2017) Effect of combined biotic and abiotic stress on some physiological aspects and antioxidant enzymatic activity in mungbean (*Vigna radiata* L.). Afr J Agric Res 12(9):700–705

Alexieva V, Sergiev I, Mapelli S, Karanov E (2001) The effect of drought and ultraviolet radiation on growth and stress markers in pea and wheat. Plant Cell Environ 24(12):1337–1344

Amede T, Schubert S, Stahr K (2003) Mechanisms of drought resistance in grain legumes I: Osmotic adjustment. SINET: Ethiopian J Sci 26(1):37–46

Andrews M, Hodge S (2010) Climate change, a challenge for cool season grain legume crop production. In: Climate change and management of cool season grain legume crops. Springer, Dordrecht, pp 1–9

Arefian M, Shafaroudi SM (2015) Physiological and gene expression analysis of extreme chickpea (*Cicer arietinum* L.) genotypes in response to salinity stress. Acta Physiol Plant 37(9):193

Ashraf M (2009) Biotechnological approach of improving plant salt tolerance using antioxidants as markers. Biotechnol Adv 27(1):84–93

Ashraf MY, Ashraf M, Arshad M (2010) Major nutrients supply in legume crops under stress environments. In: Climate change and management of cool season grain legume crops. Springer, Dordrecht, pp 155–169

Assefa T, Rao IM, Cannon SB, Wu J, Gutema Z, Blair M, Otyama P, Alemayehu F, Dagne B (2017) Improving adaptation to drought stress in white pea bean (*Phaseolus vulgaris* L.): Genotypic effects on grain yield, yield components and pod harvest index. Plant Breed 136(4):548–561

Bandeoğlu E, Eyidoğan F, Yücel M, Öktem HA (2004) Antioxidant responses of shoots and roots of lentil to NaCl-salinity stress. Plant Growth Regul 42(1):69–77

Barbieri G, Vallone S, Orsini F, Paradiso R, De Pascale S, Negre-Zakharov F, Maggio A (2012) Stomatal density and metabolic determinants mediate salt stress adaptation and water use efficiency in basil (*Ocimum basilicum* L.). J Plant Physiol 169(17):1737–1746

Bray EA, Bailey-Serres J, Weretilnyk E (2000) Responses to abiotic stresses. In: Gruissem W, Buchannan B, Jones R (eds) Biochemistry and molecular biology of plants. American Society of Plant Physiologists, Rockville, pp 1158–1249

Cattivelli L, Rizza F, Badeck FW, Mazzucotelli E, Mastrangelo AM, Francia E, Mare C, Tondelli A, Stanca AM (2008) Drought tolerance improvement in crop plants: an integrated view from breeding to genomics. Field Crop Res 105(1–2):1–14

Charlson DV, Bhatnagar S, King CA, Ray JD, Sneller CH, Carter TE, Purcell LC (2009) Polygenic inheritance of canopy wilting in soybean [*Glycine max* (L.) Merr.]. Theor Appl Genet 119(4):587–594

Chaves MM, Oliveira MM (2004) Mechanisms underlying plant resilience to water deficits: prospects for water-saving agriculture. J Exp Bot 55(407):2365–2384

Chaves MM, Flexas J, Pinheiro C (2009) Photosynthesis under drought and salt stress: regulation mechanisms from whole plant to cell. Ann Bot 103(4):551–560

Choudhary AK, Sultana R, Vales MI, Saxena KB (2018) Integrated physiological and molecular approaches to improvement of abiotic stress tolerance in two pulse crops of the semi-arid tropics. Crop J 6:99–114

Cordovilla MP, Ocana A, Ligero F, Lluch C (1995) Salinity effects on growth analysis and nutrient composition in four grain legumes-*rhizobium* symbiosis. J Plant Nutr 18(8):1595–1609

Daryanto S, Wang L, Jacinthe PA (2015) Global synthesis of drought effects on food legume production. PLoS One 10(6):0127401

Daryanto S, Wang L, Jacinthe PA (2017) Global synthesis of drought effects on cereal, legume, tuber and root crops production: a review. Agric Water Manag 179:18–33

Delgado MJ, Ligero F, Lluch C (1994) Effects of salt stress on growth and nitrogen fixation by pea, faba-bean, common bean and soybean plants. Soil Biol Biochem 26(3):371–376

Deshmukh DV, Kusalkar DV, Patil JV (2009) Evaluation of pigeonpea genotypes for morpho-physiological traits related to drought tolerance. Legume Res 32(1):46–50

Deshmukh DV, Mate SN (2013) Evaluation of pigeonpea genotypes for morpho-physiological traits related to drought tolerance. World J Agric Sci 9:17–23

Dhingra HR, Varghese TM (1993) Flowering and male reproductive functions of chickpea (*Cicer arietinum* L.) genotypes as affected by salinity. Biol Plant 35(3):447

Dita MA, Rispail N, Prats E, Rubiales D, Singh KB (2006) Biotechnology approaches to overcome biotic and abiotic stress constraints in legumes. Euphytica 147(1–2):1–24

Doering, H.W., Schulze, G., Roscher, P. 1984. Salinity effects on the micronutrient supply of plants differing in salt resistance. In 6. Colloque international pour l'optimisation de la nutrition des plantes, GERDAT, Montpellier, 2–8 Sept 1984

Du WJ, Fu SX, Yu DY (2009) Genetic analysis for the leaf pubescence density and water status traits in soybean [*Glycine max* (L.) Merr.]. Plant Breed 128(3):259–265

Dutta T, Neelapu NR, Wani SH, Challa S (2018a) Compatible solute engineering of crop plants for improved tolerance toward abiotic stresses. In: Biochemical, physiological and molecular avenues for combating abiotic stress tolerance in plants, 11:221–254

Dutta T, Neelapu NR, Wani SH, Challa S (2018b) Role and regulation of osmolytes as signaling molecules to abiotic stress tolerance in "Plant Signaling Molecules". Elsevier, ISBN 978–0–12-816451-8 (In Press)

Dagne B (2017) Improving adaptation to drought stress in white pea bean (Phaseolus vulgaris L.): Genotypic effects on grain yield, yield components and pod harvest index. Plant Breeding 136(4):548–61

Egamberdieva D, Lugtenberg B (2014) Use of plant growth-promoting rhizobacteria to alleviate salinity stress in plants. In: Use of microbes for the alleviation of soil stresses, vol 1. Springer, New York, pp 73–96

Elsheikh EAE, Wood M (1990) Effect of salinity on growth, nodulation and nitrogen yield of chickpea (*Cicer arietinum* L.). J Exp Bot 41(10):1263–1269

Elsheikh EAE, Wood M (1995) Nodulation and N_2 fixation by soybean inoculated with salt-tolerant *rhizobia* or salt-sensitive *bradyrhizobia* in saline soil. Soil Biol Biochem 27(4–5):657–661

Enéas Filho J, de Oliveira Neto OB, Prisco JT, Gomes Filho E, Nogueira CM (1995) Effects of salinity in vivo and in vitro on cotyledonary galactosidases from *Vigna unguiculata* (L.) Walp. during seed germination and seedling establishment. Revista Brasileira de Fisiologia Vegetal, Brazil

El Sayed HE (2014) Impact of magnetic water irrigation for improve the growth, chemical composition and yield production of broad bean (*Vicia faba* L.) plant. Am J Exp Agr 4(4):476–496

FAO (2010) FAO land and plant nutrition management service. FAO, Rome

FAO (2016) Food and Agriculture Organization of the United Nations. FAO, Rome

FAOSTAT (2009) Food and Agriculture Organization of the United Nations. FAOSTAT, Rome Available at: http://faostat3.fao.org/

Farooq M, Hussain M, Wakeel A, Siddique KH (2015) Salt stress in maize: effects, resistance mechanisms, and management. A review. Agron Sustain Dev 35(2):461–481

Farooq M, Gogoi N, Barthakur S, Baroowa B, Bharadwaj N, Alghamdi SS, Siddique KHM (2017) Drought stress in grain legumes during reproduction and grain filling. J Agron Crop Sci 203(2):81–102

Ferri A, Lluch C, Ocana A (2000) Effect of salt stress on carbon metabolism and bacteroid respiration in root nodules of common bean (*Phaseolus vulgaris* L.). Plant Biol 2(04):396–402

Flexas J, Bota J, Loreto F, Cornic G, Sharkey TD (2004) Diffusive and metabolic limitations to photosynthesis under drought and salinity in C3 plants. Plant Biol 6(03):269–279

Flowers TJ, Gaur PM, Gowda CL, Krishnamurthy L, Samineni S, Siddique KH, Turner NC, Vadez V, Varshney RK, Colmer TD (2010) Salt sensitivity in chickpea. Plant Cell Environ 33(4):490–509
Flowers TJ, Koyama ML, Flowers SA, Sudhakar C, Singh KP, Yeo AR (2000) QTL: their place in engineering tolerance of rice to salinity. J Exp Bot 51(342):99–106
Foolad MR, Lin GY (1997) Genetic potential for salt tolerance during germination in *Lycopersicon* species. Hortscience 32(2):296–300
Fortmeier R, Schubert S (1995) Salt tolerance of maize (*Zea mays* L.): the role of sodium exclusion. Plant Cell Environ 18(9):1041–1047
Frechilla S, Lasa B, Ibarretxe L, Lamsfus C, Aparicio-Tejo P (2001) Pea responses to saline stress is affected by the source of nitrogen nutrition (ammonium or nitrate). Plant Growth Regul 35(2):171–179
Garg N, Bhandari P (2016) Silicon nutrition and *mycorrhizal* inoculations improve growth, nutrient status, K^+/Na^+ ratio and yield of *Cicer arietinum* L. genotypes under salinity stress. Plant Growth Regul 78(3):371–387
Garg N, Manchanda G (2008) Effect of arbuscular *mycorrhizal* inoculation on salt-induced nodule senescence in *Cajanus cajan* (pigeonpea). J Plant Growth Regul 27(2):115
Garg N, Manchanda G (2009) Role of arbuscular *mycorrhizae* in the alleviation of ionic, osmotic and oxidative stresses induced by salinity in *Cajanus cajan* (L.) Millsp. (pigeonpea). J Agron Crop Sci 195(2):110–123
Ghassemi-Golezani K, Taifeh-Noori M, Oustan S, Moghaddam M (2009) Response of soybean cultivars to salinity stress. J Food Agric Environ 7(2):401–404
Ghassemi-Golezani K, Taifeh-Noori M, Oustan S, Moghaddam M, Seyyed-Rahmani S (2010) Oil and protein accumulation in soybean grains under salinity stress. Not Sci Biol 2(2):64
Ghassemi-Golezani K, Nikpour-Rashidabad N, Zehtab-Salmasi S (2012) Effect of salinity on yield and yield components of pinto bean cultivars. Inter J Plant Anim Environ Sci 2:47–51
Gholizadeh A, Dehghani H, Dvorak Y (2014) Determination of the most effective traits on wheat yield under saline stress. Agric Adv 3(4):103–110
Gilroy S, Suzuki N, Miller G, Choi WG, Toyota M, Devireddy AR, Mittler R (2014) A tidal wave of signals: calcium and ROS at the forefront of rapid systemic signaling. Trends Plant Sci 19(10):623–630
Glick BR, Cheng Z, Czarny J, Duan J (2007) Promotion of plant growth by ACC deaminase-producing soil bacteria. Eur J Plant Pathol 119(3):329–339
Hall AE, Ismail AM, Ehlers JD, Marfo KO, Cisse N, Thiaw S, Close TJ (2002) Breeding cowpea for tolerance to temperature extremes and adaptation to drought. In: Challenges and opportunities for enhancing sustainable cowpea production. IITA, Ibadan, pp 14–21
Hameed A, Dilfuza E, Abd-Allah EF, Hashem A, Kumar A, Ahmad P (2014) Salinity stress and arbuscular *mycorrhizal* symbiosis in plants. In: Use of microbes for the alleviation of soil stresses, vol 1. Springer, New York, pp 139–159
Harris BN, Sadras VO, Tester M (2010) A water-centred framework to assess the effects of salinity on the growth and yield of wheat and barley. Plant Soil 336(1–2):377–389
Hirich A, El Omari H, Jacobsen SE, Lamaddalena N, Hamdy A, Ragab R, Jelloul A, Choukr-Allah R (2014) Chickpea (‘*Cicer arietinum*’ *L.*) physiological, chemical and growth responses to irrigation with saline water. Aust J Crop Sci 8(5):646
Hussain N, Sarwar G, Schmeisky H, Al-Rawahy S, Ahmad M (2010) Salinity and drought management in legume crops. In: Climate change and management of cool season grain legume crops. Springer, Dordrecht, pp 171–191
Hussein M, Salem S, Kadambot HM (2017) Effects, tolerance mechanisms and management of salt stress in grain legumes. Plant Physiol Biochem 118:199–217
Jaleel CA, Manivannan P, Wahid A, Farooq M, Somasundaram R, Panneerselvam R (2009) Drought stress in plants: a review on morphological characteristics and pigments composition. Int J Agric Biol 11:100–105
Joshi PK, Rao PP (2017) Global pulses scenario: status and outlook. Ann N Y Acad Sci 1392(1):6–17

Kabir ME, Karim MA, Azad MAK (2004) Effect of potassium on salinity tolerance of mungbean (*Vigna radiata* L. Wilczek). J Biol Sci 4:103–110

Kapoor K, Srivastava A (2010) Assessment of salinity tolerance of Vinga mungo var. Pu-19 using ex vitro and in vitro methods. Asian J Biotechnol 2(2):73–85

Kashiwagi J, Krishnamurthy L, Crouch JH, Serraj R (2006) Variability of root length density and its contributions to seed yield in chickpea (*Cicer arietinum* L.) under terminal drought stress. Field Crop Res 95(2–3):171–181

Katerji N, van Hoorn JW, Hamdy A, Mastrorilli M (2000) Salt tolerance classification of crops according to soil salinity and to water stress day index. Agric Water Manag 43:99e109

Kaur A, Singh H, Kang JS (2017) Yield attributes and yield of pigeonpea [*Cajanus cajan* (L.) Millsp.] as influenced by growth retardants and sowing dates. Int J Curr Microbiol App Sci 6(9):1201–1205

Khan HA, Siddique KH, Munir R, Colmer TD (2015) Salt sensitivity in chickpea: growth, photosynthesis, seed yield components and tissue ion regulation in contrasting genotypes. J Plant Physiol 182:1–12

Khan HR, Paull JG, Siddique KHM, Stoddard FL (2010a) Faba bean breeding for drought-affected environments: a physiological and agronomic perspective. Field Crop Res 115(3):279–286

Khan NA, Syeed S, Masood A, Nazar R, Iqbal N (2010b) Application of salicylic acid increases contents of nutrients and antioxidative metabolism in mungbean and alleviates adverse effects of salinity stress. Int J Plant Biol 1(1):1

Kingsbury RW, Epstein E (1984) Selection for salt-resistant spring wheat1. Crop Sci 24(2):310–315

Kudapa H, Ramalingam A, Nayakoti S, Chen X, Zhuang WJ, Liang X, Kahl G, Edwards D, Varshney RK (2013) Functional genomics to study stress responses in crop legumes: progress and prospects. Funct Plant Biol 40(12):1221–1233

Kumar A, Choudhary AK, Suri VK (2016a) Influence of AM fungi, inorganic phosphorus and irrigation regimes on plant water relations and soil physical properties in okra (*Abelmoschus esculentus* L.)–pea (*Pisum sativum* L.) cropping system in Himalayan acid alfisol. J Plant Nutr 39(5):666–682

Kumar R, Singh AK, Lavania D, Siddiqui MH, Al-Whaibi MH, Grover A (2016b) Expression analysis of *ClpB/Hsp100* gene in faba bean (*Vicia faba* L.) plants in response to heat stress. Saudi J Biol Sci 23(2):243–247

Ladrera R, Marino D, Larrainzar E, González EM, Arrese-Igor C (2007) Reduced carbon availability to bacteroids and elevated ureides in nodules, but not in shoots, are involved in the nitrogen fixation response to early drought in soybean. Plant Physiol 145(2):539–546

Latef AAHA, Ahmad P (2015) Legumes and breeding under abiotic stress: an overview. In: Legumes under environmental stress: yield, improvement and adaptations. Wiley-Blackwell, Chichester, p 315

Lawlor DW, Tezara W (2009) Causes of decreased photosynthetic rate and metabolic capacity in water-deficient leaf cells: a critical evaluation of mechanisms and integration of processes. Ann Bot 103(4):561–579

Li L, Yang T, Liu R, Redden B, Maalouf F, Zong X (2017) Food legume production in China. Crop J 5(2):115–126

Li XJ, Li M, Zhou Y, Hu S, Hu R, Chen Y, Li XB (2015) Overexpression of cotton RAV1 gene in *Arabidopsis* confers transgenic plants high salinity and drought sensitivity. PLoS One 10(2):0118056

Luo GZ, Wang YJ, Xie ZM, Gai JY, Zhang JS, Chen SY (2006) The putative Ser/Thr protein kinase gene *GmAAPK* from soybean is regulated by abiotic stress. J Integr Plant Biol 48(3):327–333

Maggio A, Raimondi G, Martino A, De Pascale S (2007) Salt stress response in tomato beyond the salinity tolerance threshold. Environ Exp Bot 59(3):276–282

Mamo T, Richter C, Heiligtag B (1996) Salinity effects on the growth and ion contents of some chickpea (*Cicer arietinum* L.) and lentil (*Lens culinaris* Medic.) varieties. J Agron Crop Sci 176(4):235–247

Manchanda G, Garg N (2008) Salinity and its effects on the functional biology of legumes. Acta Physiol Plant 30(5):595–618

Mantri N, Basker N, Ford R, Pang E, Pardeshi V (2013) The role of micro-ribonucleic acids in legumes with a focus on abiotic stress response. Plant Genome 6(3):1–14

Matamoros MA, Baird LM, Escuredo PR, Dalton DA, Minchin FR, Iturbe-Ormaetxe I, Rubio MC, Moran JF, Gordon AJ, Becana M (1999) Stress-induced legume root nodule senescence. Physiological, biochemical, and structural alterations. Plant Physiol 121(1):97–112

Miyashita K, Tanakamaru S, Maitani T, Kimura K (2005) Recovery responses of photosynthesis, transpiration, and stomatal conductance in kidney bean following drought stress. Environ Exp Bot 53(2):205–214

Munns R (2002) Comparative physiology of salt and water stress. Plant Cell Environ 25(2):239–250

Munns R, Termaat A (1986) Whole-plant responses to salinity. Funct Plant Biol 13(1):143–160

Munns R, Schachtman DP, Condon AG (1995) The significance of a two-phase growth response to salinity in wheat and barley. Funct Plant Biol 22(4):561–569

Nam NH, Chauhan YS, Johansen C (2001) Effect of timing of drought stress on growth and grain yield of extra-short-duration pigeonpea lines. J Agric Sci 136(2):179–189

Nandwal AS, Godara M, Kamboj DV, Kundu BS, Mann A, Kumar B, Sharma SK (2000) Nodule functioning in trifoliate and pentafoliate mungbean genotypes as influenced by salinity. Biol Plant 43(3):459–462

Negrão S, Courtois B, Ahmadi N, Abreu I, Saibo N, Oliveira MM (2011) Recent updates on salinity stress in rice: from physiological to molecular responses. Crit Rev Plant Sci 30(4):329–377

Nishiyama Y, Murata N (2014) Revised scheme for the mechanism of photoinhibition and its application to enhance the abiotic stress tolerance of the photosynthetic machinery. Appl Microbiol Biotechnol 98(21):8777–8796

Neelapu NR, Deepak KG, Surekha C (2015) Transgenic plants for higher antioxidant. Managing Salt Tolerance in Plants: Molecular and Genomic Perspectives 5:391

Okçu G, Kaya MD, Atak M (2005) Effects of salt and drought stresses on germination and seedling growth of pea (*Pisum sativum* L.). Turk J Agric For 29(4):237–242

Pandey P, Ramegowda V, Senthil-Kumar M (2015) Shared and unique responses of plants to multiple individual stresses and stress combinations: physiological and molecular mechanisms. Front Plant Sci 6:723

Pang J, Turner NC, Du YL, Colmer TD, Siddique KH (2017) Pattern of water use and seed yield under terminal drought in chickpea genotypes. Front Plant Sci 8:1375

Pitann B, Kranz T, Zörb C, Walter A, Schurr U, Mühling KH (2011) Apoplastic pH and growth in expanding leaves of *Vicia faba* under salinity. Environ Exp Bot 74:31–36

Polania J, Poschenrieder C, Rao I, Beebe S (2016) Estimation of phenotypic variability in symbiotic nitrogen fixation ability of common bean under drought stress using 15N natural abundance in grain. Eur J Agron 79:66–73

Qados AMA (2011) Effect of salt stress on plant growth and metabolism of bean plant *Vicia faba* (L.). J Saudi Soc Agric Sci 10(1):7–15

Rabie GH, Almadini AM (2005) Role of bioinoculants in development of salt-tolerance of *Vicia faba* plants under salinity stress. Afr J Biotechnol 4(3):210

Raptan PK, Hamid A, Khaliq QA, Solaiman ARM, Ahmed JU, Karim MA (2001) Salinity tolerance of blackgram and mungbean: II. Mineral ions accumulation in different plantparts. Korean J Crop Sci 46(5):387–394

Rasool S, Ahmad A, Siddiqi TO, Ahmad P (2013) Changes in growth, lipid peroxidation and some key antioxidant enzymes in chickpea genotypes under salt stress. Acta Physiol Plant 35(4):1039–1050

Ribas-Carbo M, Taylor NL, Giles L, Busquets S, Finnegan PM, Day DA, Lambers H, Medrano H, Berry JA, Flexas J (2005) Effects of water stress on respiration in soybean leaves. Plant Physiol 139(1):466–473

Roychoudhury A, Chakraborty M (2013) Biochemical and molecular basis of varietal difference in plant salt tolerance. Annu Rev Res Biol 3:422–454

Rao GS, Reddy NN, Surekha C (2015) Induction of plant systemic resistance in Legumes cajanus cajan, Vigna radiata, Vigna mungo against plant pathogens Fusarium oxysporum and alternaria alternata–a Trichoderma viride mediated reprogramming of plant defense mechanism. Int J Recent Sci Res 6(5):4270–4280

Saha P, Chatterjee P, Biswas AK (2010) NaCl pretreatment alleviates salt stress by enhancement of antioxidant defense system and osmolyte accumulation in mungbean (*Vigna radiata* L. Wilczek). Indian J Exp Biol 48:593–600

Sehrawat N, Yadav M, Bhat KV, Sairam RK, Jaiwal PK (2015) Effect of salinity stress on mungbean [*Vigna radiata* (L.) wilczek] during consecutive summer and spring seasons. J Agric Sci Belg 60(1):23–32

Sekeroglu N, Kara MS, Dede O, Askin T (1999) Effect of salinity on germination, early seedling growth, Na and K constituents of chickpea. Turk J Field Crops 4(79):84

Serraj R, Krishnamurthy L, Kashiwagi J, Kumar J, Chandra S, Crouch JH (2004) Variation in root traits of chickpea (*Cicer arietinum* L.) grown under terminal drought. Field Crop Res 88(2–3):115–127

Sharma HC, Srivastava CP, Durairaj C, Gowda CLL (2010) Pest management in grain legumes and climate change. In: Climate change and management of cool season grain legume crops. Springer, Dordrecht, pp 115–139

Singh AK, Singh SS, Prakash V, Kumar S, Dwivedi SK (2015) Pulses production in India: present status, bottleneck and way forward. J Agri Search 2(2):75–83

Soussi M, Ocana A, Lluch C (1998) Effects of salt stress on growth, photosynthesis and nitrogen fixation in chick-pea (*Cicer arietinum* L.). J Exp Bot 49(325):1329–1337

Stoddard FL, Balko C, Erskine W, Khan HR, Link W, Sarker A (2006) Screening techniques and sources of resistance to abiotic stresses in cool-season food legumes. Euphytica 147(1–2):167–186

Surekha C, Aruna L, Hossain MA, Wani SH, Neelapu NRR (2015) Present status and future prospects of transgenic approaches for salt tolerance in plants/crop plants. In: Wani SH, Hossain MA (eds) Managing Salt Tolerance in Plants. CRC Press, New York, pp 329–352

Swarnalakshmi K, Yadav V, Senthilkumar M, Dhar DW (2016) Biofertilizers for higher pulse production in India: scope, accessibility and challenges. Indian J Agron 61:173–181

Swathi L, Reddy DM, Sudhakar P, Vineela V (2017) Screening of mungbean (*Vigna radiata* L. Wilczek) genotypes against water stress mediated through polyethylene glycol. Int J Curr Microbiol App Sci 6(10):2524–2531

Surekha CH, Neelapu NR, Prasad BS, Ganesh PS (2014a) Induction of defense enzymes and phenolic content by Trichoderma viride in Vigna mungo infested with Fusarium oxysporum and Alternaria alternata. Int J Agric Sci Res 4(4):31–40

Surekha C, Kumari KN, Aruna LV, Suneetha G, Arundhati A, Kavi Kishor PB (2014b) Expression of the Vigna aconitifolia P5CSF129A gene in transgenic pigeonpea enhances proline accumulation and salt tolerance. Plant Cell, Tissue and Organ Cult (PCTOC) 116(1):27–36

Surekha CH, Neelapu NR, Kamala G, Prasad BS, Ganesh PS (2013) Efficacy of Trichoderma viride to induce disease resistance and antioxidant responses in legume Vigna Mungo infested by Fusarium oxysporum and Alternaria alternata. Int J Agr Sci Res 3(2):285–294

Srivastava A (2010) Assessment of salinity tolerance of Vinga mungo var. Pu-19 using ex vitro and in vitro methods. Asian J Biotechnol 2(2):73–85

Taiz L, Zeiger E (2006) Plant physiology, 4th edn. Sinauer Associate, Sunderland

Talebi R, Ensafi MH, Baghebani N, Karami E, Mohammadi K (2013) Physiological responses of chickpea (Cicer arietinum) genotypes to drought stress. Environ Exp Biol 11:9–15

Tardieu F, Parent B, Simonneau T (2010) Control of leaf growth by abscisic acid: hydraulic or non-hydraulic processes? Plant Cell Environ 33(4):636–647

This D, Comstock J, Courtois B, Xu Y, Ahmadi N, Vonhof WM, Fleet C, Setter T, McCouch S (2010) Genetic analysis of water use efficiency in rice (*Oryza sativa* L.) at the leaf level. Rice 3(1):72–86

Toker C, Yadav SS (2010) Legumes cultivars for stress environments. In: Climate change and management of cool season grain legume crops. Springer, Dordrecht, pp 351–376

Toker C, Canci H, Yildirim T (2007) Evaluation of perennial wild cicer species for drought resistance. Genet Resour Crop Evol 54(8):1781–1786

Toker C, Mutlu N, Pratap A, Kumar J (2011) Breeding for abiotic stress. In: Pratap A, Kumar J (eds) Biology and breeding of food legumes. CABI, Cambridge, pp 241–260

Tuberosa R, Salvi S (2006) Genomics-based approaches to improve drought tolerance of crops. Trends Plant Sci 11(8):405–412

Turner NC, Colmer TD, Quealy J, Pushpavalli R, Krishnamurthy L, Kaur J, Singh G, Siddique KH, Vadez V (2013) Salinity tolerance and ion accumulation in chickpea (*Cicer arietinum* L.) subjected to salt stress. Plant Soil 365(1–2):347–361

Ullah SM, Soja G, Gerzabek MH (1994) Ion uptake, osmoregolation and plant-water relations in faba beans (*Vicia faba*) under salt stress. Bodenkultur 44:291–301

Vadez V, Sinclair TR, Serraj R (2000) Asparagine and ureide accumulation in nodules and shoots as feedback inhibitors of N_2 fixation in soybean. Physiol Plant 110(2):215–223

Valentine AJ, Benedito VA, Kang Y (2010) Legume nitrogen fixation and soil abiotic stress: from physiology to genomics and beyond. In: Annual plant reviews volume 42: nitrogen metabolism in plants in the post-genomic era. Wiley-Blackwell, Oxford/Hoboken, pp 207–248

Van Hoorn JW, Katerji N, Hamdy A, Mastrorilli M (2001) Effect of salinity on yield and nitrogen uptake of four grain legumes and on biological nitrogen contribution from the soil. Agric Water Manag 51(2):87–98

Varshney RK, Mohan SM, Gaur PM, Gangarao NVPR, Pandey MK, Bohra A, Sawargaonkar SL, Chitikineni A, Kimurto PK, Janila P, Saxena KB (2013) Achievements and prospects of genomics-assisted breeding in three legume crops of the semi-arid tropics. Biotechnol Adv 31(8):1120–1134

Wakabayashi K, Hoson T, Kamisaka S (1997) Osmotic stress suppresses cell wall stiffening and the increase in cell wall-bound ferulic and diferulic acids in wheat coleoptiles. Plant Physiol 113(3):967–973

Wani SH, Dutta T, Neelapu NRR, Surekha C (2017) Transgenic approaches to enhance salt and drought tolerance in plants. Plant Gene 11:219–231

Wild A (2003) Soils, land and food: managing the land during the twenty-first century. Cambridge University Press, New York

Yadav HD, Yadav OP, Dhankar OP, Oswal MC (1989) Effect of chloride salinity and boron on germination, growth and mineral-composition of chickpea (*Cicer arietinum* L.). Ann Arid Zone 28(1–2):63–67

Yin Z, Liang J, Hao X, Lu H, Hao J, Yin F (2015) Physiological response of mung bean to polyethylene glycol drought stress at flowering period. Am J Plant Sci 6(05):785

Yeo AR (2000) QTL: their place in engineering tolerance of rice to salinity. Journal of Experimental Botany 51(342):99–106

Zegaoui Z, Planchais S, Cabassa C, Djebbar R, Belbachir OA, Carol P (2017) Variation in relative water content, proline accumulation and stress gene expression in two cowpea landraces under drought. J Plant Physiol 218:26–34

Chapter 5
Salt Stress Responses in Pigeon Pea (*Cajanus cajan* L.)

Aditya Banerjee, Puja Ghosh, and Aryadeep Roychoudhury

5.1 Introduction

The crop plants belonging to the leguminosae family are a popular food choice because of their high protein content and efficient bio-productivity in marginal soils. Pigeon pea (*Cajanus cajan* L.) is a major food legume consumed as split pulse in the tropics and sub-tropics (Varshney et al. 2012). The crop is popular in more than 50 countries of Africa, Latin America, Carribean and parts of Asia. The Indian subcontinent is a major producer of pigeon pea accounting for about 90% of the production (Waheed et al. 2006). The green pods are prepared as vegetables, whereas the leaves are treated as animal fodder. The crop has high edible, nutritional and ethno-botanical values (Qasim et al. 2014). Extracts of the plant parts have been used as medicinal remedies for diabetes, stomach ailments, pain, ulcers and inflammation (Upadhyay et al. 2010). Bioactive compounds isolated from the crop exhibited antibacterial, anti-plasmodial, anti-inflammatory, antioxidant, hypocholesterolemic, hepato-protective and anthelmintic activities (Pal et al. 2011; Banerjee and Roychoudhury 2018a).

Subbarao et al. (2000) initially proposed that pigeon pea crops have a tendency to tolerate drought stress. Under nitrogen deficient conditions, these crops establish a symbiotic relation with nitrogen fixing microorganisms to improve soil fertility and sustain even in marginal soils (Fujita et al. 2004). Thus, cultivation of pigeon pea in arid and semi-arid regions can improve the overall fertility and nitrogen content of the soil. Due to such potential impacts of pigeon pea cultivation, focussed studies investigating the physiology of pigeon pea plants alone or in interaction with other companion intercrops have been performed under field conditions (Tayyab et al. 2015).

A. Banerjee · P. Ghosh · A. Roychoudhury (✉)
Department of Biotechnology, St. Xavier's College (Autonomous),
Kolkata, West Bengal, India

S. H. Wani, M. Jain (eds.), *Pulse Improvement*,
https://doi.org/10.1007/978-3-030-01743-9_5

Pigeon pea, being a rainfed water-intensive crop experiences different moisture-associated stresses throughout its life cycle. Salt stress is the most prevalent kind of abiotic stress which inflicts such vastly cultivated crops. This is because it has been found that > 6% of the total land and 30% of the irrigated land areas are afflicted with salt concentrations unsuitable for agricultural pursuits (Banerjee and Roychoudhury 2018b; Banerjee and Roychoudhury 2017). Such toxic concentration of salt in the soil directly deteriorates crop productivity, as it interferes with mineral nutrient uptake and reduces crucial agronomic traits like plant height, leaf area, crop growth rate, total dry matter, net assimilation rate and seed development (Paul et al. 2017). This chapter presents a comprehensive compilation of the less known effects of salt stress on the developmental physiology of pigeon pea plants.

5.2 Agronomic Properties of Pigeon Pea

Pigeon pea is a high yielding, hardy tropical and sub-tropical perennial shrub popularly cultivated for its edible seeds encapsulated within flat, straight, pubescent pods. Each pod (5–9 cm long and 12–13 mm wide) contains around 2–9 small coated seeds which might be brown, red or black in colour (FAO 2016). The dry seeds of pigeon pea are commonly used as pulses in Indian and Indonesian cuisines. The pods and leaves are often used as palatable protein-rich fodder instead of alfalfa. Breeding programs have successfully created pigeon pea varieties well adapted to low-moisture conditions. These cultivars were found to be disease resistant and well suited to different production systems and cropping cycles. The fast growing habit along with drought tolerance due to deep tap root system has established pigeon pea as a rainfed crop of the semi-arid regions. A wide variety of soil support the cultivation of these crops, though best suited growth occurs within the pH range of 5–7. Pigeon pea is highly susceptible to salt stress and waterlogging. However, some reported varieties tolerated 6–12 dS m^{-1} (Duke 1983). Maximum cultivation area of pigeon pea lies in Asia covering around 4.3 million hectare (m ha). India is the largest producer with a pigeon pea coverage area of 3.6 m ha followed by Myanmar (560,000 ha), Kenya (196,000 ha), China (150,000 ha), Malawi (123,000 ha), Uganda (86,000 ha), Mozambique (84,000 ha), Tanzania (68,000 ha) and Nepal (21,000 ha) (Heuze et al. 2017). Pigeon pea cultivation mainly stretches throughout southern India and across some pockets in central India (Fig. 5.1a). In a significant comparison, it was observed that large stretches of south Indian soil contain high levels of total dissolved solids (TDS) like Na^+, Cl^-, K^+ etc. (Fig. 5.1b) and thus pigeon pea crops grown in these areas are exposed to variable edaphic stresses including salinity.

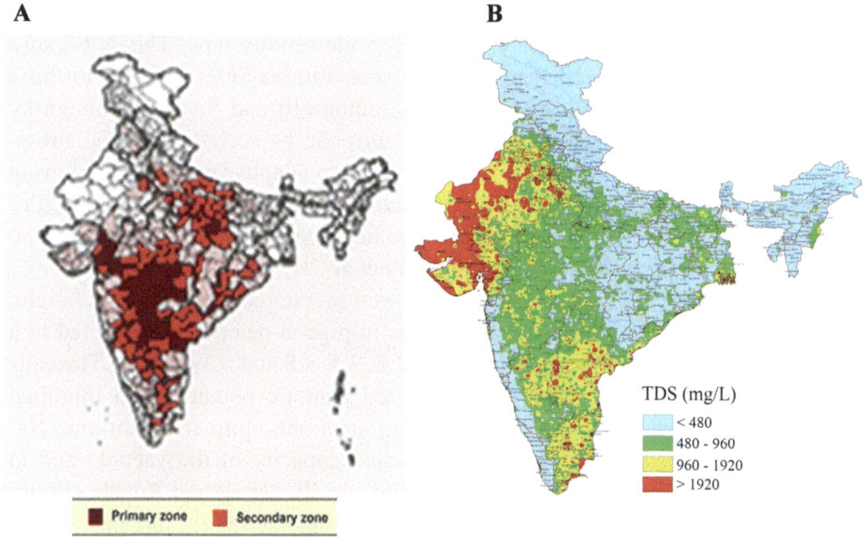

Fig. 5.1 Comparing the geographical distribution of pigeon pea cultivation and soil contamination in India, (**a**): The primary and secondary zones of pigeon pea cultivation across India. Pigeon pea is mainly cultivated throughout South and South-West India with some pockets of Central India. (Source: http://www.icrisat.org/what-we-do/satrends/feb2006.htm), (**b**): The distribution of the total dissolved solids (TDS) shows moderately-high content in South Indian soils. Source: (http://news.mit.edu/sites/mit.edu.newsoffice/files/images/2014/MIT-Desalination-India-02.jpg)

5.3 Salinity Affects Physiological Parameters in Pigeon Pea

Pigeon pea plants exposed to 50 mM and 100 mM NaCl solutions exhibited visible phytotoxic symptoms like browning and drying of leaves. Growth of the root system was considerably inhibited. Analysis of the basic biometric parameters showed that toxicity appeared initially in the older leaves followed by the younger ones at the plant tips. Fresh and dry weights were reduced by 17–23%. When grown under 100 mM NaCl stress, the ratio of dry weight of roots/ dry weight of shoots reduced by 30–40% compared to the control plants. Leaf area was reduced by 64–67% due to direct ion toxicity and nutritional imbalance. The effects of salinity using Na_2SO_4 exhibited similar degeneration of the basic physiological and anatomical indices in pigeon pea (Amuthavalli and Sivasankaramoorthy 2012).

Salinity triggers the accumulation of a vast group of compatible solutes like proline, polyamines, glycine-betaine, sugars, etc. These act as essential osmolytes and maintain the cellular osmoticum (Roychoudhury and Banerjee 2016). It is well established that Pro acts as a crucial physiological index for determining the extent of plant response to salt stress (Roychoudhury et al. 2015). Amuthavalli and Sivasankaramoorthy (2012) determined that the levels of Pro concomitantly increased with the concentration of salt stress applied in pigeon pea plants.

Interestingly, it was found that Pro accumulation in pigeon pea was associated with both alterations in the water potential and even the salinity type. This is because higher Pro levels were recorded in plants subjected to Na_2SO_4 compared to those treated with high concentrations of NaCl (Amuthavalli and Sivasankaramoorthy 2012). Such increase in the concentration of Pro can be accredited to the stress-induced metabolic interruption accompanied with an adaptive response conferring cellular protection. Pro levels can increase due to elevated rate of proteolysis. Pro maintains the osmotic potential within tissues and confers enzyme protection and stability to membrane systems (Roychoudhury et al. 2015) (Fig. 5.2).

In another study, Tayyab et al. (2016) showed severe reduction in plant height, relative growth rate, fresh and dry biomasses in pigeon pea plants subjected to a gradient of sea salt concentrations (0.5, 1.6, 2.8, 3.5, 3.8 and 4.3 dS m^{-1}). The cells exhibited ion imbalances resulting in lowered osmotic potential and inhibited physiological growth. This is because during such sub-optimal conditions, Na^+ concentration exceeds the compartmentalization capacity of the vacuole and in turn disrupts cellular metabolism and cell division (Banerjee and Roychoudhury 2016a). It is reported that several members of the legume family like pea, chick-pea, faba and mungbean show stunted growth during salt stress (Elsheikh and Wood 1990; Delgado et al. 1994; Akhtar et al. 2013). Like in pigeon pea, fresh and

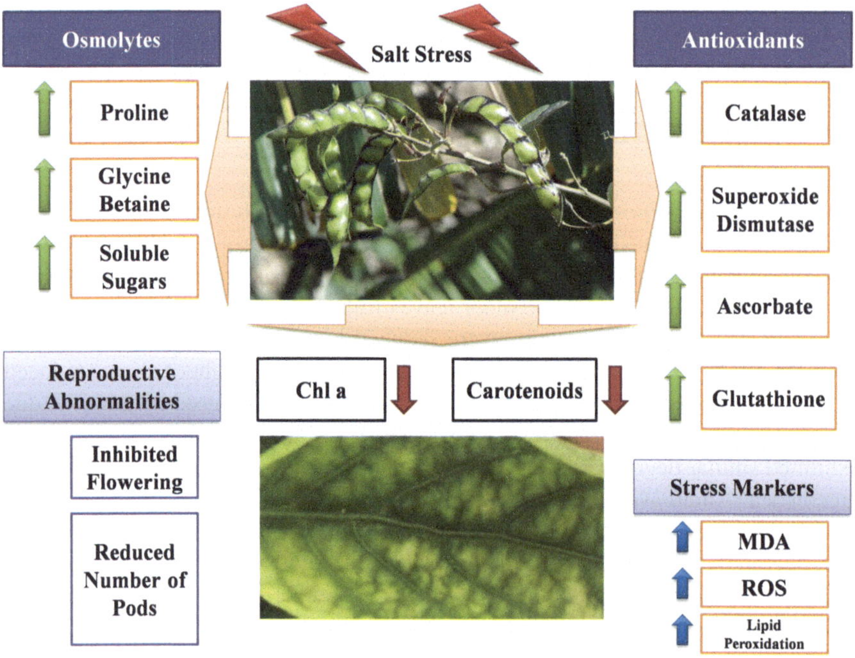

Fig. 5.2 Salt stress in pigeon pea triggers the accumulation of compatible solutes like proline, glycine betaine and soluble sugars. The antioxidant system is triggered to scavenge toxic ROS. Prolonged salt stress however deteriorates photosynthetic efficiency due to degradation of Chl a and carotenoids

dry biomass significantly decreased in pea, common bean and sesbania grown in saline medium (Hernandez et al. 1999; Mahmood et al. 2008).

Salt stress also affected reproductive growth in pigeon pea. Significant reduction in the number of flowers and pods was recorded. High concentration of sea salts stimulated increased flower shedding which ultimately reduced the effective number of pods (Vadez et al. 2007). The number and weight of seeds also deteriorated. Inappropriate ovule fertilization caused due to inhibited development of flowers, pollen grains and embryos reduced the yield of pigeon pea during salt stress (Torabi et al. 2013).

Salt stress induced loss of turgor in pigeon pea plants by reducing the moisture content and tissue succulence. As a result the cells experience low water potential throughout the stress period (Abideen et al. 2014). To counteract such dehydration, plants accumulate carbohydrates, amino acids, proteins and ammonium compounds for equilibrating plant-water relations at the cellular level (Roychoudhury et al. 2015). Pigeon pea plants exposed to salinity accumulated soluble sugars acting as organic osmotica (Tayyab et al. 2016). Elevated sugar content promotes osmoprotection, osmoregulation, carbon storage and scavenging of reactive oxygen species (ROS) (Banerjee and Roychoudhury 2018c). Excess accumulation of Na^+ during salt stress disrupts protein and nucleic acid structure by interfering with hydrogen bonding and polar contacts. Hence, the content of total soluble proteins was found to be greatly diminished in stressed pigeon pea plants (Tayyab et al. 2016). Pigeon pea plants survived 3.5 dS m^{-1} salinity of sea water by diminishing relative growth rate and by maintaining low water potential and leaf succulence. The entire metabolic machinery was channelized towards the production of osmolytes like soluble sugars (Tayyab et al. 2016).

5.4 Salinity Affects Photosynthesis in Pigeon Pea

Photosynthesis is the basic physiological process of autotrophic organisms like plants. Abiotic stresses like salinity, drought, heavy metal toxicity, etc. severely deteriorate photosynthesis by degrading the photosynthetic pigments and machinery. Such degradation is moderated via the uncontrolled production of ROS like hydroxyl, superoxide radicals, hydrogen peroxide (H_2O_2), etc. (Banerjee and Roychoudhury 2018b). Low salt concentration increased the total chlorophyll content and the chlorophyll a/b ratio, whereas higher concentrations drastically degraded chlorophyll pigments due to excessive Na^+ accumulation in the cytosol (Yang et al. 2011). Tayyab et al. (2016) showed chlorophyll a (Chl a) to be more sensitive to salt stress compared to chlorophyll b (Chl b) in pigeon pea. The authors suggested that Chl b could get converted to Chl a during stress. Thus, due to rapid degradation of Chl a in pigeon pea, the total Chl content decreased. High activity of the chlorophyllase also triggers Chl breakdown during salt stress (Eckardt 2009).

Interestingly, the carotenoid content of the cell was unaffected during low salinity (1.6 dS m^{-1}). It was suggested that these carotenoids protect the photosynthetic machinery in pigeon pea resulting in increased Chl content during low salt stress (Tayyab et al. 2016). However, under high salt stress, β-carotene gets converted to zeaxanthin for promoting protection from photoinhibition (Banerjee and Roychoudhury 2016b). Under such situations, carotenoid content drastically decreased in pigeon pea (Tayyab et al. 2016) (Fig. 5.2). In another study, it was observed that 100 mM NaCl and Na_2SO_4 reduced the Chl a content in chickpea by 75% and 77% respectively. The reduction could be due to overall mineral deficiency, leading to inhibited synthesis of Chl coupled to rapid Chl a degradation (Amuthavalli and Sivasankaramoorthy 2012).

5.5 Treatments for Salt Stress Tolerance in Pigeon Pea

Reports regarding the use of exogenous biotic and abiotic factors in generating salt tolerance in pigeon pea are scarce. However, the available excerpts show lowered salt susceptibility in the plants. Inoculation with arbuscular mycorrhizal fungi (AMFs) significantly ameliorated salt stress in the salt-sensitive pigeon pea cultivar, Paras (Pandey and Garg 2017). Mycorrhization with AMFs like *Rhizophagus irregularis* individually or together with *Funneliformis mosseae* promoted high biomass accumulation, improved redox stability and activated the antioxidant system in salt-stressed Paras. This resulted in effective scavenging of ROS and reduced cellular and molecular damages under saline environments (Pandey and Garg 2017). Low doses of gamma irradiation (0.0025 and 0.005 kGy) on pigeon pea seeds ameliorated salt susceptibility in the genetically diverse pigeon pea varieties, Pusa-991 and Pusa-992 (Kumar et al. 2017). Irradiation prior to sowing positively affected the carbon flow dynamics throughout the plants and elevated the concentration of osmolytes like glycine betaine. The protease activities and the Na^+/K^+ ratios were significantly reduced in the treated sets. Among the two selected varieties, Pusa-992 exhibited better salt tolerance compared to Pusa-991 (Kumar et al. 2017).

5.6 Transgenic Approaches for Generating Salt Tolerance in Pigeon Pea

Genetic engineering in pigeon pea to derive salt-tolerant lines have not been exhaustively investigated. Only one instance is available where overexpression of the mutated Δ^1-pyrroline-5-carboxylate synthetase gene (*P5CSF129A*) from *Vigna aconitifolia* conferred salt tolerance in transgenic pigeon pea plants (Surekha et al. 2014). The mutated gene was not subjected to feedback regulation and thus its overexpression markedly elevated the endogenous concentration of Pro. The transgenics exhibited better growth performance, improved levels of Chl and relative water content, along with lower lipid peroxidation even under 200 mM NaCl-mediated stress

(Surekha et al. 2014). Thus, enhanced accumulation of Pro ameliorated salt shock in pigeon pea by stabilising the cellular homeostasis.

Heterologous expression of the cold and drought regulatory protein (CDR) of pigeon pea conferred abiotic stress tolerance in the transgenic rice lines (Sunitha et al. 2017). The genetically modified indica varieties exhibited improved germination, rates of seedling survival, biomass, shoot and root length under different stresses like salt, drought and cold. Pronounced activities of superoxide dismutases and catalase, along with the accumulation of Pro and reducing sugar were observed in the transgenics. The abscisic acid (ABA)-dependent genes were up-regulated in the rice plants which also formed larger panicles and higher number of filled grains, compared to the control plants under stress (Sunitha et al. 2017). Similar results have been shown by Tamirisa et al. (2014) where the overexpression of *CDR* from pigeon pea conferred salt tolerance in transgenic *Arabidopsis* plants. Microarray analyses confirmed the up-regulation of 1780 genes in the transgenic plants compared to the control sets. Increased levels of antioxidants, Pro and reducing sugars were also observed in genetically modified *Arabidopsis* exposed to stress (Tamirisa et al. 2014). This effectively shows that CDR acts as a novel regulator in stress ameliorative pathways which involve antioxidant and ABA-mediated signaling among plant species. Recently a cyclin-dependent kinase regulatory subunit gene (*CKS*) from pigeon pea was overexpressed in *Arabidopsis* to produce salt and drought tolerant plants (Tamirisa et al. 2017). The transgenics exhibited higher biomass, Pro and GSH levels, along with decreased water loss and malondialdehyde content. Under salt stress, the genetically modified *Arabidopsis* lines showed reduced stomatal conductance, lower transpiration and higher photosynthetic rates compared to the non-transformed plants (Tamirisa et al. 2017). Gene expression analyses revealed that the transgenic lines maintained up-regulated levels of both ABA-dependent and ABA-independent genes involved in conferring multiple abiotic stress tolerance (Tamirisa et al. 2017). Thus *CKS* from pigeon pea can be a potential molecular target for generating multiple stress tolerance among crop plants.

5.7 Conclusion and Future Perspectives

Salinity is the most pronounced type of edaphic abiotic stress, responsible for massive crop losses throughout the world. Pigeon pea is a pulse crop which can be grown on marginal soils. Thus, it is very crucial to understand the biochemical dynamics operating at the systemic level during salt stress. Salinity induces unregulated accumulation of ROS in the tissues of pigeon pea plants. This triggers lipid peroxidation and damages to protein and enzyme structures. To counteract such degenerative effects, the plants accumulate high levels of osmolytes like Pro, glycine betaine, etc. These compatible solutes scavenge the toxic ROS and help to maintain the cellular osmotica during severe oxidative stress. The photosynthetic efficiency is also greatly reduced due to overall damage to the photosynthetic machinery. Chl a and carotenoid contents drastically decreased in the pigeon pea plants experiencing long term salt stress. Under such situations, the application of some exogenous biotic and abiotic factors may help in mitigating salt-mediated toxicity. Mycorrhization with two fungal strains and gamma irradiation, prior to sowing helped the treated plants to overcome salt susceptibility. A genotype-dependent salt response was also visualized under such circumstances. Transgenic pigeon pea overexpressing a mutated version of the *P5CS* gene accumulated elevated levels of Pro without any feedback inhibition. These lines were significantly tolerant to salt stress. Heterologous expression of the *CDR* and *CKS* genes in model plants like rice and *Arabidopsis* confirmed the involvement of these genes in regulating the ABA-dependent and ABA-independent stress responses under saline environments.

The association of salt stress with the transcriptomics and proteomics has been rarely investigated in pigeon pea. Eucidation of these aspects can be novel future perspectives. Genome-wide association studies identifying gene expression profiles in salt-stressed pigeon pea would definitely aid in the identification of new molecular targets which can be put to use via transgenic technologies. Genomics-assisted stringent breeding might also be practised to identify quantitative trait loci and potential markers present in salt-tolerant cultivars of pigeon pea. Preparation of a blueprint in this regard would facilitate easy screening of stress- tolerant cultivars by standardised laboratory protocols. Though high salinity led to compromised reproductive development, low saline concentrations did not affect pigeon pea yields (Tayyab et al. 2016). Thus, pigeon pea plants can be promoted as a protein-rich edible seed crop and a source of nutritious green fodder which can be grown in marginal soil. This would improve the socio-economic condition of farmers from theoretically unproductive soil of the world.

Acknowledgements Financial assistance from Council of Scientific and Industrial Research (CSIR), Government of India through research grant [38(1387)/14/EMR-II] is gratefully acknowledged.

References

Abideen Z, Koyro HW, Huchzermeyer B et al (2014) Moderate salinity stimulates growth and photosynthesis of *Phragmites karka* by water relations and tissue specific ion regulation. Environ Exp Bot 105:70–76

Akhtar J, Ahmad R, Ashraf MY et al (2013) Influence of exogenous application of salicylic acid on salt-stressed mungbean (*Vigna radiata*): growth and nitrogen metabolism. Pak J Bot 45:119–125

Amuthavalli P, Sivasankaramoorthy S (2012) Effect of salt stress on the growth and photosynthetic pigments of pigeon pea (*Cajanus cajan*). J Appl Pharma Sci 2:131–133

Banerjee A, Roychoudhury A (2016a) Group II late embryogenesis abundant (LEA) proteins: structural and functional aspects in plant abiotic stress. Plant Growth Regul 79:1–17

Banerjee A, Roychoudhury A (2016b) Plant responses to light stress: oxidative damages, photoprotection and role of phytohormones. In: Ahammed GJ, Yu J-Q (eds) Plant hormones under challenging environmental factors. Springer, Dordrecht, pp 181–213

Banerjee A, Roychoudhury A (2017) Epigenetic regulation during salinity and drought stress in plants: histone modifications and DNA methylation. Plant Gene 11:199–204
Banerjee A, Roychoudhury A (2018a) Effect of salinity stress on growth and physiology of medicinal plants. In: Ghorbanpour M et al (eds) Medicinal plants and environmental challenges. Springer International Publishing AG, Cham, pp 177–188
Banerjee A, Roychoudhury A (2018b) Regulation of photosynthesis under salinity and drought stress. In: Singh VP, Singh S, Singh R, Prasad SM (eds) Environment and photosynthesis: a future prospect. Studium Press (India), New Delhi, pp 134–144
Banerjee A, Roychoudhury A (2018c) Abiotic stress, generation of reactive oxygen species, and their consequences: an overview. In: Singh VP, Singh S, Tripathi D, Mohan Prasad S, Chauhan DK (eds) Revisiting the role of reactive oxygen species (ROS) in plants: ROS boon or bane for plants? John Wiley & Sons, Inc., Hoboken, pp 23–50
Delgado M, Ligero F, Lluch C (1994) Effects of salt stress on growth and nitrogen fixation by pea, Faba-bean, common bean and soybean plants. Soil Biol Biochem 26:371–376
Duke JA (1983) Handbook of energy crops. NewCROPS web site. http://www.hort.purdue.edu/newcrop/duke_energy/dukeindex.html
Eckardt NA (2009) A new chlorophyll degradation pathway. Plant Cell Onln 21:700–700
Elsheikh E, Wood M (1990) Effect of salinity on growth, nodulation and nitrogen yield of chickpea (*Cicer arietinum* L.). J Exp Bot 41:1263–1269
FAO (2016) Grassland index. A searchable catalogue of grass and forage legumes. FAO, Rome http://www.fao.org/ag/AGP/AGPC/doc/GBASE/Default.htm
Fujita K, Kai Y, Takayanagi M et al (2004) Genotypic variability of Pigeonpea in distribution of photosynthetic carbon at low phosphorus level. Plant Sci 166:641–649
Hernandez J, Campillo A, Jimenez A et al (1999) Response of antioxidant systems and leaf water relations to NaCl stress in pea plants. New Phytol 141:241–251
Heuze V, Thiollet H, Tran G et al. (2017) Pigeon pea (*Cajanus cajan*) seeds. Feedipedia, a program by INRA, CIRAD, AFZ and FAO. https://www.feedipedia.org/node/329
Kumar P, Sharma V, Atmaram CK, Singh B (2017) Regulated partitioning of fixed carbon (^{14}C), sodium (Na^+), potassium (K^+) and glycine betaine determined salinity stress tolerance of gamma irradiated pigeonpea [*Cajanus cajan* (L.) Millsp]. Environ Sci Pollut Res Int 24:7285–7297
Mahmood A, Athar M, Qadri R, Mahmood N (2008) Effect of NaCl salinity on growth, nodulation and total nitrogen content in *Sesbania sesban*. Agric Conspectus Sci 73:137–141
Pal D, Mishra P, Sachan N, Ghosh AK (2011) Biological activities and medicinal properties of *Cajanus cajan* (L) Millsp. J Adv Pharm Technol Res 2:207
Pandey R, Garg N (2017) High effectiveness of *Rhizophagus irregularis* is linked to superior modulation of antioxidant defence mechanisms in *Cajanus cajan* (L.) Millsp. genotypes grown under salinity stress. Mycorrhiza 27:669. https://doi.org/10.1007/s00572-017-0778-8
Paul S, Roychoudhury A, Banerjee A et al (2017) Seed pre-treatment with spermidine alleviates oxidative damages to different extent in the salt (NaCl)-stressed seedlings of three indica rice cultivars with contrasting level of salt tolerance. Plant Gene 11:112–123
Qasim M, Abideen Z, Adnan MY et al (2014) Traditional ethnobotanical uses of medicinal plants from coastal areas of Pakistan. J Coast Life Med 2:22–30
Roychoudhury A, Banerjee A (2016) Endogenous glycine betaine accumulation mediates abiotic stress tolerance in plants. Trop Plant Res 3:105–111
Roychoudhury A, Banerjee A, Lahiri V (2015) Metabolic and molecular-genetic regulation of proline signaling and its cross-talk with major effectors mediates abiotic stress tolerance in plants. Turk J Bot 39:887–910
Subbarao G, Chauhan Y, Johansen C (2000) Patterns of osmotic adjustment in Pigeonpea—its importance as a mechanism of drought resistance. Eur J Agron 12:239–249
Sunitha M, Srinath T, Reddy VD, Rao KV (2017) Expression of cold and drought regulatory protein (CcCDR) of pigeonpea imparts enhanced tolerance to major abiotic stresses in transgenic rice plants. Planta 245:1137–1148

Surekha CH, Kumari KN, Aruna LV, Suneetha G, Arundhati A, Kavi Kishor PB (2014) Expression of the *Vigna aconitifolia P5CSF129A*gene in transgenic pigeonpea enhances proline accumulation and salt tolerance. Plant Cell Tissue Organ Cult 116:27–36

Tamirisa S, Vudem DR, Khareedu VR (2014) Overexpression of pigeonpea stress-induced cold and drought regulatory gene (*CcCDR*) confers drought, salt, and cold tolerance in *Arabidopsis*. J Exp Bot 65:4769–4781

Tamirisa S, Vudem DR, Khareedu VR (2017) A cyclin dependent kinase regulatory subunit (CKS) gene of pigeonpea imparts abiotic stress tolerance and regulates plant growth and development in *Arabidopsis*. Front Plant Sci 8:165

Tayyab AR, Azeem M, Ahmed N (2015) Seed germination and seedling growth of Pigeon pea (*Cajanus cajan* (L.) Millspaugh) at different salinity regimes. Int J Biol Biotechnol 12:155–160

Tayyab AM, Qasim M et al (2016) Salt stress responses of pigeon pea (*Cajanus cajan*) on growth, yield and some biochemical attributes. Pak J Bot 48:1353–1360

Torabi F, Ahmad M, Enteshari S et al (2013) Effects of salinity on the development of hydroponically grown borage (*Borago officinalis* L.) male gametophyte. Not Bot Horti Agrobot Cluj-Napoca 41:65–72

Upadhyay B, Dhaker AK, Kumar A (2010) Ethnomedicinal and ethnopharmaco-statistical studies of eastern Rajasthan, India. J Ethnopharmacol 129:64–86

Vadez V, Krishnamurthy L, Serraj R et al (2007) Large variation in salinity tolerance in chickpea is explained by differences in sensitivity at the reproductive stage. Field Crop Res 104:123–129

Varshney RK, Chen W, Li Y et al (2012) Draft genome sequence of pigeonpea (*Cajanus cajan*), an orphan legume crop of resource-poor farmers. Nat Biotechnol 30:83–89

Waheed A, Hafiz IA, Qadir G et al (2006) Effect of salinity on germination, growth, yield, ionic balance and solute composition of Pigeon pea (*Cajanus cajan* (L.) Millsp). Pak J Bot 38:1103

Yang J, Zheng W, Tian Y et al (2011) Effects of various mixed salt-alkaline stresses on growth, photosynthesis, and photosynthetic pigment concentrations of *Medicago ruthenica* seedlings. Photosynthetica 49:275–284

Chapter 6
Pisum Improvement Against Biotic Stress: Current Status and Future Prospects

Reetika Mahajan, Aejaz Ahmad Dar, Shazia Mukthar, Sajad Majeed Zargar, and Susheel Sharma

6.1 Introduction

Pea (*Pisum sativum* L.) is a major cool-season pulse crop, an important part of sustainable cropping systems that belongs to the Leguminosae family (genus, *Pisum*; subfamily, Faboideae) (Nemecek et al. 2008; Duc et al. 2010; Jensen et al. 2012). Leguminosae family which accounts around 27% of the world crop production is the second most important crop plant family after Poaceae (Graham and Vance 2003). Pea is mainly cultivated in temperate regions of the world. China stands first in the production of vegetable peas (10.60 Mt) followed by India (4 Mt), whereas in case of dry pea production, Canada holds the first rank (3.85 Mt) followed by China (1.6 Mt). India produced 0.60 Mt of dry pea annually (FAOSTAT 2013). Worldwide, vegetable pea production is around 17.43 Mt, whereas dry pea production is around 11.16 Mt which represents the second most important legume after common bean (FAOSTAT 2013). Pea seeds are an important source of starch (18.6–54.1%), proteins (15.8–32.1%), fibers (5.9–12.7%), sucrose (1.3–2.1%), and oil (0.6–5.5%) along with minerals, vitamins, and micronutrients (Bastianelli et al. 1998; Arnoldi et al. 2015). Pea can be consumed by human in various forms like fresh seedlings, immature pods, and seeds (green vegetable or dry seeds). Pea hay can be used as fodder for animals (Bastida-Garcia et al. 2011). Nitrogen-fixing ability of pea has made this crop important for development of low-input farming system. Moreover, pea plays an important role as a break crop for reducing pest and pathogen pressure

R. Mahajan (✉) · A. A. Dar · S. Mukthar · S. Sharma
School of Biotechnology, Sher-e-Kashmir University of Agricultural Science and Technology of Jammu, Chatha, Jammu, J&K, India

S. M. Zargar
Division of Biotechnology, Sher-e-Kashmir University of Agricultural Science and Technology of Kashmir, Shalimar, Srinagar, J&K, India

S. H. Wani, M. Jain (eds.), *Pulse Improvement*,
https://doi.org/10.1007/978-3-030-01743-9_6

(Nemecek and Kägi 2007; Hayer et al. 2010; Macwilliam et al. 2014). In the eighteenth century, the fundamental discoveries which are now the scientific basis of genetics were established by Gregor Mendel by using pea as a model plant system (Ellis et al. 2011). Since then pea breeding for improvement of its yield and production is practiced. Various environmental factors like biotic and abiotic conditions affect the pea production in the world. Moreover, with the increase in population by 2050, there is a need to increase the crop production to feed the mankind. Both the abovementioned factors are of major concern globally. Over the last decade, advancement in molecular biology and biotechnology has made various methods available for crop improvement program, but the gap between the plant researchers and breeders is one of the major constraints for the improvement program. Thus, there is a need to narrow down the gap between the two important plant communities across the world for improvement of crops and development of smart crops which could withstand the harsh conditions and have higher yield.

6.2 Genetic and Genomic Resources Available in *Pisum*

6.2.1 Pea Genome and Karyotype

Pea (*Pisum sativum* L.) is an imperative vegetable crop cultivated globally and a valuable model plant in genetics since the days of Gregor Johann Mendel (Bar and Ori 2015). In spite of the phylogenetic similarity of legumes, they differ much in their ploidy level, chromosome number, genome size, and reproductive biology. Unlike the major crops of legumes, *M. truncatula* and *Lotus japonicus* have a small genome size, open to forward and reverse genetic analyses, and consequently act as model research plants of legume species. The pea has a large genome of 4300 Mb structured in seven chromosome pairs ($2n = 2x = 14$) (Praca-Fontes et al. 2014). The pea has about 37.7% GC content, and 30% of cytosine are 5methyl-cytosine (5meC), and 50% of 5meC have the sequence of C(A/T)G (Murray et al. 1978; Salinas et al. 1988). DNA melting studies and reassociation kinetics have shown that about 75–97% genome is made of repetitive sequences. Recently next-generation sequencing (NGS) established the presence of 50–60% highly to moderately repeated sequences (Flavell et al. 1974; Novák et al. 2010). The clustering of sequence reads found that *Ty3/gypsy* LTR-retrotransposons are the major components of the pea repeats. Ogre elements represented 20–33% of the pea genome, whereas *Ty1/copia* and other kinds of repeats were found at low proportion (Macas et al. 2007). Pea repeats have become the hot subject of many studies targeting mainly on LTR-retrotransposons (Cyclops), MITE elements (Zaba), PIGY, Angela, PDR, Stowaway, and centromeric retrotransposons (Chavanne et al. 1998; Neumann et al. 2011). Such diverse repeats of pea provide valuable markers for allowing cytogenetic chromosomal discrimination within its karyotype. The satellite repeat PisTR-B is the example of the most convenient cytogenetic marker of all pea chromosomes (Smýkal et al. 2012) (Fig. 6.1).

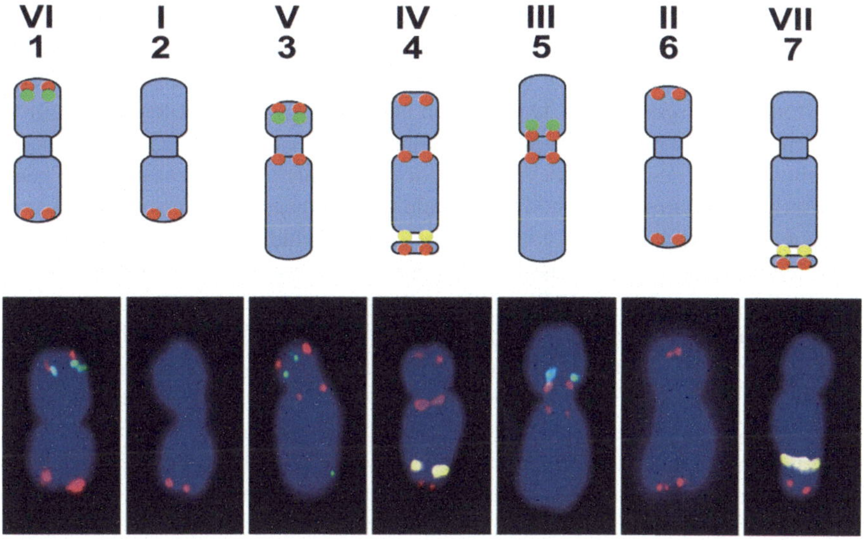

Fig. 6.1 The upper panel shows a scheme of the pea karyotype with the loci for PisTR-B (red), 5S rDNA (green), and 45S rDNA (yellow). The Arabic and Roman numerals refer to chromosome type and linkage group, respectively. The bottom panel shows the same loci detected by FISH on isolated metaphase chromosomes. Bar = 5 μm. (Smýkal et al. 2012)

The pea karyotype consists of seven chromosomes, of which five are acrocentric and two with a secondary constriction corresponding to the 45S rRNA. The chromosomes in pea are recognizable based on morphology and in situ hybridization, and these distinguished chromosomes can be identified with linkage groups. The chromosome and linkage group numbers are alluded to utilizing Arabic and Roman numerals, separately (1 = VI, 2 = I, 3 = V, 4 = IV, 5 = III, 6 = II, and 7 = VII) (Fuchs et al. 1998; Ellis and Poyser 2002) (Fig. 6.1).

6.2.2 *Genetic and Genomic Resources*

The large genome size and high transposable content of the genome of garden pea obstruct its gene sequencing and discovery (Tayeh et al. 2015a).Therefore, transcriptomes of different tissues of pea serve the complete source of available gene sequences (Kulaeva et al. 2017). Transcriptome analysis can provide extensive information about expressed genes. The high quantity and quality of transcriptomic data obtained can be used for genome map construction, molecular marker designing, and characterization of host-pathogen interactions. All these transcriptomic data have been uploaded to the NCBI database, and it allows users to perform BLAST searching and to study gene polymorphism (Kaur et al. 2012; Zhuang et al.

2012). Zhukov et al. (2015) studied RNA sequencing from pea nodules and root tips by Illumina GAIIx system, followed by de novo transcriptome assembly using Trinity program. They reported about 58,000 and 37,000 contigs from "Nodules" and "Root Tips," respectively. Out of them about 13,000 contigs (nodule-specific) were annotated to known plant protein-coding sequences. Of these, 581 sequences have shown full CDSs and were thus identified as nodule-specific transcripts of pea. The useful information obtained from nodule-specific gene sequences of pea can be used for designing gene-based markers, polymorphic studies, and real-time PCR.

Different types of markers used for detecting polymorphism were morphological markers, isozymes, RFLP, RAPD, SSR, and high-throughput parallel genotyping. Many significant progresses were gained with the usage of SSR and SNP markers. Among them, genomic and EST-based SSRs were widely used for studying genetic diversity and bridging different genetic maps (Mishra et al. 2012; Sun et al. 2014; Tayeh et al. 2015b). SNP markers are the choice of today because of their abundant frequency, easy way of scoring, and responsive to high-throughput genotyping. Consensus maps were built to produce higher mapping resolution and better genome coverage and have shown close relationship between pea and *Medicago truncatula*, *Lotus japonicus*, soybean, pigeon pea, chickpea, and lentil (Bordat et al. 2011; Leonforte et al. 2013b; Sindhu et al. 2014).

Development of next-generation sequencing allowed identification of thousands of SNPs across the genome, as established by polymorphism studies and genetic map construction. In 1912, the first genetic linkage map of pea was reported, and the first genetic map was constructed in 1925. The complete genetic map constructed in the early 2000s which consisted of seven linkage groups was found reliable with that of pea karyotype. The first genetic maps were built with the development of new sequencing technologies and the appearance of EST databases (Kulaeva et al. 2017). The high proportion of genomic synteny found between *P. sativum* and *Medicago truncatula* Gaertn. (a model legume whose genome was sequenced and annotated) has opened up chance of comparative study between pea linkage groups and chromosomes of *M. truncatula* (Fig. 6.2). This approach in pea can determine the order of nucleotide sequence of specific genes by studying gene mapping and hunting of candidate genes present in *M. truncatula* (Kalo et al. 2004). The approach of comparative mapping has allowed analysis of the paleo-history of the pea genome by proposing evolution of the seven pea chromosomes from the hexaploid ancestor of Eudicot (Bordat et al. 2011; Smýkal et al. 2012).

The Pea Marker Database (PMD) was intended to squeeze the valuable information about pea markers. Version 1 (PMD1) contains information of 2484 genic markers including their chromosomal locations in linkage groups and the sequences of their corresponding transcripts. Version 2 (PMD2) is an updated version of PMD1 and consists of 15,944 markers with several advanced characteristic features. The information present in PMD1 and 2 is easily accessible at www.peamarker.arriam.ru. The transcriptome assemblies of pea are not only useful in searching specific genes but also play a crucial part in construction of high-density genetic maps. Thus they act as the vital tools for the identification of markers and loci associated with

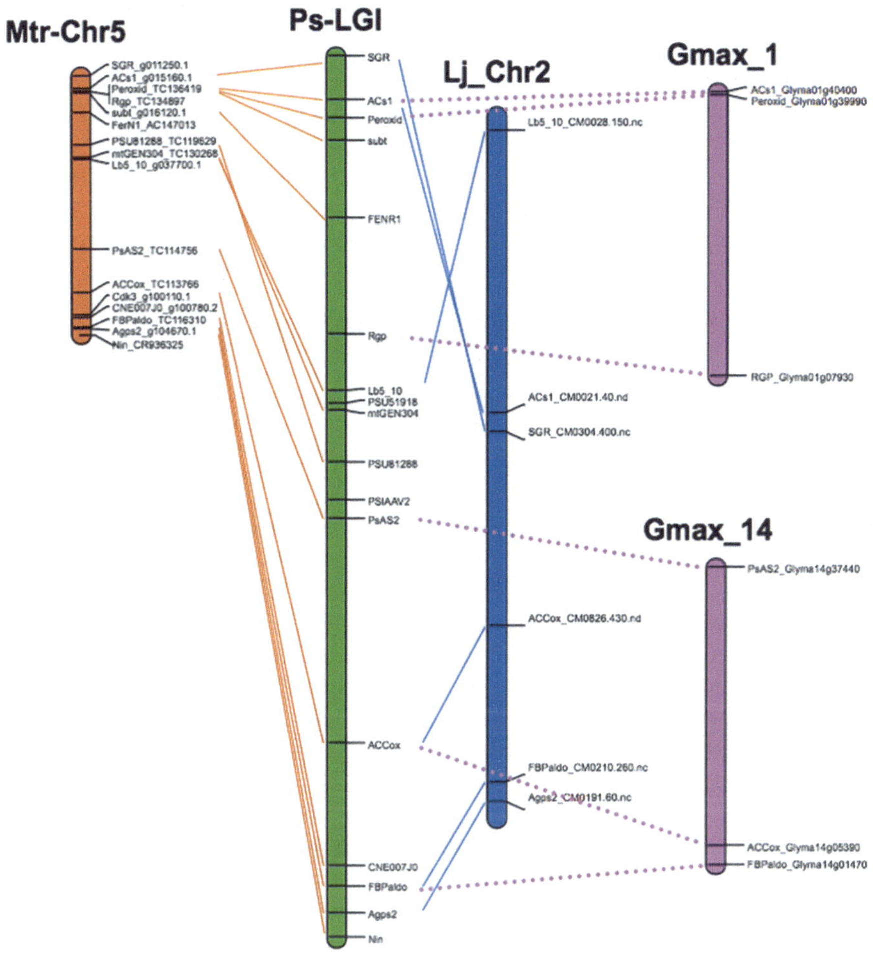

Fig. 6.2 Comparative maps of *P. sativum* and *M. truncatula*, *L. japonicus*, and *Glycine max* for linkage group LGI, as an example of synteny conservation among legume species. (Smýkal et al. 2012)

their traits of interest (Kulaeva et al. 2017). The valuable genomic information available in different programs have determined the pea community to make speedy progress toward targeted and most efficient molecular breeding programs of pea by utilizing the rich diversity of germplasm and their wild relatives (Tayeh et al. 2015a).

The genomics approaches, for instance, fast neutron and TILLING method, were used for reverse genetics study. The TILLING method induces point mutations with mutagens such as ethyl methane sulfonate and involves mutational screening systems to find out induced mutations in genomic DNA targets. The characterization data of a large TILLING population of pea is available in an online database, UTILLdb, which contains phenotypic information in addition to gene sequence of mutant genes (Wang et al. 2008; Dalmais et al. 2008). The commercial pea variety 'Cameor' was used to develop the TILLING population and BAC library for genome sequencing and positional cloning. The BAC library developed from PI 269818 was found useful for the isolation and identification of disease resistance genes (*Fw*, for *Fusarium wilt* resistance) (Coyne et al. 2007). In the future, the availability of pea chloroplast genome sequence can be very useful in transgenic and evolutionary applications (Smýkal et al. 2012).

6.3 Biotic Stress in Crops Especially in Legumes

Any external factor negatively influencing the physiology, growth, productivity, metabolism, and survival of plants is known as stress. Stress in plants is classified into two categories – abiotic and biotic. Abiotic stress is a broad-spectrum stress caused primarily by nonliving organisms which include drought, salinity, heat, chilling, waterlogging, extreme changes in soil pH, mechanical stress (such as wind, hail, wounding, etc.), effect of herbicides and weedicides, and exposure to other heavy metals, whereas biotic stress is the stress caused mainly by the living organisms which mostly include pathogen (bacteria, fungi, viruses, nematodes, etc.), insects, and weeds limiting crop productivity to a great extent.

Like other domesticated crops, legumes are the most indispensable crops which include chickpea (*Cicer arietinum*), soybean (*Glycine max*), groundnut (*Arachis hypogaea*), common bean (*Phaseolus vulgaris*), cowpea (*Vigna unguiculata*), pigeon pea (*Cajanus cajan*), lentil (*Lens culinaris*), mung bean (*Vigna radiata*), pea (*Pisum sativum*), etc. These crops are categorized into two papilionoid clades, namely, galegoid and phaseoloid, which are also known as cool- and warm-season legumes, respectively (Lewis et al. 2005). Legumes playing a pivotal role in human nutrition are suffering from considerable yield loss nowadays because of their unprecedented exposure to biotic factors such as bacteria, fungi, nematodes, insects, and weeds reducing legume production drastically. These diseases are described as follows:

6.3.1 Fungal Diseases in Legumes

Fungal diseases have gained importance from several decades as they cause huge loss of yield to the legumes. Aerial fungal diseases are having greater influence on yield varying among years and cropping regions. However, some of these diseases impact vast areas of cultivated legumes causing huge losses in quantity and quality. Biotrophic pathogens causing foliar diseases such as powdery mildews, rusts, and downy mildews are major constraints in production of legumes (Sillero et al. 2006). Powdery mildew is a serious disease influencing several legumes. Its causal organism is *Erysiphe* species which is an important fungi causing about 40% of yield loss in legumes (Sillero et al. 2006; Bari 1987). This disease is common in pea, black gram, and mung bean because of its widespread propagation in climates like cool nights and dry and warm days which affects the quality and yield of legumes. Various rust species are known to infect forage and grain legumes, most of which belong to genus *Uromyces*, such as *U. ciceris-arietini* infecting chickpea; *U. appendiculatus* causing disease in common bean; *U. pisi* infecting pea; *U. viciae-fabae* on lentil, common vetch, and faba bean; *U. vignae* on cowpea; and *U. striatus* infecting alfalfa. Other rust species belonging to different genera can be major constraint on legumes such as *P. meibomiae* and *Phakopsora pachyrhizi* infecting soybean or *Puccinia arachidis* infecting groundnut (Rubiales et al. 2002). Asian rust caused by *Phakopsora pachyrhizi* is a serious disease that is having deleterious effects on yield in soybean, and the disease is spreading in the world at an alarming pace (Carmona et al. 2005; du Preez et al. 2005). Ascochyta blight, chocolate spot, and anthracnose are known as important necrotrophic fungal diseases attacking various grain legumes (pea, chickpea), faba bean, lentil, and lupin, respectively (Tivoli et al. 2006). In chickpea, the most common aerial or foliar diseases of major importance are ascochyta blight and Botrytis gray mold (BGM) caused by *Ascochyta rabiei* and *Botrytis cinerea* (seedborne disease), respectively, causing huge loss to the yield accounting about 100% in certain conditions (Nene and Reddy 1987). Chocolate spot which is caused by *Botrytis fabae* is a devastating disease of leaf in faba bean which leads to considerable decrease in yield accounting to about 60% (Hanounik 1981), particularly in humid regions. Anthracnose is another devastating disease in legumes caused by *Colletotrichum* species (e.g., in lupin caused by *Colletotrichum lupine* and in soybean and pea by *Colletotrichum truncatum*), followed by phomopsis and brown spot caused by *Diaporthe toxica* and *Pleiochaeta setosa*, respectively (Sweetingham et al. 1998).Various soilborne pathogens are known to attack legume crops most of which attack at seedling stage (Infantino et al. 2006). Such diseases are known as damping-off caused by various pathogens including *Fusarium*, *Pythium* spp., and *Rhizoctonia*. Fusarium root rot and Fusarium wilt (caused by *Fusarium* spp. and *F. oxysporum)* are the most important obstacles for the enhanced production of legumes causing drastic which affect the adult plants and seedlings particularly in lentil, common bean, and chickpea (Hamwieh et al. 2005) causing chlorosis or yellowing of leaf, wilting, and death of crop. Pythium root rot is another constraint in legumes caused by oomycete of pythium species also called as water mold which is responsible for poor development of roots and thus deprives the plant from nutrition and renders it unhealthy. Rhizoctonia root rot is a common soilborne disease caused by *Rhizoctonia solani* in soybean, in chickpea, and in lentils. *Aphanomyces* is another type of fungal pathogen present in soil causing root rot in leguminous plants. Its symptoms are similar to pythium. For example, *Aphanomyces euteiches* damages the roots of pea, thus influencing the growth of crop and ultimately the yield.

6.3.2 *Bacterial Diseases in Legumes*

Bacterial diseases are of minor importance in legumes than that of fungi and viruses. Among the bacterial diseases, bacterial blight, bacterial pustule, and bacterial wilt are the most important diseases in legumes. The causal organism of bacterial blight and bacterial pustule includes *Pseudomonas syringae* pv. *pisi* in pea and pv. *glycinea* in *soybean, Xanthomonas axonopodis* pv. *phaseoli* causing bean blight in common bean, and *Xanthomonas campestris* pv. *cassiae* in chickpea and pv. *glycines* in soybean, respectively (Allen 1983; Baltrus et al. 2012; Khare and khare 2012). Bacterial blight and pustule are mostly disseminated in the high-altitude, cool, humid, and lowland tropical regions, respectively. Both pathogens are seedborne and can occur simultaneously in the environment. Bacterial blight is usually confused with ascochyta blight which is a fungal disease. Symptom of bacterial blight includes angular lesions that are initially water-soaked and is confused with physical damage due to hail at earlier stage. Bacterial blight lesions are restricted by leaf veins which makes the shape of lesion angular. In contrary to bacterial blight lesions, circular lesions of ascochyta blight expand beyond the leaf veins. Bacterial wilt is another constraint in groundnut caused by *Pseudomonas solanacearum*, an aerobic rod-shaped gram-negative bacterium (van der Linden et al. 2013). The disease is seedborne and is manifested by causing damping-off and killing of seedlings.

6.3.3 *Viral Disease of Legumes*

Viral diseases encompass the major biotic stress restricting production of legumes particularly growing in subtropical and tropical regions (Loebenstein and Thottappilly 2003; Rao et al. 2008; Sastry and Zitter 2014). About 150 viruses which belong to different genera are responsible for causing natural infection to cultivated food legumes (ICTV 2006). Viral diseases are transmitted either through seed or through vectors. Examples of some seed-transmitted viruses which are widespread include Bean common mosaic virus, Cucumber mosaic virus (CMV), Alfalfa mosaic virus, Soybean mosaic virus, Peanut mottle virus (PeMoV), Peanut stripe virus (PStV), Pea seed-borne mosaic virus (PSbMV), and Bean yellow mosaic virus (BYMV). In addition to above mentioned viruses, Bean golden mosaic virus (BGMV) has been regarded as the most vital disease limiting bean production in lowlands of the Caribbean, in parts of Central America accounting yield losses up to production and the, with yield losses 100% (Coyne et al. 2003). Complex viral diseases like chickpea stunt are endemic to Africa and Asia, while groundnut rosette is prevalent in Africa only (Alegbejo and Abo 2002; Kumar et al. 2008c). Other complex viruses include Pigeonpea sterility mosaic virus (PPSMV), Mungbean yellow mosaic India virus (MYMIV), Mungbean yellow mosaic virus (MYMV), and Pea bud necrosis virus (PBNV) which are limited to Southeast Asia (Kumar et al. 2008a; Malathi and John 2008; Mandal et al. 2012). Tobacco streak virus (TSV) is another most important biotic constraint in groundnut (Kumar et al. 2008b).

6.3.4 Biotic Stress Due to Insects and Nematodes

Another biotic stress is due to the insect pests present in agricultural fields posing serious threat to leguminous crops. Insects cause severe damage by feeding on plants directly, providing site for pathogen attack, and acting as vectors for various viruses such as hairy caterpillar, white fly, thrips, and aphids which act as vectors for yellow mosaic virus (Xu et al. 2010; Hodgson et al. 2012; Musser et al. 2011; Viteri et al. 2009). Insects show clear symptoms after attack in all parts of plant including leaf, stem, pods, and flowers. Some of the major insects causing significant financial loss to legumes include greenbug, armyworm, and leafhopper (Mehrkhou et al. 2012; Kumar et al. 2012). Nematodes on the contrary do not show conspicuous symptoms after attack, but their infestation is manifested with development retardation, inability to fix nitrogen, and delay in normal growth characteristics. Among various nematodes, the most important ones infesting legumes are *Heterodera* species (cyst nematode), *Meloidogyne* sp. (root knot nematode), *Rotylenchulus* species (reniform nematode), and *Ditylenchus* species (stem nematode) causing severe loss to farmers (Vovlas et al. 2011; Lombardo et al. 2011; Leach et al. 2012; Ahmad and Prasad 2012).

6.3.5 Biotic Stress Due to Weeds

Various parasitic plants which have evolved as weeds are known to pose severe threat to major crops including legumes (Rubiales et al. 2006). Weeds unlike other biotic stress factors affect legumes indirectly by competing with the nutrients available to the cultivated plant which results in growth, agroprocessing, and lesser yield of plant causing crop loss up to 90%. Weed infestation is severe in leguminous plants such as soybean, pigeon pea, mung bean, groundnut chickpea, and black gram particularly when it takes place in early stages of their development. Various important weeds affecting legumes include various species of *Avena*, *Convolvulus*, *Sonchus*, *Phaleria*, *Chenopodium*, *Orobanche*, and *Striga* (Rashid et al. 2009; Doring et al. 2012; Smitchger et al. 2012).

6.4 Impact of Biotic Stress on *Pisum*

Pea crop is known for farmers by vast diversity between areas, between fields in small area, and between years. Despite this diversity, pea crop suffers from various biotic stresses including the diseases such as bacterial, fungal, viral, pest, and nematode infestation and weeds. Among all these biotic stresses, pea suffers significant crop loss due to the presence of fungus *Aphanomyces euteiches*, causing root disease during wet years. Root disease can be evaded by winter pea during

Table 6.1 Biotic stresses and their source of resistance in garden pea

Diseases resistant	Sources
Powdery mildew	P-388, T 6587, 6588, JP-83, JP-71, JP-4, PRS-4, T-56, P-185, 6583, 7588, T-10
Fusarium wilt	Sylvia, Selection-1, Alaska, Kalanagni, S-17, S-45, Early Giant, JM-1, JM-2, Sel. 123-3-2, GC-468, JP 501/2, Grey *Ba*dger
Pea rust	JP-4, FC-1, Pant P 11, HUDP 16, JPBB-3, HUP 14
Pea seed-borne mosaic virus	American Wonder, Little Marvel PI 193586, PI 193835, X78123, X78126, X78127
Pea enation mosaic virus	OSU-547-29, OSU-559-6, OSU-546-3, OSU-584-16
Ascochyta blight	Australian Winter, Kinnauri, JI-9, JI-143, JI-370, JI-96, PI-173052, PI-174922
Common root rot	Minnesota 494–All, PI 175227, MN 144, MN 313, MN 314
Fusarium root rot	PI-792022, PI-792024

Source: Dhall (2015)

reproductive phase of the crop cycle, but during mild winters, frost acclimation occurs which renders available cultivars susceptible to various aerial diseases such as bacterial blight (*Pseudomonas syringae* pv. pisi) and ascochyta blight leading to heavier yield loss, thereby reducing thousand grain weight leading to change in plant architecture by increasing fragility (Bénézit et al. 2005). It was found that *E. baeumleri* and *E. trifolii* can overcome er1 gene-induced resistance to *E. pisi* (Ondřej et al. 2005; Attanayake et al. 2010). Moreover *E. trifolii* could also overcome resistance gene Er3 (Fondevilla et al. 2013). List of biotic stress in garden pea and their sources of resistance is given in Table 6.1.

6.5 Approaches to Tackle Biotic Stress

Biotic stresses have drastic effect on crop production. From the last two decades, a number of studies have been done to improve the production of pea. These studies include both breeding and biotechnology approaches. With the goal of providing sufficient and nutritive food material to growing population, scientists are searching a number of ways to develop biotic stress-tolerant pea crop. Here in this section, we will be discussing the approaches to tackle biotic stress in pea.

6.5.1 Breeding Approaches

Plant breeding is one of the oldest approaches which is used by the farmers since the start of agriculture practices. To make crops tolerant to both biotic and abiotic stress, breeders have been using various approaches like classical breeding which includes selection of desirable plant based on morphology and its performance in the field, crossing and modern breeding which includes marker-assisted selection (MAS), reverse breeding and doubled haploids (DH), and gene pyramiding for altering the genetic makeup of crops. Conventional breeding has increased the yield gains by approximately 2% per year (Warkentin et al. 2015). In pea, certain diseases like powdery mildew, rust, ascochyta blight, white rot, wilt, root rot, and collar rot have decreased the yield and production rate of pea crop across the globe. Powdery mildew disease which is mainly caused by *Erysiphe pisi* and newly identified causative agents *Erysiphe trifolii* and *Erysiphe baeumleri* has led to 25–50% yield losses across the world (Munjal et al. 1963; Warkentin et al. 1996). This disease has affected the seed number per pod, pod number per plant, plant height, and number of nodes (Gritton and Ebert 1975). Powdery mildew resistance in pea carries a single recessive gene (Harland 1948). Till date only two recessive (er1 and er2) and one dominant (Er3) genes for powdery mildew resistance have been identified in *Pisum* germplasm (Harland 1948; Heringa et al. 1969; Fondevilla et al. 2007a, b). Gene pyramiding approach was used to develop moderate resistance to ascochyta blight in pea. Only few pea lines with moderate level of ascochyta blight disease resistance were identified through evaluating 3500 cultivated pea accessions (Kraft et al. 1998; Zhang et al. 2006). However, various studies revealed that *P. fulvum* (wild pea) accessions have higher resistance to ascochyta blight (Clulow et al. 1991a; Wroth 1998; Fondevilla et al. 2005; Jha et al. 2012). It was found that wild pea accession P651 (*P. fulvum*) along with two accessions, namely, P670 (*P. sativum* ssp. *elatius*) and P665 (*P. sativum* ssp. *syriacum*), has higher level of ascochyta blight resistance than other wild pea accessions (Fondevilla et al. (2005). Moreover wild pea accession P651 is being mostly used for breeding purpose (Sindhu et al. 2014; Jha et al. 2016). Modern breeding approaches like QTL identification to a particular trait have also been studied in pea for identification of various QTLs related to various biotic and abiotic stresses. More than 30 QTLs related to ascochyta blight disease have been identified on all linkage group of pea (Timmerman-Vaughan et al. 2002, Timmerman-Vaughan et al. 2004; Tar'an et al. 2003; Prioul et al. 2004, Fondevilla et al. 2008a, b, 2011a, b; Jha et al. 2016). Backcross approach for development of weevil resistance (*Bruchus pisorum* L.) pea variety by transferring identified pea weevil resistance from *P. fulvum* (Clement et al. 2002; Clement et al. 2009; Aryamanesh et al. 2012). Variation for resistance to root rot has been identified and introgressed into pea accessions for development of root rot-resistant pea variety, and QTLs linked with partial resistance to *Aphanomyces* root rot were also identified (Pilet-Nayel et al. 2002, 2005; McGee et al. 2012). Potyviruses and seedborne mosaic virus are the major virus threat to pea production across the world. Thus breeding approaches are being used to select major gene resistance to these viruses for the improvement of pea production (van Leur et al. 2007). Some of the biparental crosses which were developed for improvement of pea production from both biotic and abiotic stresses are enlisted in Table 6.2.

Marker-assisted selection (MAS) is one of the important and successful modern breeding approaches which is being used for the development of resistance crops. In pea, powdery mildew resistance has been provided by introgression of er1 and er2

Table 6.2 Showing pea biparental cross with responsive traits (biotic and abiotic)

Original cross	Mapped traits	References
Erygel × 661	Plant height, flowering time, number of nodes, resistance to *Ascochyta blight*	Dirlewanger et al. (1994)
JI1089 × JI296	Leaf resistance to *Mycosphaerella pinodes*	Clulow et al. (1991a, b)
Vinco × Hurst's Greenshaft	Resistance to *Pseudomonas syringae* pv. pisi	Hunter et al. (2001)
Partridge × Early Onward	Resistance to *Pseudomonas syringae* pv. pisi	Hunter et al. (2001)
A26 × Rovar	*Ascochyta blight* resistance	Timmerman-Vaughan et al. (2004)
IFPI3260 × IFPI3251	Resistance to *Uromyces pisi* (Pers.) Wint.	Barilli et al. (2010)
Puget × 90–2079	Partial resistance to *Aphanomyces euteiches*	Pilet-Nayel et al. (2002, 2005); Loridon et al. (2005); Hamon et al. (2013)
A88 × Rovar 133	Resistance to *Ascochyta blight* field epidemics, plant reproductive maturity	Timmerman-Vaughan et al. (2002)
Carneval × MP1401	Lodging reaction, plant height, resistance to *Mycosphaerella blight*, grain yield, seed protein concentration, days to maturity	Tar'an et al. (2003, 2004)
JI296 × DP	Resistance to *Mycosphaerella pinodes*, plant height, flowering date	Prioul et al. (2004); Loridon et al. (2005); Tayeh et al. (2015a)
DSP × 90–2131	Earliness at flowering, plant height, *Aphanomyces* root rot resistance	Hamon et al. (2013)
Shawnee × Bohatyr	Seed mineral content, partial resistance to *Fusarium wilt* race2	Loridon et al. (2005), McPhee et al. (2012)
Baccara × PI180693	*Aphanomyces* root rot resistance, earliness at flowering	Hamon et al. (2011, 2013), Duarte et al. (2014), Tayeh et al. (2015a)
Kaspa × Yarrum	Powdery mildew resistance	Sudheesh et al. (2014)
JI281 × JI399	Resistance genes(Ppi2) to *Pseudomonas syringae* pv pisi	Ellis et al. (1992), Hall et al. 1997, Hunter et al. (2001)
JI15 × JI399	Resistance genes (Ppi1 and Ppi2) to *Pseudomonas syringae* pv pisi	Ellis et al. (1992), Hall et al. (1997), Hunter et al. (2001)
Kaspa × ps1771	Powdery mildew resistance, boron tolerance	Sudheesh et al. (2014)

(continued)

Table 6.2 (continued)

Original cross	Mapped traits	References
"Afghanistan"(sym2) × A1078–239	Tolerance to *Fusarium root rot*	Weeden and Porter (2007)
CMG × PI220174	Tolerance to *Fusarium* root rot	Weeden and Porter (2007)
Carman × Rewarde	Resistance to *Fusarium* root rot	Feng et al. (2011)
HUVP1 × FC1	Resistance to *Uromyces fabae* (Pers.) de-Bary	Rai et al. (2011)
MN313 × OSU1026	Tolerance to *Aphanomyces euteiches*	Weeden et al. (2000)
Baccara × 552f	*Aphanomyces root rot* resistance, earliness at flowering	Hamon et al. (2011, 2013)
P665 × Messire	Resistance to *Mycosphaerella pinodes, Orobanche crenata, Pseudomonas syringae*, earliness, root length, aerial biomass, drought tolerance	Valderrama et al. (2004), Fondevilla et al. (2008a, b), (2011a, b), (2012), Carrillo et al. (2012), Iglesias-García et al. (2015)
JI1794 × Slow	Tolerance to *Fusarium solani* f. sp. pisi, pod dehiscence, dry seed weight	Timmerman-Vaughan et al. (1996), Weeden et al. (1998, 2002), Hance et al. (2004)
Pennant × ATC113	*Bruchus pisorum* resistance	Byrne et al. (2008), Aryamanesh et al. (2014)

Source: Tayeh et al. (2015a)

gene in susceptible cultivar via marker-assisted selection. For this, molecular markers like SSR, SCAR, CAPs, and AFLP markers have been used. For er1 gene, SSR and SCAR markers (Timmerman et al. 1994; Tiwari et al. 1998; Janila and Sharma 2004; Ek et al. 2005; Pereira and Leitão 2010); for er2, AFLP and SCAR markers (Tiwari et al. 1999; Katoch et al. 2010); and, for Er3, SCAR markers have been identified (Fondevilla et al. 2008a; b). Pea accessions (wild and cultivated) with uncharacterized moderate resistance to powdery mildew have also been identified with accessions showing higher resistance to er1, er2, and Er3 (Pal et al. 1980; Sharma 1992; Dang et al. 1994; Thakur et al. 1996; Fondevilla et al. 2007a). Breeding approaches have also been utilized for the development of abiotic stress tolerance pea. It is well known that abiotic stresses like drought, salinity, frost, heat, and waterlogging have decreased global pea production. Hr allele which is responsible for delaying flower initiation has been introgressed in pea for improved production in winter season (Lejeune-Hénaut et al. 2008). Efforts have been taken to produce improved seed nutrient concentration, seed quality, seed yield (Ubayasena et al. 2011; Jha et al. 2013), boron toxicity tolerance (Bagheri et al. 1994), heat tolerance during flowering (Petkova et al. 2009), iron deficiency (Kabir et al. 2012), and salinity (Leonforte et al. 2013a) for meeting increased food demand. Identifying and selecting unique stress resistant source that will enhance the persistence of existing resistance genes is important in pea breeding. Drawback of breeding approaches like season dependency, time consumption for the generation of population, lot of labor for field analysis, and transfer of undesirable traits with desirable ones has led to the utilization of other advanced approaches for tackling the stress.

6.5.2 Biotechnology Approach

Biotechnology is gaining more attention in the scientific community as most of the approaches used in this are not season dependent and less time-consuming. In these approaches the alteration is done in genetic level by utilizing high-throughput technologies. Approaches like genomics, transcriptomics, proteomics, metabolomics, and genetic transformation are in frequent use for the improvement of various crops. Here in this section, we will be discussing about the various approaches which are used by the scientist to understand the mechanism of biotic stress tolerance in pea crop. Genetic diversity is one of the important and foremost steps for studying the variation in a large number of germplasm. Studies revealed that desirable genotypes for a particular trait can be selected by evaluating the germplasm either by morphological or molecular bases. Molecular markers from low throughput to high throughput have been used in evaluating diversity for a particular trait in pea. A novel er1 allele (er1-6) has been identified in Chinese pea landraces which confer resistance to powdery mildew disease (Sun et al. 2016). SNP1121 marker specific to powdery mildew disease was also developed which can be used for Chinese pea landraces improvement program by marker-assisted selection (Sun et al. 2016). *Agrobacterium*-mediated gene transfer has been done in pea for the development of transgenic pea (Puonti-Kaerlas et al. 1990). Bean α-amylase inhibitor1 from *Phaseolus vulgaris* which showed resistance to pea bruchid has been transferred to pea cultivar for providing resistance to pea weevil (*Bruchus pisorum*) under field conditions in Australia and over several seasons (Morton et al. 2000). Resistance to herbicides in pea has been achieved via *Agrobacterium*-mediated genetic transformation containing mutant ahas/als gene (Nifantova et al. 2005). Transgenic pea has been developed by stacking two antifungal genes (chitinase and glucanase) for resistance against fungal diseases (Amian et al. 2011). Genetic transformation in pea is slightly difficult than other legume crops due to low rate of transformation and difficulty in plant regeneration (Svabova et al. 2005; Warkentin et al. 2015). Various transcriptome studies have been carried out by using a pea 6k oligo-array (Ps6kOLI1) to study the effect of mutations in genes involved in primary metabolism or hormone deficiency on the seed transcriptome (Weigelt et al. 2008, 2009; Riebeseel et al. 2010; Radchuk et al. 2010). Microarray approach has been utilized for identification of differential gene expressed in resistance to *Mycosphaerella pinodes* in pea (Fondevilla et al. 2011a; b). RNA-seq approach has been used to study the expression of 20 differential genes in pea accession infected with powdery mildew (Barilli et al. 2014). Another important approach next-generation sequencing which has made it possible to generate transcriptome data without the information of sequenced genome has been utilized in pea improvement program. For this a library of 450 Mb has been prepared from various parts of pea like flowers, leaves, cotyledons, epicotyl and hypocotyl, and light-treated etiolated seedlings. These were assembled into 324,428 unigenes and annotated with *A. thaliana*, *M. truncatula*, *G. max*, and other databases. This information will expedite molecular approaches for pea improvement program (Franssen et al. 2011).

Proteomics, an important approach which facilitates the understanding of biological as well as physiological mechanism in a plant, has been utilized in pea improvement program against both biotic and abiotic stresses. Leaf proteome analysis of powdery mildew resistant and susceptible pea cultivar resulted in identification of 108 spots including defense proteins more in the susceptible cultivar (Curto et al. 2006). Response to crenate broomrape (*Orobanche crenata*) has been studied in pea by using proteomic approach which leads to identification of proteins involved in carbohydrate and nitrogen metabolism and electron transport and stress-related proteins (Castillejo et al. 2004). Similarly, response to *Mycosphaerella pinodes* was also studied in pea by using a proteomic approach resulting in identification of 31 proteins comprised of photosynthesis, metabolism, transcription/translation and defense, and stress-related proteins by using a combination of peptide mass fingerprinting (PMF) and MSMS fragmentation (Castillejo et al. 2010). A compatible interaction between pea and downy mildew pathogen *Peronospora viciae* has been studied via using proteomics (Amey et al. 2008). Comparative proteomic approach which has been used to study the effect of beneficial microbes in pea reveals that change in protein network in treated pea provides resistance to pea against *Sclerotinia sclerotiorum* (Jain et al. 2015). Metabolomic approach has also been employed in pea improvement program (Charlton et al. 2004, 2008).

6.6 Current Scenario for *Pisum* Improvement

Many researchers and scientists are working currently for the *Pisum* improvement by tackling abiotic and biotic stresses, identifying components and factors for increasing pea production, and mapping genes for involvement in many traits. The discussion about *Pisum* improvement is mentioned already in above sections. But here we will discuss and focus more about the crop improvement from last few years. The study of Basher et al. (2017) revealed the existence of significant variation among the pea genotypes. They found that genetic coefficients of variation were slightly lower than their corresponding phenotypic coefficients because of the impact of environment. The path coefficient analysis has shown that pod size, seed weight, and number of seeds per pod had maximum effect on the grain yield per plant. Similarly Khan et al. (2017) and Deepika et al. (2017) observed considerable genetic variability and reported that focus should be given on pod size, seeds per pod, seeds per plant, and 100-seed weight for improving the seed yield during breeding programs.

Powdery mildew (PM) is one of the most destructive diseases of pea caused by the ascomycete fungus *Erysiphe pisi* Syd., which grows on the leaves and develops majority on the pods and stem with enormous impact on seed production (Warkentin et al. 1996). The naturally occurring genetic resistances to PM are monogenic recessive and are conferred by two independent loci, namely, er1 and er2. A third type of genetic resistance, also monogenic but dominant (Er3), is identified in *P. fulvum* and can be introgressed into *P. sativum* by interspecific crossing (Fondevilla et al. 2008a; b). The er2 locus confers powdery mildew resistance to leaves, whereas er1 locus confers resistance toward a broad spectrum of pathotypes found on all plant tissues (Katoch et al. 2010; Pereira and Leitão 2010). The er1 gene was recognized as a homolog of the barley powdery mildew resistance gene MLO and is also called as PsMLO1. SSRs based on the highly polymorphic microsatellite markers involving the er1/PsMLO1 gene resistance can be used in multiple pea powdery mildew resistance breeding programs (Cardoso et al. 2017).

Development of novel and reliable SSR markers is helpful in both fundamental and applied genomics of the crop. The Illumina HiSeq 2500 System was used to reveal 8899 putative SSR, and out of which 3275 nonredundant primers were designed for amplification. Among that 1644 SSRs were chosen for primer validation, and 841 produced obvious polymorphisms among genotypes of cultivated pea and their wild relatives (*P. fulvum* Sm.). The number of alleles per locus varied from 2 to 10, and their polymorphism information content (PIC) ranged from 0.08 to 0.82 with an average value of 0.38 (Yang et al. 2015). Genetic structure, diversity, and interrelationships were also studied in a collection of 151 pea genotypes using 21 morphological characters and 20 SSR primers (Rana et al. 2016). The pod size, seed weight, and seed yield per plant showed significant amount of variation among quantitative traits. The SSR primers amplified 179 alleles with mean value of 8.95 alleles per primer, exposing a high level of genetic diversity. The mean polymorphism information content was 0.72. The observed heterozygosity varied from 0.10 to 0.99, with a mean value of 0.46, and expected heterozygosity varied from 0.47 to 0.94 with a mean of 0.75. The STRUCTURE analysis showed three gene pools belonged to diverse geographic regions indicating germplasm transfer and exchange during the domestication. Such type of study can be very helpful in understanding the genetic configuration and selection of appropriate diverse genotypes of pea for future improvement programs.

Marker-assisted breeding is often used nowadays in major crops for efficient crop improvement by utilizing next-generation sequencing technology in identifying many loci and markers linked with the traits of interest. In this regard, Ma et al. (2017) mapped 1683 markers including 1608 SNPs and generated a linkage map of 1310.1 cM. The comparative mapping of pea with other legumes revealed the high level of synteny between the genome of pea and *Medicago truncatula*. The QTL analysis of the RIL population showed one QTL for each of the nutrient traits. The genome-wide SNPs together with the genetic linkage map have identified QTLs associated with the nutrients of pea, and thus it can facilitate marker-assisted selection for high-quality nutritional traits in breeding programs.

Salt stress is one of the harsh environmental factors limiting productivity and yield of crops with a total loss of 12 billion US dollars (Shabala 2013). The productivity losses in crops because of salt stress are due to mainly imbalanced nutrition,

ion toxicity, osmotic stress, and oxidative stress (Hussain et al. 2012). The high content of sodium in plants damages many biomembranes and subcellular organelles, thus affecting normal growth and development (Quintero et al. 2007). Pea is the most salt-tolerant vegetable crop among legumes as 50% of its yield decline was found at 100 mM NaCl (Subbarao and Johansen 1994). Ghezal et al. (2016) reported that salt stress had adversely affected chlorophyll formation, seed germination, growth, metabolism, membrane structure, and mineral composition (potassium and phosphorus). However, it increased the accumulation of proline, soluble sugars, Na^+, and secondary metabolites. It was found that seed priming showed a good response to salt stress by protecting membrane integrity, maintaining high levels of osmotica (sugars, proline, K^+, and P), and improving chlorophyll and carotenoid contents. The bio-priming of pea seeds had thus increased the pea salinity tolerance which can have a good impact on the seed yield.

Plants have evolved an adaptation to withstand cold stress induced by low, nonfreezing temperatures. However, frost damage can cause irreversible injury such as destruction of cell membrane system and loss of photosynthetically active tissue (Xin and browse 2000; Menon et al. 2015). For legume species, frost stress is also one of the principal limiting abiotic factors affecting their production by damaging seedlings and limiting plant growth (Maqbool et al. 2010). Therefore, in-depth research on plant frost tolerance is needed for exploring the molecular basis of cold acclimation and increased production of most crops. The studies on frost tolerance and breeding for winter-hardy cultivars play fundamental roles in the stable increase of pea production. The frost tolerance of pea was evaluated and conducted on 672 diverse pea accessions with 267 informative SSR markers. Sixteen accessions were identified as the most winter-hardy for their ability to survive in all nine field experiments with a mean survival rate of 0.57, ranging from 0.41 to 0.75. Association analysis detected seven markers that repeatedly had associations with frost tolerance in at least two different environments. One of the markers was found to be involved in the metabolism of glycoproteins in response to chilling stress and may provide a novel mechanism of frost tolerance in pea (Liu et al. 2017). The frost tolerance associated markers in winter-hardy germplasm and will play a vital role in marker-assisted breeding for winter-hardy pea cultivar.

6.7 Our Vision and Future Prospects

Various qualities of pea like short life cycle, self-pollinating crop, and easily manageable make it the model system for Mendel's experiments in 1866. Since then this plant system has been studied. Till date various research have been done for pea improvement. Now, pea breeding should include study related to *Rhizobia*, *Mycorrhiza*, and other beneficial microorganism interactions with pea for the development of smart crop against stresses, study phenology, and morphology of the pea plant to have better and unique cropping systems. On the other hand, techniques of gene mapping, cloning, gene silencing, CRISPR, and genetic transformation are important to create novel gene combinations to overcome biotic stresses in pea. Colinearity and candidate gene approach combination has resulted in identification of genes underlying agronomically important traits including virus resistances and plant architecture. The high-throughput omics approaches could be utilized for the development of novel, highly accurate selective breeding tools for improved pea genotypes that will help the breeders to overcome critical stress problems under current and future climates.

6.8 Conclusion

It is clear that biotic agents like bacteria, fungi, virus, insect, and pest have adverse impact on the crop production which ultimately leads to decrease in agricultural output. The decrease crop production and increased population is a threat to the mankind. Adequate food diet is the key to successful and better life. Providing the adequate and balanced food diet to ever-growing population is important in the present. Thus, with the help of breeding and biotechnological interventions, researchers and breeders are able to improve the production of crop to a certain extent. However, there is a need to have better understanding of various mechanisms for development of smart climate resilience crops.

References

Ahmad P, Prasad MNV (2012) Environmental adaptations and stress tolerance in plants in the era of climate change. Springer, New York

Alegbejo MD, Abo ME (2002) Etiology, ecology, epidemiology and control of groundnut rosette disease in Africa. J Sustain Agric 20:17–29

Allen DJ (1983) The pathology of tropical food legumes: disease resistance in crop improvement. Wiley, New York, 413 pp

Amey RC, Schleicher T, Slinn J, Lewis M, Macdonald H, Neill SJ, Spencer-Phillips PTN (2008) Proteomic analysis of a compatible interaction between *Pisum sativum* (pea) and the downy mildew pathogen *Peronospora viciae*. Eur J Plant Pathol 122:41–55

Amian AA, Papenbrock J, Jacobsen HJ, Hassan F (2011) Enhancing transgenic pea (*Pisum sativum* L.) resistance against fungal diseases through stacking of two antifungal genes (chitinase and glucanase). GM Crops 2(2):104–109. https://doi.org/10.4161/gmcr.2.2.16125

Arnoldi A, Zanoni C, Lammi C, Boschin G (2015) The role of grain legumes in the prevention of hypercholesterolemia and hypertension. Crit Rev Plant Sci 34:144–168. https://doi.org/10.1080/07352689.2014.897908

Aryamanesh N, Byrne O, Hardie DC, Khan T, Siddique KHM, Yan G (2012) Large-scale density-based screening for pea weevil resistance in advanced backcross lines derived from cultivated field pea (*Pisum sativum*) and *Pisum fulvum*. Crop Pasture Sci 63:612–618. https://doi.org/10.1071/CP12225

Aryamanesh N, Zeng Y, Byrne O, Hardie DC, Al-Subhi AM, Khan T et al (2014) Identification of genome regions controlling cotyledon, pod wall/seed coat and pod wall resistance to pea weevil through QTL mapping. Theor Appl Genet 127:489–497. https://doi.org/10.1007/s00122-013-2234-2

Attanayake RN, Glawe DA, McPhee KE, Dugan FM, Chen W (2010) Erysiphe trifolii—a newly recognized powdery mildew pathogen of pea. Plant Pathol 59:712–720. https://doi.org/10.1111/j.1365-3059.2010.02306.x

Bagheri A, Paull JG, Rathjen AJ (1994) The response of *Pisum sativum* L. germplasm to high concentrations of soil boron. Euphytica 75:9–17. https://doi.org/10.1007/BF00024526

Baltrus DA, Nishimura MT, Dougherty KM et al (2012) The molecular basis of host specialization in bean pathovars of *Pseudomonas syringae*. Mol Plant Microbe Interact 25(877):888

Bar M, Ori N (2015) Compound leaf development in model plant species. Curr Opin Plant Biol 23:61–69

BARI (Bangladesh Agricultural Research Institute) (1987) Yield losses assessment of mungbean due to powdery mildew. In: Plant Pathology Division (ed) Annual report, 1983/84. BARI, Joydebpur, Gazipur, pp 81–82

Barilli E, Satovic Z, Rubiales D, Torres AM (2010) Mapping of quantitative trait loci controlling partial resistance against rust incited by Uromyces pisi (Pers.) Wint. in a Pisum fulvum L. intraspecific cross. Euphytica 175:151–159. https://doi.org/10.1007/s10681-010-0141-z

Barilli E, Rubiales D, Gjetting T, Lyngkjær MF (2014) Differential gene transcript accumulation in peas in response to powdery mildew (*Erysiphe pisi*) attack. Euphytica 198(1). https://doi.org/10.1007/s10681-014-1062-z

Basher I, Ishtiaq S, Fiaz S, Sajjad M (2017) Association of yield attributing traits in pea (*Pisum sativum* L.) germplasm. Banat's J Biotechnol VIII(15). https://doi.org/10.7904/2068–4738–VIII(15)–43

Bastianelli D, Grosjean F, Peyronnet C, Duparque M, Regnier JM (1998) Feeding value of pea (Pisum sativum, L.) - 1 chemical composition of different categories of pea. Anim Sci 67:609–619. https://doi.org/10.1017/S1357729800033051

Bastida-Garcia JL, González-Ronquillo M, Domínguez Vara IA, Romero-Bernal J, Castelán Ortega O (2011) Effect of field pea (Pisum sativum L.) level on intake, digestion, ruminal fermentation and in vitro gas production in sheep fed maintenance diets. Anim Sci J 82:654–662. https://doi.org/10.1111/j.1740-0929.2011.00884.x

Bénézit M, Biarnès V, Jeuffroy M, Carmona MA, Gally ME, Lopez SE (2005) Asian soybean rust: incidence, severity, and morphological characterization of *Phakopsora pachyrhizi* (uredinia and telia) in Argentina. Plant Dis 89:109–109

Bordat A, Savois V, Nicolas M, Salse J, Chauveau A, Bourgeois M et al (2011) Translational genomics in legumes allowed placing in silico 5460 unigenes on the pea functional map and identified candidate genes in *Pisum sativum* L. G3 1:93–103. https://doi.org/10.1534/g3.111.000349

Byrne OM, Hardie DC, Khan TN, Speijers J, Yan G (2008) Genetic analysis of pod and seed resistance to pea weevil in a Pisum sativum x P. fulvum interspecific cross. Aust J Agric Res 59:854–862. https://doi.org/10.1071/AR07353

Cardoso A, Pereira R, Fonseca M, Leitão J (2017) A microsatellite sequence in the fifth intron provides a broad-spectrum SSR marker for multiple alleles of the er1/ PsMLO1 powdery mildew resistance gene in *Pisum sativum* L. Mol Breed 37:84. https://doi.org/10.1007/s11032-017-0685-x

Carmona MA, Gally ME, Lopez SE (2005) Asian soybean rust: incidence, severity, and morphological characterization of *Phakopsora pachyrhizi* (uredinia and telia) in Argentina. Plant Dis 89:109–109

Carrillo E, Rubiales D, Pérez-De-Luque A, Fondevilla S (2012) Characterization of mechanisms of resistance against Didymella pinodes in Pisum spp. Eur J Plant Pathol 135:761–769. https://doi.org/10.1007/s10658-012-0116-0

Castillejo MA, Amiour N, Gaudot ED, Rubiales D, Jorrín JV (2004) A proteomic approach to studying plant response to crenate broomrape (*Orobanche crenata*) in pea (*Pisum sativum*). Phytochemistry 65:1817–1828

Castillejo MA, Curto M, Fondevilla S, Rubiales D, Jorrin JV (2010) Two-dimensional electrophoresis based proteomic analysis of the pea (*Pisum sativum*) in response to *Mycosphaerella pinodes*. J Agric Food Chem 58:12822–12832

Charlton A, Allnutt T, Holmes S, Chisholm J, Bean S, Ellis N, Mullineaux P, Oehlschlager S (2004) NMR profiling of transgenic peas. Plant Biotechnol J 2:27–35

Charlton A, Donarski JA, Harrison M, Jones SA, Godward J, Oehlschlager S, Arques JL, Ambrose M, Chinoy C, Mullineaux PM et al (2008) Responses of the pea (*Pisum sativum* L.) leaf metabolome to drought stress assessed by nuclear magnetic resonance spectroscopy. Metabolomics 4:312–327

Chavanne F, Zhang DX, Liaud MF, Cerff R (1998) Structure and evolution of *Cyclops*: a novel giant retrotransposon of the *Ty3/Gypsy* family highly amplified in pea and other legume species. Plant Mol Biol 37:363–375

Clement SL, Hardie DC, Elberson LR (2002) Variation among accessions of *Pisum fulvum* for resistance to pea weevil. Crop Sci 42:2167–2173. https://doi.org/10.2135/cropsci2002.2167

Clement SL, McPhee KE, Elberson LR, Evans MA (2009) Pea weevil, Bruchus pisorum L. (Coleoptera: Bruchidae), resistance in Pisum sativum x Pisum fulvum interspecific crosses. Plant Breed 128:478–485. https://doi.org/10.1111/j.1439-0523.2008.01603.x

Clulow SA, Lewis BG, Matthews P (1991a) A pathotype classification for Ascochyta pinodes. J Phytopathol 131:322–332. https://doi.org/10.1111/j.1439-0434.1991.tb01203.x

Clulow SA, Matthews P, Lewis BG (1991b) Genetic analysis of resistance to Mycosphaerella pinodes in pea seedlings. Euphytica 58:183–189. https://doi.org/10.1007/BF00022819

Coyne DP, Steadman JR, Godoy-Lutz G, Gilbertson R, Arnaud-Santana E, Beaver JS, Myers JR (2003) Contributions of the Bean/Cowpea CRSP to management of bean diseases. Field Crop Res 82:155–168

Coyne CJ, McClendon MT, Walling JG, Timmerman-Vaughan GM, Murray S et al (2007) Construction and characterization of two bacterial artificial chromosome libraries of pea (*Pisum sativum* L.) for the isolation of economically important genes. Genome 50:871–875

Curto M, Camafeita E, Lopez JA, Maldonado AM, Rubiales D, Jorrín JV (2006) A proteomic approach to study pea (*Pisum sativum*) responses to powdery mildew (*Erysiphe pisi*). Proteomics *6*:S163–S174

Dalmais M, Schmidt J, Le Signor C, Moussy F, Burstin J et al (2008) UTILLdb, a *Pisum sativum* in silico forward and reverse genetics tool. Genome Biol 9. https://doi.org/10.1186/gb-2008-9-2-r43

Dang JK, Sangwan MS, Yadar RK, Chowdhury RK (1994) Sources of resistance against powdery mildew in pea. Legum Res 17:231–232

Deepika, Mahajan R, Lay P, Sharma P, Salgotra RK, Sharma S (2017) Assessment of genetic variability, heritability and genetic advance among *Pisum* accessions. Vegetos Int J Plant Res 30:482–488

Dhall RK (2015) Breeding for biotic stresses resistance in vegetable crops: a review. Res & Rev: J Crop Sci Technol 4(1) ISSN: 2319-3395(online)

Dirlewanger E, Isaac PG, Ranade S, Belajouza M, Cousin R, de Vienne D (1994) Restriction fragment length polymorphism analysis of loci associated with disease resistance genes and developmental traits in Pisum sativum L. Theor Appl Genet 88:17–27. https://doi.org/10.1007/BF00222388

Doring TF, Storkey J, Baddeley J et al (2012) Legume based plant mixtures for delivery of multiple ecosystem services: weed diversity and weed control. In: McCracken K (ed) SAC SEPA biennial conference: valuing ecosystems: policy, economic and management interactions. Edinburgh, 3–4 Apr 2012, pp 163–168

du Preez ED, van Rij NC, Lawrance KF (2005) First report of soybean rust caused by *Phakopsora pachyrhizi* on dry beans in South Africa. Plant Dis 89:206–206

Duarte J, Rivière N, Baranger A, Aubert G, Burstin J, Cornet L et al (2014) Transcriptome sequencing for high throughput SNP development and genetic mapping in pea. BMC Genomics 15:126. https://doi.org/10.1186/1471-2164-15-126

Duc G, Blancard S, Hénault C, Lecomte C, Petit M, Bernicot M-H, et al (2010) Potentiels et leviers pour développer la production et l'utilisation des protéagineux dans le cadre d'une agriculture durable en Bourgogne. Innov Agron 11:157–171

Ek M, Eklund M, Von Post R, Dayteg C, Henriksson T, Weibull P, Ceplitis A, Isaac P, Tuvesson S (2005) Microsatellite markers for powdery mildew resistance in pea (Pisum sativum L.). Hereditas 142:86–91

Ellis THN, Poyser SJ (2002) An integrated and comparative view of pea genetic and cytogenetic maps. New Phytol 153:17–25

Ellis THN, Turner L, Hellens RP, Lee D, Harker CL, Enard C et al (1992) Linkage maps in pea. Genetics 130:649–663

Ellis THN, Hofer JI, Timmerman-Vaughan GM, Coyne CJ, Hellens RP (2011) Mendel, 150 years on. Trends Plant Sci 16:590–596

FAOSTAT (2013) http://faostat3.fao.org/

Feng J, Hwang R, Chang KF, Conner RL, Hwang SF, Strelkov SE et al (2011) Identification of microsatellite markers linked to quantitative trait loci controlling resistance to Fusarium root rot in field pea. Can J Plant Sci 91:199–204. https://doi.org/10.4141/cjps09176

Flavell RB, Bennett MD, Smith JB, Smith DB (1974) Genome size and the proportion of repeated nucleotide sequence DNA in plants. Biochem Genet 12:257–269

Fondevilla S, Avila CM, Cubero JI, Rubiales D (2005) Response to *Ascochyta pinodes* in a germplasm collection of Pisum spp. Plant Breed 124:313–315. https://doi.org/10.1111/j.1439-0523.2005.01104.x

Fondevilla S, Carver TLW, Moreno MT, Rubiales D (2007a) Identification and characterisation of sources of resistance to Erysiphe pisi Syd. in Pisum spp. Plant Breed 126:113–119. https://doi.org/10.1111/j.1439-0523.2006.01312.x

Fondevilla S, Torres AM, Moreno MT, Rubiales D (2007b) Identification of a new gene for resistance to powdery mildew in Pisum fulvum, a wild relative of pea. Breed Sci 57:181–184

Fondevilla S, Rubiales D, Moreno MT, Torres AM (2008a) Identification and validation of RAPD and SCAR markers linked to the gene Er3 conferring resistance to Erysiphe pisi DC in pea. Mol Breed 22:193–200. https://doi.org/10.1007/s11032-008-9166-6

Fondevilla S, Satovic Z, Rubiales D, Moreno MT, Torres AM (2008b) Mapping of quantitative trait loci for resistance to *Ascochyta pinodes* in Pisum sativum subsp. syriacum. Mol Breed 21:439–454. https://doi.org/10.1007/s11032-007-9144-4

Fondevilla S, Almeida NF, Satovic Z, Rubiales D, Patto MCV, Cubero JI et al (2011a) Identification of common genomic regions controlling resistance to *Mycosphaerella pinodes*, earliness and architectural traits in different pea genetic backgrounds. Euphytica 182:43–52. https://doi.org/10.1007/s10681-011-0460-8

Fondevilla S, Küster H, Krajinski F, Cubero JI, Rubiales D (2011b) Identification of genes differentially expressed in a resistant reaction to *Mycosphaerella pinodes* in pea using microarray technology. BMC Genomics 12. https://doi.org/10.1186/1471-2164-12-28

Fondevilla S, Martin-Sanz A, Satovic Z, Dolores Fernandez-Romero M, Rubiales D, Caminero C (2012) Identification of quantitative trait loci involved in resistance to Pseudomonas syringae pv. syringae in pea (Pisum sativum L.). Euphytica 186:805–812. https://doi.org/10.1007/s10681-011-0592-x

Fondevilla S, Rotter B, Krezdorn N, Jüngling R, Winter P, Rubiales D (2013) Identification of genes involved in resistance to Didymella pinodes in pea by deepSuperSAGE genome-wide transcriptome profiling. Book of abstracts of First Legume Society Conference, pp 148

Franssen SU, Shrestha RP, Bräutigam A, Bornberg-Bauer E, Weber APM (2011) Comprehensive transcriptome analysis of the highly complex Pisum sativum genome using next generation sequencing. BMC Genomics 12. https://doi.org/10.1186/1471-2164-12-227

Fuchs J, Kühne M, Schubert I (1998) Assignment of linkage groups to pea chromosomes after karyotyping and gene mapping by fluorescent *in situ* hybridization. Chromosoma 107:272–276

Ghezal N, Rinez I, Sbai H, Saad I, Farooq M et al (2016) Improvement of *Pisum sativum* salt stress tolerance by bio-priming their seeds using *Typha angustifolia* leaves aqueous extract. S Afr J Bot 105:240–250

Graham PH, Vance CP (2003) Legumes: importance and constraints to greater use. Plant Physiol 131:872–877
Gritton ET, Ebert RD (1975) Interaction of planting date and powdery mildew on pea plant performance. Am Soc Horticult Sci 100:137–142
Hall KJ, Parker JS, Ellis THN, Turner L, Knox MR, Hofer JMI et al (1997) The relationship between genetic and cytogenetic maps of pea. II. Physical maps of linkage mapping populations. Genome 40:755–769. https://doi.org/10.1139/g97-798
Hamon C, Baranger A, Coyne CJ, McGee RJ, Le Goff I, L'Anthoëne V et al (2011) New consistent QTL in pea associated with partial resistance to Aphanomyces euteiches in multiple French and American environments. Theor Appl Genet 123:261–281. https://doi.org/10.1007/s00122-011-1582-z
Hamon C, Coyne CJ, McGee RJ, Lesné A, Esnault R, Mangin P et al (2013) QTL meta-analysis provides a comprehensive view of loci controlling partial resistance to Aphanomyces euteiches in four sources of resistance in pea. BMC Plant Biol 13:45. https://doi.org/10.1186/1471-2229-13-45
Hamwieh A, Udupa SM, Choumane W, Sarker A, Dreyer F, Jung C, Baum M (2005) A genetic linkage map of *Lens* sp. based on microsatellite and AFLP markers and the localization of fusarium vascular wilt resistance. Theor Appl Genet 110:669–677
Hance ST, Grey W, Weeden NF (2004) Identification of tolerance to Fusarium solani in Pisum sativum ssp. elatius. Pisum Genet 36:9–13
Hanounik S (1981) Influence of Ronilan on the severity of chocolate spot and yield of faba bean. FABIS Newslett 3:50–51
Harland SC (1948) Inheritance of immunity to mildew in Peruvian forms of Pisum sativum. Heredity 2:263–269
Hayer F, Bonnin E, Carrouee B, Gaillard G, Nemecek T, Schneider A et al (2010) Designing sustainable crop rotations using life cycle assessment of crop combinations in 9th European IFSA symposium, Vienna
Heringa RJ, Van Norel A, Tazelaar MF (1969) Resistance to powdery mildew (Erysiphe polygoni D.C.) in peas (Pisum sativum L.). Euphytica 18:163–169
Hodgson EW, Kemis M, Geisinger B (2012) Assessment of Iowa soybean growers for insect pest management practices. J Ext 50:4RIB6
Hunter PJ, Ellis N, Taylor JD (2001) Association of dominant loci for resistance to Pseudomonas syringae pv. pisi with linkage groups II, VI and VII of Pisum sativum. Theor Appl Genet 103:129–135. https://doi.org/10.1007/s001220100566
Hussain M, Jang KH, Farooq M, Lee DJ (2012) Morphological and physiological evaluation of Korean rice genotypes for salt resistance. Int J Agric Biol 14:970–974
ICTV dB Management (2006) ICTVdB–the universal virus database, version 4. Columbia University, New York
Iglesias-García R, Prats E, Fondevilla S, Satovic Z, Rubiales D (2015) Quantitative trait loci associated to drought adaptation in pea (Pisum sativum L.). Plant Mol Biol Report:1–11. https://doi.org/10.1007/s11105-015-0872-z
Infantino A, Kharrat M, Riccioni L, Coyne CJ, McPhee KE, Grunwald NJ (2006) Screening techniques and sources of resistance to root diseases in cool season food legumes. Euphytica 147:201–221
Jain A, Singh A, Singh S, Singh V, Singh HB (2015) Comparative proteomic analysis in pea treated with microbial consortia of beneficial microbes reveals changes in the protein network to enhance resistance against Sclerotinia sclerotiorum. J Plant Physiol 182:79–94
Janila P, Sharma B (2004) RAPD and SCAR markers for powdery mildew resistance gene er in pea. Plant Breed 123:271–274
Jensen ES, Peoples MB, Boddey RM, Gresshoff PM, Hauggaard-Nielsen H, Alves BJR et al (2012) Legumes for mitigation of climate change and the provision of feedstock for biofuels and biorefineries. A review. Agron Sustain Dev 32:329–364. https://doi.org/10.1007/s13593-011-0056-7

Jha AB, Warkentin TD, Gurusamy V, Tar'an B, Banniza S (2012) Identification of ascochyta blight resistance in wild Pisum species for use in pea breeding. Crop Sci 52:2462–2468. https://doi.org/10.2135/cropsci2012.04.0242

Jha AB, Arganosa G, Tar'An B, Diederichsen A, Warkentin TD (2013) Characterization of 169 diverse pea germplasm accessions for agronomic performance, *Mycosphaerella blight* resistance and nutritional profile. Genet Resour Crop Evol 60:747–761. https://doi.org/10.1007/s10722-012-9871-1

Jha AB, Tar'an B, Stonehouse R, Warkentin TD (2016) Identification of QTLs associated with improved resistance to ascochyta blight in an interspecific pea recombinant inbred line population. Crop Sci 56:2926–2939. https://doi.org/10.2135/cropsci2016.01.0001

Kabir AH, Paltridge NG, Able AJ, Paull JG, Stangoulis JCR (2012) Natural variation for Fe-efficiency is associated with upregulation of Strategy I mechanisms and enhanced citrate and ethylene synthesis in Pisum sativum L. Planta 235:1409–1419. https://doi.org/10.1007/s00425-011-1583-9

Kalo P, Seres A, Taylor SA, Jakab J, Kevei Z, Kereszt A et al (2004) Comparative mapping between Medicago sativa and *Pisum sativum*. Mol Gen Genomics 272:235–246. https://doi.org/10.1007/s00438-004-1055-z PMID: 15340836

Katoch V, Sharma S, Pathania S, Banayal DK, Sharma SK, Rathour R (2010) Molecular mapping of pea powdery mildew resistance gene er2 to pea linkage group III. Mol Breed 25:229–237. https://doi.org/10.1007/s11032-009-9322-7

Kaur S, Pembleton LW, Cogan NOI et al (2012) Transcriptome sequencing of field pea and faba bean for discovery and validation of SSR genetic markers. BMC Genomics 13:104

Khan MRA, Mahmud F, Reza MA, Mahbub MM, Shirazy BJ, Rahman MM (2017) Genetic diversity, correlation and path analysis for yield and yield components of pea (*Pisum sativum* L.). World J Agric Sci 13:11–16

Khare M, Khare U (2012) Studies on survival of *Xanthomonas campestris* pv. glycines. Ind Phytopathol 48:180–181

Kraft JM, Dunne B, Goulden D, Armstrong S (1998) A search for resistance in peas to *Mycosphaerella pinodes*. Plant Dis 82:251–253. https://doi.org/10.1094/PDIS.1998.82.2.251

Kulaeva OA, Zhernakov AI, Afonin AM, Boikov SS, Sulima AS et al (2017) Pea Marker Database (PMD): a new online database combining known pea (*Pisum sativum* L.) gene-based markers. PLoS One 12:e0186713. https://doi.org/10.1371/journal.pone.0186713

Kumar PL, Kumari SMG, Waliyar F (2008a) Virus disease Studies on survival of *Xanthomonas campestris* pv. glycines. Ind Phytopathol 48:180–181

Kumar PL, Prasada Rao RDVJ, Reddy AS, Madhavi KJ, Anitha K, Waliyar F (2008b) Emergence and spread of Tobacco streak virus menace in Indian and control strategies. Indian J Plant Protect 36:1–8

Kumar PL, Jones AT, Waliyar F (2008c) Virus diseases of pegionpea. In: Rao GP, Kumar PL, Holguin-Peña RJ (eds) Vegetable and pulse crops: Vol. 3. Characterization, diagnosis and management of plant viruses. Studium Press LLC, Hardcover, pp 235–258

Kumar JK, Prasad A, Richard SA (2012) In vitro antioxidant activity and preliminary phytochemical analysis of medicinal legumes. J Pharm Res 5:3059–3062

Leach M, Agudelo P, Lawton-Rauh A (2012) Genetic variability of *Rotylenchulus reniformis*. Plant Dis 96:30–36

Lejeune-Hénaut I, Hanocq E, Bethencourt L, Fontaine V, Delbreil B, Morin J et al (2008) The flowering locus Hr colocalizes with a major QTL affecting winter frost tolerance in Pisum sativum L. Theor Appl Genet 116:1105–1116. https://doi.org/10.1007/s00122-008-0739-x

Leonforte A, Forster JW, Redden RJ, ME N, Salisbury PA (2013a) Sources of high tolerance to salinity in pea (Pisum sativum L.). Euphytica 189:203 216. https://doi.org/10.1007/s10681-012-0771-4

Leonforte A, Sudheesh S, Cogan NOI, Salisbury PA, Nicolas ME, Materne M et al (2013b) SNP marker discovery, linkage map construction and identification of QTLs for enhanced

salinity tolerance in field pea (*Pisum sativum* L.). BMC Plant Biol 13:161. https://doi.org/10.1186/1471-2229-13-161
Lewis G, Schrire B, Mackind B, Lock M (2005) Legumes of the world. Royal Botanic Gardens, Kew
Liu R, Fang L, Yang T, Zhang X, Hu J et al (2017) Marker-trait association analysis of frost tolerance of 672 worldwide pea (*Pisum sativum* L.) collections. Sci Rep 7:5919. https://doi.org/10.1038/s41598-017-06222-y
Loebenstein G, Thottappilly G (eds) (2003) Virus and virus-like diseases of major crops in developing countries, vol 800. Kluwer Academic Publishers, Dordrecht
Lombardo S, Colombo A, Rapisarda C (2011) Severe damage caused by *Meloidogyne artiellia* on cereals and leguminous in Sicily. Redia 94:149–151
Loridon K, McPhee K, Morin J, Dubreuil P, Pilet-Nayel ML, Aubert G et al (2005) Microsatellite marker polymorphism and mapping in pea (Pisum sativum L.). Theor Appl Genet 111:1022–1031. https://doi.org/10.1007/s00122-005-0014-3
Ma Y, Coyne CJ, Grusak MA, Mazourek M, Cheng P et al (2017) Genome-wide SNP identification, linkage map construction and QTL mapping for seed mineral concentrations and contents in pea (*Pisum sativum* L.). BMC Plant Biol 17:43. https://doi.org/10.1186/s12870-016-0956-4
Macas J, Neumann P, Navrátilová A (2007) Repetitive DNA in the pea (*Pisum sativum* L.) genome: Comprehensive characterization using 454 sequencing and comparison to soybean and *Medicago truncatula*. BMC Genomics 8. https://doi.org/10.1186/1471-2164-8-427
Macwilliam S, Wismer M, Kulshreshtha S (2014) Life cycle an economic assessment of Western Canadian pulse systems: the inclusion of pulses in crop rotations. Agric Syst 123:43–53. https://doi.org/10.1016/j.agsy.2013.08.009
Malathi VG, John P (2008) Gemini viruses infecting legumes. In: Rao GP, Kumar PL, Holguin-Peña RJ (eds) Vegetable and pulse crops: Vol. 3. Characterization, diagnosis and management of plant viruses. Studium Press LLC, Hardcover, pp 97–123
Mandal B, Jain RK, Krishnareddy M, Krishna Kumar NK, Ravi KS, Pappu HR (2012) Emerging problems of Tospoviruses (*Bunyaviridae*) and their management in the Indian Subcontinent. Plant Dis 96:468–479
Maqbool A, Shafiq S, Lake L (2010) Radiant frost tolerance in pulse crops-a review. Euphytica 172:1–12
McGee RJ, Coyne CJ, Pilet-Nayel ML, Moussart A, Tivoli B, Baranger A et al (2012) Registration of pea germplasm lines partially resistant to aphanomyces root rot for breeding fresh or freezer pea and dry pea types. J Plant Regist 6:203–207. https://doi.org/10.3198/jpr2011.03.0139crg
McPhee KE, Inglis DA, Gundersen B, Coyne CJ (2012) Mapping QTL for Fusarium wilt Race 2 partial resistance in pea (Pisum sativum). Plant Breed 131:300–306. https://doi.org/10.1111/j.1439-0523.2011.01938.x
Mehrkhou F, Talebi AA, Moharramipour S, Naveh VH (2012) Demographic parameters of Spodoptera exigua (Lepidoptera: Noctuidae) on different soybean cultivars. Environ Entomol 41:326–332
Menon M, Barnes WJ, Olson MS (2015) Population genetics of freeze tolerance among natural populations of *Populus balsamifera* across the growing season. New Phytol 207:710–722
Mishra RK, Gangadhar BH, Nookaraju A, Kumar S, Park SW (2012) Development of EST-derived SSR markers in pea (*Pisum sativum*) and their potential utility for genetic mapping and transferability. Plant Breed 131:118–124. https://doi.org/10.1111/j.1439-0523.2011.01926.x
Morton RL, Schroeder HE, Bateman KS, Chrispeels MJ, Armstrong E, Higgins TJV (2000) Bean a-amylase inhibitor 1 in transgenic peas (*Pisum sativum*) provides complete protection from pea weevil (*Bruchus pisorum*) under field conditions. Proc Natl Acad Sci U S A 97(8):3820–3825 10.1073ypnas.070054597
Munjal RL, Chenulu VV, Hora TS (1963) Assessment of losses due to powdery mildew (*Erysiphe polygoni*) on pea. Indian Phytopathol 19:260–267
Murray MG, Cuellar RE, Thompson WF (1978) DNA sequence organization in the pea genome. Biochemistry 17:5781–5790

Musser FR, Catchot AL, Gibson BK, Knighten KS (2011) Economic injury levels for southern green stink bugs (Hemiptera: Pentatomidae) in R7 growth stage soybeans. Crop Prot 30:63–69

Nemecek T, Kägi T (2007) Life cycle inventories of Swiss and European agricultural production systems. In: Final Report Ecoinvent V2. 0 No 15a. Agroscope Reckenholz-Tänikon Research Station ART. Swiss Centre for Life Cycle Inventories, Zurich/Dübendorf. SimaPro Ph.D, 7.

Nemecek T, Von Richthofen JS, Dubois G Casta P, Charles R, Pahl H (2008) Environmental impacts of introducing grain legumes into European crop rotations. Eur J Agron 28:380–393. https://doi.org/10.1016/j.eja.2007.11.004

Nene YL, Reddy MV (1987) Chickpea diseases and their control. In: Saxena MC, Singh KB (eds) The Chickpea. CAB International, Oxon, UK, pp 233–270

Neumann P, Navratilova A, Koblizkova A, Kejnovsky E, Hribova E, Hobza R, Widmer A, Dolezel J, Macas J (2011) Plant centromeric retrotransposons: A structural and cytogenetic perspective. Mob DNA 2. https://doi.org/10.1186/1759-8753-2-4

Nifantova SN, Simonenko IV, Komarnitskiĭ IK, Kuchuk NV (2005) Production of transgenic pea (*Pisum sativum* L.) plants resistant to the herbicide pursuit. Tsitol Genet 39(2):16–21

Novák P, Neumann P, Macas J (2010) Graph-based clustering and characterization of repetitive sequences in next-generation sequencing data. BMC Bioinforma 11. https://doi.org/10.1186/1471-2105-11-378

Ondřej M, Dostálová R, Odstrčilová L (2005) Response of Pisum sativum germplasm resistant to Erysiphe pisi to inoculation with Erysiphe baeumleri, a new pathogen of peas. Plant Prot Sci 41:95–103

Pal AB, Brahmappa RDR, Ullasa BA (1980) Field resistance of pea germplasm to powdery mildew (Erysiphe polygoni) and rust (Uromyces fabae). Plant Dis 64:1085–1086

Pereira G, Leitão J (2010) Two powdery mildew resistance mutations induced by ENU in Pisum sativum L. affect the locus er1. Euphytica 171(3):345–354

Petkova V, Nikolova V, Kalapchieva SH, Stoeva V, Topalova E, Angelova S (2009) Physiological response and pollen viability of Pisum sativum genotypes under high temperature influence. Iv Balkan Symp Veg Potato 830:665–671. https://doi.org/10.17660/actahortic.2009.830.96

Pilet-Nayel L, Muehlbauer FJ, McGee RJ, Kraft JM, Baranger A, Coyne CJ (2002) Quantitative trait loci for partial resistance to Aphanomyces root rot in pea. Theor Appl Genet 106:28–39

Pilet-Nayel ML, Muehlbauer FJ, McGee RJ, Kraft JM, Baranger A, Coyne CJ (2005) Consistent quantitative trait loci in pea for partial resistance to Aphanomyces euteiches isolates from the United States and France. Phytopathology 95:1287–1293. https://doi.org/10.1094/PHYTO-95-1287

Prioul S, Frankewitz A, Deniot G, Morin G, Baranger A (2004) Mapping of quantitative trait loci for partial resistance to Ascochyta pinodes in pea (Pisum sativum L.) at the seedling and adult plant stages. Theor Appl Genet 108:1322–1334. https://doi.org/10.1007/s00122-003-1543-2

Puonti-Kaerlas J, Eriksson T, Engström P (1990) Production of transgenic pea (*Pisum sativum* L.) plants by Agrobacterium tumefaciens – mediated gene transfer. Theor Appl Genet 80:246–252

Quintero JM, Fournier JM, Benlloch M (2007) Na^+ accumulation in shoot is related to water transport in K^+-starved sunflower plants but not in plants with a normal K^+ status. J Plant Physiol 164:60–67

Radchuk R, Emery NJR, Weier D, Vigeolas H, Geigenberger P, Lunn JE, Feil R, Weschke W, Weber H (2010) Sucrose non-fermenting kinase 1 (SnRK1) coordinates metabolic and hormonal signals during pea cotyledon growth and differentiation. Plant J 61(324):338

Rai R, Singh AK, Singh BD, Joshi AK, Chand R, Srivastava CP (2011) Molecular mapping for resistance to pea rust caused by Uromyces fabae (Pers.) de-Bary. Theor Appl Genet 123:803–813. https://doi.org/10.1007/s00122-011-1628-2

Rana JC, Rana M, Sharma V, Nag A, Rakesh K, Chahota RK, Sharma TR (2016) Genetic diversity and structure of pea (*Pisum sativum* L.) germplasm based on morphological and SSR markers. Plant Mol Biol Report. https://doi.org/10.1007/s11105-016-1006-y

Rashid A, Khan RU, Marwat SK (2009) Importance of weed control in chickpea under rainfed condition. Am Eurasian J Agric Environ Sci 5:456–459

Riebeseel E, Häusler RE, Radchuk R, Meitzel T, Hajirezaei MR, Emery RJN, Küster H, Nunes-Nesi A, Fernie AR, Weschke W et al (2010) The 2-oxoglutarate/malate translocator mediates amino acid and storage protein biosynthesis in pea embryos. Plant J 61:350–363

Rubiales D, Emeran AA, Sillero JC (2002) Rusts on legumes in Europe and North Africa. Grain Legumes 37:8–9

Rubiales D, P'erez-de-Luque A, Sillero JC, Rom´an B, Kharrat M, Khalil S, Joel DM, Riches C (2006) Screening techniques and sources of resistance against parasitic weeds in grain legumes. Euphytica 147:187–199

Salinas J, Matassi G, Montero LM, Bernardi G (1988) Compositional compartmentalization and compositional patterns in the nuclear genomes of plants. Nucleic Acids Res 16:4269–4285

Sastry KS, Zitter TA (2014) Management of virus and viroid diseases of crops in the tropics, Chapter 2. In: Plant virus and viroid diseases in the tropics, vol 1. Springer Science+Business Media B.V., pp 149–480

Shabala S (2013) Learning from halophytes: physiological basis and strategies to improve abiotic stress tolerance in crops. Ann Bot 112:1209–1221

Sharma N (1992) Evaluation of varietal susceptibility in pea to Erysiphe polygoni. Ann Appl Biol 120:110–111

Sillero JC, Fondevilla S, Davidson J, Vaz Patto MC, Warkentin TD, Thomas J, Rubiales D (2006) Screening techniques and sources of resistance to rusts and mildews in grain legumes. Euphytica 147:255–272

Sindhu A, Ramsay L, Sanderson LA, Stonehouse R, Li R, Condie J et al (2014) Gene-based SNP discovery and genetic mapping in pea. Theor Appl Genet 127:2225–2241. https://doi.org/10.1007/s00122-014-2375-y

Smitchger JA, Burke IC, Yenish JP (2012) The critical period of weed control in lentil (*Lens culinaris*) in the pacific Northwest. Weed Sci 60:81–85

Smýkal P, Aubert G, Burstin J, Coyne CJ, Ellis NTH et al (2012) Pea (*Pisum sativum* L.) in the genomic era. Agronomy 2:74–115. https://doi.org/10.3390/agronomy2020074

Subbarao GV, Johansen C (1994) Strategies and scope for improving salinity tolerance in crop plants. In: Pessarakli M (ed) Handbook of plant and crop stress. Marcel Dekker, New York, pp 559–579

Sudheesh S, Lombardi M, Leonforte A, Cogan NOI, Materne M, Forster JW et al (2014) Consensus genetic map construction for field pea (Pisum sativum L.), trait dissection of biotic and abiotic stress tolerance and development of a diagnostic marker for the er1 powdery mildew resistance gene. Plant Mol Biol Report 33:1391–1403. https://doi.org/10.1007/s11105-014-0837-7

Sun X, Yang T, Hao J, Zhang X, Ford R, Jiang J et al (2014) SSR genetic linkage map construction of pea (*Pisum sativum* L.) based on Chinese native varieties. Crop J 2:170–174. https://doi.org/10.1016/j.cj.2014.03.004

Sun S, Fu H, Wang Z, Duan C, Zong X, Zhu Z (2016) Discovery of a novel er1 allele conferring powdery mildew resistance in Chinese pea (Pisum sativum L.) Landraces. PLoS One 11(1):e0147624. https://doi.org/10.1371/journal.pone.0147624

Svabova L, Smýkal P, Griga M, Ondrej V (2005) *Agrobacterium*-mediated transformation of *Pisum sativum* in vitro and in vivo. Biol Plant 49:361–370

Sweetingham MW, Jones RAC, Brown AGP (1998) Diseases and pests. In: Gladstones JS, Atkins C, Hamblin J (eds) Lupins as crop plants: biology, production and utilization. CAB International, Wallingford, pp 263–290

Tar'an B, Warkentin T, Somers DJ, Miranda D, Vandenberg A, Balde S et al (2003) Quantitative trait loci for lodging resistance, plant height and partial resistance to ascochyta blight in field pea (Pisum sativum L.). Theor Appl Genet 107:1482–1491. https://doi.org/10.1007/s00122-003-1379-9

Tar'an B, Warkentin T, Somers DJ, Miranda D, Vandenberg A, Blade S et al (2004) Identification of quantitative trait loci for grain yield, seed protein concentration and maturity in field pea (Pisum sativum L.). Euphytica 136:297–306. https://doi.org/10.1023/B:EUPH.0000032721.03075.a0

Tayeh N, Aubert G, Pilet-Nayel M-L, Lejeune-Hénaut I, Warkentin TD, Burstin J (2015a) Genomic tools in pea breeding programs: status and perspectives. Front Plant Sci 6:1037

Tayeh N, Klein A, Le Paslier MC, Jacquin F, Houtin H, Rond C et al (2015b) Genomic prediction in pea: effect of marker density and training population size and composition on prediction accuracy. Front Plant Sci 6:941. https://doi.org/10.3389/fpls.2015.00941

Thakur BR, Kapoor AS, Jamwal RS (1996) Varietal resistance of pea to powdery mildew in dry temperature zone of Himachal Pradesh. Indian Phytopathol 49:92–93

Timmerman GM, Frew TJ, Weeden NF, Miller AL, Goulden DS (1994) Linkage analysis of er-1, a recessive Pisum sativum gene for resistance to powdery mildew fungus (Erysiphe pisi D.C.). Theor Appl Genet 88:1050–1055

Timmerman-Vaughan GM, McCallum JA, Frew TJ, Weeden NF, Russell AC (1996) Linkage mapping of quantitative trait loci controlling seed weight in pea (Pisum sativum L). Theor Appl Genet 93:431–439. https://doi.org/10.1007/BF00223187

Timmerman-Vaughan GM, Frew TJ, Russell AC, Khan T, Butler R, Gilpin M et al (2002) QTL mapping of partial resistance to field epidemics of ascochyta blight of pea. Crop Sci 42:2100–2111. https://doi.org/10.2135/cropsci2002.2100

Timmerman-Vaughan GM, Frew TJ, Butler R, Murray S, Gilpin M, Falloon K et al (2004) Validation of quantitative trait loci for Ascochyta blight resistance in pea (Pisum sativum L.), using populations from two crosses. Theor Appl Genet 109:1620–1631. https://doi.org/10.1007/s00122-004-1779-5

Tivoli B, Baranger A, Avila CM, Banniza S, Barbetti M, Chen W, Davidson J, Lindeck K, Kharrat M, Rubiales D, Sadiki M, Sillero JC, Sweetingham M, Muehlbauer FJ (2006) Screening techniques and sources of resistance to foliar diseases caused by major necrotrophic fungi in grain legumes. Euphytica 147:223–253

Tiwari KR, Penner GA, Warkentin TD (1998) Identification of coupling and repulsion phase markers for powdery mildew resistance genes er1 in pea. Genome 41:440–444

Tiwari KR, Penner GA, Warkentin TD (1999) Identification of AFLP markers for powdery mildew resistance gene er2 in pea. Pisum Genet 31:27–29

Ubayasena L, Bett K, Tar'an B, Warkentin T (2011) Genetic control and identification of QTLs associated with visual quality traits of field pea (Pisum sativum L.). Genome 54:261–272. https://doi.org/10.1139/g10-117

Valderrama MR, Roman B, Satovic Z, Rubiales D, Cubero JI, Torres AM (2004) Locating quantitative trait loci associated with Orobanche crenata resistance in pea. Weed Res 44:323–328. https://doi.org/10.1111/j.1365-3180.2004.00406.x

Van der Linden L, Bredenkamp J, Naidoo S et al (2013) Gene for- gene tolerance to bacterial wilt in Arabidopsis. Mol Plant-Microbe Interact 26:398–406

van Leur JAG, Aftab M, Leonforte B, Moore S, Freeman AJ (2007) Control of pea seed borne mosaic virus in field pea through resistance breeding. In: Proceedings 16th Biennial Conference of the Australasian Plant Pathology Society, Adelaide, 2007.

Viteri D, Cabrera I, De E, Jensen C (2009) New record of *Thrips* species associated with soybeans in Puerto Rico. Florida Entomol 92:181–185

Vovlas N, Troccoli A, Palomares-Rius J et al (2011) *Ditylenchus gigas* spp. parasitizing broad bean: a new stem nematode singled out from the *Ditylenchus dipsaci* species complex using apolyphasic approach with molecular phylogeny. Plant Pathol 60:762–777

Wang Z, Luo Y, Li X, Wang L, Xu S, Yang J et al (2008) Genetic control of floral zygomorphy in pea (*Pisum sativum* L.). Proc Natl Acad Sci U S A 105:10414–10419

Warkentin TD, Rashid KY, Xue AG (1996) Fungicidal control of powdery mildew in field pea. Can J Plant Sci 76:933–935

Warkentin TD, Smykal P, Coyne CJ, Weeden N, Domoney C, Bing D et al (2015) Pea (Pisum sativum L.). In: De Ron AM (ed) Grain Legumes, Handbook of Plant Breeding. Springer Science+Business Media, New York, pp 37–83

Weeden NF, Porter L (2007) The genetic basis of Fusarium root rot tolerance in the 'Afghanistan' pea. Pisum Genet 39:35–36

Weeden NF, Ellis THN, Timmerman-Vaughan GM, Swiecicki WK, Rozov SM, Berdnikov VA (1998) A consensus linkage map for Pisum sativum. Pisum Genet 30:1–4

Weeden NF, McGee R, Grau CR, Muehlbauer FJ (2000) A gene influencing tolerance to common root rot is located on linkage group IV. Pisum Genet 32:53–55

Weeden NF, Brauner S, Przyborowski JA (2002) Genetic analysis of pod dehiscence in pea (*Pisum sativum* L.). Cell Mol Biol Lett 7:657–663

Weigelt K, Kuster H, Radchuk R, Muller M, Weichert H, Fait A, Fernie AR, Saalbach I, Weber H (2008) Increasing amino acid supply in pea embryos reveals specific interactions of N and C metabolism, and highlights the importance of mitochondrial metabolism. Plant J 55:909–926

Weigelt K, Küster H, Rutten T, Fait A, Fernie AR, Miersch O, Wasternack C, Emery RJN, Desel C, Hosein F et al (2009) ADP-glucose pyrophosphorylase-deficient pea embryos reveal specific transcriptional and metabolic changes of carbon-nitrogen metabolism and stress responses. Plant Physiol 149:395–411

Wroth JM (1998) Possible role for wild genotypes of Pisum spp. to enhance ascochyta blight resistance in pea. Aust J Exp Agric 38:469–479. https://doi.org/10.1071/EA98024

Xin Z, Browse J (2000) Cold comfort farm: the acclimation of plants to freezing temperatures. Plant Cell Environ 23:893–902

Xu R, Li W, Zhang LF, Lin YH, Qi B, Xing H (2010) A study on the inheritance of resistance to whitefly in soybean. Sci Agric Sin 43:72–78

Yang T, Fang L, Zhang X, Hu J, Bao S, Hao J et al (2015) High-throughput development of SSR markers from pea (*Pisum sativum* L.) based on next generation sequencing of a purified Chinese commercial variety. PLoS One 10:e0139775. https://doi.org/10.1371/journal.pone.0139775

Zhang R, Hwang SF, Chang KF, Gossen BD, Strelkov SE, Turnbull GD et al (2006) Genetic resistance to Ascochyta pinodes in 558 field pea accessions. Crop Sci 46:2409–2414. https://doi.org/10.2135/cropsci2006.02.0089

Zhuang X, McPhee KE, Coram TE, Peever TL, Chilvers MI (2012) Rapid transcriptome characterization and parsing of sequences in a non-model host-pathogen interaction; pea- *Sclerotinia sclerotiorum*. BMC Genomics 13:668

Zhukov VA, Zhernakov AI, Kulaeva OA, Ershov NI, Borisov AY, Tikhonovich IA. (2015) De Novo Assembly of the Pea (*Pisum sativum* L.) Nodule Transcriptome. International Journal of Genomics. 695947. https://doi.org/10.1155/2015/695947

Chapter 7
Insights into Insect Resistance in Pulse Crops: Problems and Preventions

Santisree Parankusam, Sricindhuri Katamreddy, Pradeep Reddy Bommineni, Pooja Bhatnagar-Mathur, and Kiran K. Sharma

7.1 Introduction

7.1.1 What are Pulses and Their Importance?

Pulses are the dry seeds of food legumes, known as smart foods due to their high protein and essential amino acid content. They play a key role in crop rotation by improving soil fertility due to their ability to establish symbiotic relationship with soil microbes that can fix atmospheric nitrogen (Sharma et al. 2016). Pulses contribute significantly to the nutritional security and economy of majority of the population by complementing to cereals in crop rotation and avoiding protein-calorie malnutrition. Because pulses are the affordable source of proteins and minerals, thus they play a key role in alleviating undernourishment-related issues especially in developing countries. Being highly water efficient, pulses can be grown in all three seasons and thus constitute an important source of income for millions of poor farmers in the semiarid and subtropical regions. Moreover, at some regions farmers grow pulse crops either in monoculture or intercropping with other food and commercial crop plants including cotton, soybean, groundnut, etc. Hence, UN declared 2016 as International Year of Pulses to acknowledge the potential of pulses in addressing future global food security, nutrition, and environmental sustainability. Because of increasing demand and advent of the new agriculture technology during mid-1960s, the productivity of pulses has been doubled from the level of 441 kg/ha to about 800 kg/ha. The major pulse crops under global cultivation include pigeon pea (*Cajanus cajan*), chickpea (*Cicer arietinum*), rajamash (*Phaseolus vulgaris*), mung bean (*Vigna radiate*), cowpea (*Vigna unguiculata*), ricebeans

S. Parankusam (✉) · S. Katamreddy · P. R. Bommineni · P. Bhatnagar-Mathur · K. K. Sharma
International Crops Research Institute for the Semi-Arid Tropics (ICRISAT), Hyderabad, Telangana, India
e-mail: S.Parankusam@cgiar.org

S. H. Wani, M. Jain (eds.), *Pulse Improvement*,
https://doi.org/10.1007/978-3-030-01743-9_7

(*Vigna umbellata*), urd bean (*Vigna mungo*), lentil (*Lens culinaris*), faba bean (*Vicia faba*), and lathyrus (*Lathyrus sativus*).

7.1.2 Global Status of Pulse Production and Consumption

The global area under pulse production, as recorded during 2013, is about 807.54 lakh ha producing 730.07 lakh tones with an average production of 904 kg/ha. The percentage of global cultivated area and production of major pulse crops is given in Fig. 7.1. According to the FAO statistics, the average pulse production in the world was about 63.9 million tons during the last decade (2003–2012), with Asia being the major (45.62%) contributor of the global production (http://faostat3.fao.org/home/E, accessed in 2014). Approximately, 2.6 million pounds of pulses have been harvested every year in the USA, of which 75% is exported to the developing countries (Asif et al. 2013). Global statistical data reveal that the production of pulses cannot be sufficient to meet the requirements of the growing population. In spite of that, an average marginal global annual production increase of 0.77% has only witnessed in the last five decades. Pulses are grown approximately by 171 countries in the world, of these, beans cultivated by 120 countries accounting for 37.5% of global area under pulse production. Similarly, chickpea is being grown by 52 countries contributing about 16.77%, cowpea by 33 countries contributing 15.05%, followed by peas grown in 96 countries contributing 8.50%, pigeon pea by 21 counties contributing 7.70%, lentils grown by 51 countries contributing 5.38%, and others contributing 11.1% to the global pulse cultivation. The global production of dry beans was 33.50% followed by dry peas 15.91%, chickpea 17.95%, cowpea 8.45%, pigeon pea 6.50%, lentil 6.78%, and others 17.69%. India has been ranked first in the world with 35% of cultivated area and 25% of global production of pulses. Given the growing demand from increasing population, production shortages due to climatic changes and urbanization that consequent price rise have substantially reduced the per capita consumption of pulses round the globe which is expected to go further down in future (Kumar et al. 2011).

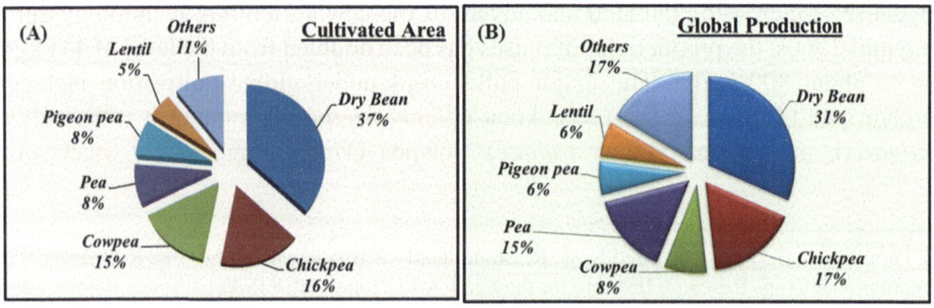

Fig. 7.1 The pie charts depict the global cultivated area (**a**) and production (**b**) of the major pulse crops

7.1.3 Major Pests of Pulses Crops

Pulses are constantly challenged by a number of insect pests, diseases, and weeds (Shanower et al. 1999). Although about 200 species found to attack pulse crops, only 1 dozen has been a major cause of damage to pulses (Keneni et al. 2011). Insect pests feeding on flowers, pods, and seeds are the most important biotic constraints affecting yields both in field and storage. For example, insect pests such as pod borers *Maruca vitrata*, *Helicoverpa armigera*, aphids *Aphis craccivora* Koch, whitefly *Bemisia tabaci* Genn, tobacco caterpillar *Spodoptera* sp., thrips *Megaleurothrips distalis*, leaf hopper *Empoasca* sp., and *Caliothrips indicus* cause extensive damage under field conditions while bruchids *Callosobruchus* sp. damage the grain in storage (Reddy 2009). Seeds attacked by storage insect pests particularly bruchid beetles including *Callosobruchus chinensis*, *C. maculatus*, *C. analis*, and *Bruchus incarnates* suffer poor germination and are no longer suitable for food (Keneni et al. 2011). Insect pests cause severe damage to pulse crops either by sucking sap or chewing plant parts like leaves, stems, roots, or fruits and by acting as a vector for viral, bacterial, or fungal transmission. The direct damage due to insect pests includes fouling, deformations, necrosis of plant tissues, and dissemination of plant pathogens, whereas indirect damage involves the loss of harvest quality, thus elevating the overall cost of crop production (Bardner and Fletcher 1974). The approximate pulse crop loss from insect pests has been up to 2–2.1 million ton accounting approximately to 6000 crore annually. However, the extent of loss may vary with climatic conditions and crop and pest type (Oerke 2006). Unfortunately, pulse crops are infested with number of pests throughout the cropping season due to favorable climatic conditions.

A wide variety of insect pests cause a variety of damage symptoms in pulses. A large number of beetles and pod-sucking bugs cause severe foliar damage resulting in stunted growth, late flowering and fruiting (Singh and van Emden 1979). Another major class, the sap-sucking insects such as aphids, whiteflies, and mirids deplete the assimilates by removal of sap and thus act as vectors for virus transmission. The best examples of aphids include soybean aphid (*Aphis glycine*), cowpea aphid (*A. craccivora*), spotted alfalfa (*Therioaphis trifolii*), bluegreen aphid (*Acyrthosiphon kondoi*), and pea aphid (*Acyrthosiphon pisum*) (Kamphuis et al. 2013a). Of late, the most notorious pests, pod borers, belonging to the order Lepidoptera cause great damage to foliage, flower, and pods. The pod borers of species *Helicoverpa* and *Spodoptera* have gained a great deal of attention in recent years due to their wide host range, severity of damage, and resistance to commonly used chemical insecticides. They often suck away sap from developing grains rendering them to shriveled condition (War et al. 2013; Higgins et al. 2012; Acharjee et al. 2010). If the incidence of Hubner is not controlled in key pulse growing areas, it can even lead to complete crop failure especially in long-duration varieties, and hence farmers often discourage to grow pulses in these areas (Gopali et al. 2013; Sharma et al. 2010).

Here we describe few most important groups of pests common for majority of cultivated pulses.

7.1.3.1 Gram Pod Borer (*Helicoverpa armigera*)

Among the variety of insects feeding on pulse, the pod borer, *Helicoverpa armigera*, is the most damaging pest worldwide, and its frequent occurrence often results in complete crop failure. Pod borer can invade the crop from seedling to maturity. Normally larvae remain hidden in the foliage of crop unnoticed till the formation of pods. After pod formation, they devour the seed inside by making holes in the pod walls. Because of its complex nature and non-availability of good resistance sources in cultivated gene pool, breeding for resistance to *Helicoverpa* has remained a serious challenge (Sharma et al. 2008). However, the use of moderate resistant cultivars, early sowing, growing short-duration varieties, intercropping, and using pheromone traps, bird perches, and spraying chemicals such as monocrotophos, endosulfan, and indoxacarb help to reduce pod borer damage.

7.1.3.2 Bruchids (*Callosobruchus chinensis*)

Bruchids (*Callosobruchus* spp.) are cosmopolitan and most dangerous storage pest of pulses grown in tropics and subtropics (Mishra et al. 2017). The adult and grub feed on the grain by making holes. Although crops are infested in the field, infestations are often too low to detect at harvest. Bruchids breed rapidly in storage and make the infested grain unmarketable before detection. Cleaning out old seed residues in storage places before harvest, treating infected or high-risk seeds or area with fumigation, and slowing bruchid development by reducing storage temperatures are few strategies to counter bruchid damage.

7.1.3.3 Aphids (*Aphis craccivora*)

The black aphid (*Aphis craccivora*) is the main aphid pest of pulse crops with wide distribution across many parts of the world. Aphid nymphs and adults suck sap from the leaves, stem, and pods, causing depletion of the photosynthates. The aphids are also capable of transmitting several viral diseases such as leaf roll luteovirus, chickpea chlorotic dwarf virus, and chickpea distortion mosaic virus (Kamphuis et al. 2013a). Severe infestation of aphids leads to deformation of the leaves and shoots and stunted appearance and in extreme cases even causes the plant death. Using natural aphid killers, neem oil or insecticidal soap wash may help to reduce crop damage.

7.1.3.4 Pulse Beetle (*Callosobruchus chinensis* and *C. maculatus*)

These are the most destructive pests of legume seeds and causes significant economic loss during storage or preservation (Ranga Rao and Rameswar Rao 2010). While *C. chinensis* is Asian in origin, *C. maculatus* has African origin. However, both the species are widely distributed throughout the tropics and warm temperate parts of the world. Infested stored seed can be recognized by the white eggs on the seed surface and the round holes in the seed coat. Coating the seed with small quantities of vegetable oil or neem oil, fumigation with aluminum phosphide, and periodic drying of the seeds are often practiced to reduce damage.

7.1.3.5 Cutworm (*Agrotis ipsilon*)

Cutworm larvae mostly feed on leaves, stems, and roots causing mortality up to 30% in chickpea. The larvae are gray-black in color. The pest is active during night time while hide themselves under the clods during day time. The caterpillar cut the plants or branches during night. Deep ploughing, crop rotation, intercropping with wheat or linseed or mustard, and endosulfan or aldrin spraying minimize the crop damage from this pest.

7.2 Defense Mechanisms in Pulse Crops Against Insect Pests

It is essential to understand various morphological, biochemical, and molecular defense responses to insect feeding to improve insect resistance in pulse crops. In general pulse crops either resist or tolerate the damage caused by insect pests depending on the species and developmental stage of plant and insect pest. Plants produce various toxins and defense proteins that target physiological processes in the insect (Mithofer and Boland 2012). The biochemical and molecular mechanisms of defense against the insect pests are diverse and dynamic (Howe and Jander 2008; War et al. 2012). Evidently, pulses have developed multidimensional direct or indirect strategies and specific plant resistance traits that reduce insect feeding broadly at three stages, i.e., (1) restricting pest landing and feeding, (2) reducing palatability to insect, and (3) interfering with the metabolism of the attacking insect pest (War et al. 2012).

The direct defense mechanisms are mediated by certain structural/chemical traits that limit insect landing on the crop. Typical examples of direct defense are the morphological features such as increased thorns, prickles, trichome density or lignification, or specific secondary metabolites that restrict insect attack and damage (Kaplan et al. 2009). Few trichomes not only act as mechanical barriers but also harbor secretory structures that release toxins. Alternatively, pulses defend insect pests by some other chemical and physical plant traits that attract natural enemies from an additional trophic level of the attacking herbivores.

The other important mechanism is releasing specialized metabolites which are toxic, antidigestive, or, at least, unpalatable to the attaching insect. At times, the injured tissues upon infestation release certain volatile organic compounds, terpenoids, or few aromatic compounds to attract parasitoids or predators of the feeding insect (De Moraes et al. 1998; Kessler and Baldwin 2001). Allomones such as arcelins, L-canavanine, polyhydroxy alkaloids, and saponins have been reported to confer resistance to insect pests in grain legumes (Dhaliwal and Dhaliwal 1993).

Plant resistance to insects can be constitutively present throughout life cycle or induced upon insect attack (Schoonhoven et al. 2005). The mechanical injury caused by insect infestation or some of the oral secretions of insects trigger inducible defenses through the stimulation of defense chemical synthesis, protease inhibitors, chitinases, and polyphenol oxidases which in turn reduce the insect damage (Pandey et al. 2017). It is hypothesized that the response to insect attack starts with the insect recognition through oral secretions or through insect mouth parts followed by the signals generated from wounded plant cells that eventually turn into signal transduction events. Calcium ion fluxes, phosphorylation cascades, and hormonal cross-talk may relay the signal in the cell to reprogramming the transcriptome, proteome, and metabolome to generate defense compounds and genotypic and/or phenotypic responses (Bruce 2014). Some responses of host plants to different insect herbivores are very general and provide protection against a variety of invaders, whereas others are more specific and target particular types of attacker (Kusnierczyk et al. 2007; Ali and Agrawal 2012; Stam et al. 2014). Combining direct and indirect defenses in crops may be a valuable contribution to the future development of integrated pest management (IPM) strategies (Dicke 1999).

7.2.1 Structural Defense

Several physiochemical characteristics contribute to insect resistance in grain legumes (Clement et al. 1993). Trichromes seem to be responsible for preventing heavy aphid colonization of chickpeas, as aphids feed successfully on areas of the plant where trichomes are not found (Edwards 2001). In case of soybean, trichome length has also been implicated in resistance to beanflies, whiteflies, and pod borers (Chiang and Norris 1983; Lam and Pedigo 2001). For instance, the resistance to leafhoppers in soybeans has been shown to be related to the orientation and size of trichomes more than their density (Lam and Pedigo 2001). Resistance of *Cajanus scarabaeoides* to *H. armigera* has attributed to high density of non-glandular trichomes on pods (Kaplan et al. 2009). A positive correlation has been reported between pod length, basal girth of stem, and pod borer damage (Nanda et al. 1996). Varieties with brown seeds and green pods having streaks have been reported to be least susceptible to pod borer damage (Nanda et al. 1996). Besides, there is evidence on the reduced aphid damage on leafless (*af*) and semi-leafless (*st*) phenotypes of peas due to the reduction in feeding space (Kareiva and Sahakian 1990). Damage due to *H. armigera*, *S. litura*, and jassids in groundnut exhibits significant correlation with morphological characteristics including stem thickness, leaflet shape, leaf length, leaf hairiness, standard petal length, stipule, and peg length. Some of these traits also serve as phenotypic markers for pest resistance in groundnut (Sharma et al. 2003). Nonetheless, pod and seed morphology have also been implicated in resistance to insects. In cowpea, resistance to weevils strongly related to the seed coat thickness (Koona et al. 2002), while resistance to hemipteran and lepidopteran pod pests has been linked to a variety of morphological characters including pod toughness, hull thickness, peduncle length, and seed location in the pod (Koona et al. 2002). Lima bean secretes extra floral nectar to attract ants that can predate upon attacking insects and herbivores (Heil 2004). Interestingly, this secretion can also be induced by jasmonic acid treatment suggesting the existence of insect-induced signaling pathway.

7.2.2 *Chemical Defense*

Plants produce a number of metabolites with defensive properties against insect pests, yet there is little information on what specific phytochemicals contribute to insect resistance (Shimoda et al. 1997). They belong to various chemical classes including isoprene-derived terpenoids, steroids N-containing alkaloids, and phenolic compounds such as flavonoids, cyanogenic glycosides, and proteinase inhibitors (PIs). As mentioned earlier, volatile terpenoids released from lima beans in response to insect feeding attract predators to help control the attacked insects (Petitt and Wietlisbach 1992; Shimoda et al. 1997). Quercetin has been shown to play an important role in food selection behavior of *H. armigera* larvae, while high levels of stilbene—a phytoalexin—have been recorded in *H. armigera*-resistant pigeon pea cultivars (Green et al. 2002). Similarly, oxalic and malic acid exudates from leaf trichomes of chickpea have been shown to contribute to resistance from foliage and pod-feeding insects including *H. armigera* and *H. punctigera* (Yoshida et al. 1997). Proteins such as chitinases, cysteine proteases, lectins, and leucine aminopeptidases are toxic to the insect gut depending on their survival in the alkaline gut conditions. Moreover, anti-insect activity of these toxic plant proteins is also protected by the inducible protease inhibitors by preventing the degradation of these antinutritional or toxic proteins. The most common PIs in pulse crops are Kunitz type, having inhibitory activity against trypsin and chymotrypsin. Besides, these protease inhibitors also affect digestion in the insect gut resulting into depleted nutrient utilization (Zhu-Salzman et al. 2008). Amylase and protease inhibitors have been shown to have adverse effects on larval growth and development of *H. armigera* in pigeon pea (Giri and Kachole 1998). However, some of the insects such as *H. armigera* has coevolved to overcome the effect of host plant PIs either by producing a different set of proteases that are not susceptible or overshooting the proteases beyond the amount of host plant PIs (Ranjekar et al. 2003). In such cases, PIs from the nonhost plants have been found to be an effective alternative against the attacking pest. In support of this notion, a study showed that protease inhibitors from wild relatives including *Cajanus albicans*, *C. cajanifolius*, *C. sericeus*, *Flemingia bracteata*, and *Rhynchosia bracteata* showed complete inhibition of *H. armigera* gut proteinases (HaGPs) (Parde et al. 2012). These wild relatives have been shown to contain amylases, phytolectins, and a range of secondary metabolites which can help in plant defense mechanism against insects. It is generally assumed that secondary metabolites play a role in legume defense against various stress conditions, although their role in insect defense is not clear. For instance, phloem alkaloids from narrow-leafed lupins, *L. angustifolius* L., confer aphid resistance through antibiosis (Edwards et al. 2003). Moreover, leaf alkaloid concentration is also negatively correlated with the red-legged earth mite infestation (Wang et al. 2000). In addition, several wild relatives of groundnut having a different lipid composition, especially more of n-alkanes, have shown a resistant reaction to *S. jrugiperda* and thrips (Yang et al. 1993).

7.2.3 Molecular Defense

In order to control the pulse crop losses due to insect attack, it is imperative to study the molecular events associated with infestation and defense. The knowledge on the activation or repression of insect responsive genes after injury is also key to breeding and genetic manipulation efforts targeted to enhance plant resistance to insect pests. A recent comprehensive high-throughput transcriptomic study in chickpea described the early insect wound-responsive genes in leaves. Leaf transcriptome simulated with oral secretions of *H. armigera* followed by mechanical injury identified a total of 1835 differentially expressed genes of which 1334 were induced and 501 genes were repressed significantly. The gene expression of several ERFs, calcium-binding protein genes, cell wall modifying genes, WRKYs, and defense-related genes were shouted up, while F-box genes and the NAC/MYB transcription factors were present among the significantly repressed genes. In addition, genes coding for kinases and hormone signaling display both up- and downregulation during early phases of injury. Interestingly, a rapid regulation of various hormonal networks has been observed even within 20 minutes of simulation by *H. armigera* that resulted in the suppression of growth pathways and activation of defense pathways (Pandey et al. 2017).

7.3 Strategies to Combat Insect Attack in Pulse Crops

7.3.1 Breeding and Molecular Breeding Approaches for Insect Resistance in Pulses

In nature, variation in plant resistance to insects among individual species may occur due to the difference in their genetic makeup and differential natural and artificial selection pressures (Schoonhoven et al. 2005). Plants have evolved diverse ways to cope with individual insect attack that has resulted in natural variation for resistance toward insect pests. Studying the molecular genetics of this variation has been important to facilitate the identification of resistance genes and the process of resistance against insects. In order to develop insect-resistant varieties, it is imperative to identify, characterize, and categorize effective sources of resistance. However, elaborative studies on host plant resistance to insect pest in pulse crops are restricted to few important pulse crops including pigeon pea, cowpea, chickpea, and soybean only. However, few attempts have been made in mung bean and urdbean. This is due to the difficulties involved in screening and selection of the test material under uniform insect infestation across seasons and locations. In addition, it is difficult to rear some of the insect pests on artificial diets to ensure screening and selection under optimum levels of insect infestation. Most importantly, due to the ease with which insects can be controlled with help of insecticides, there has been an insufficient focus on developing cultivars with resistance to insect pests. Elite germplasm with resistance to insect pests has been identified in pulse crops, but the insect resistance-breeding programs are still underway and that to confine to few crop pests only. For example, in lentil, genotypic differences for susceptibility have been observed only to aphid, pod borer, and seed weevil, but not many efforts have been made to breed for resistance to other insects. Moreover the sources of resistance have not been used extensively in breeding programs partially due to fertility barriers (Clement et al. 1993; Sharma and Ortiz 2000).

7.3.1.1 Host Plant Resistance Studies in Major Pulse Crops

Although successful breeding of insect resistance into commercial cultivars has proven difficult in many legume crops, screening of legume genetic resources for insect resistance has showed some promise (Clement and Quisenberry 1999). Although varieties with resistance to target insect pests have been developed and released for pigeon pea, chickpea, cowpea, mung bean, urdbean, field pea, soybean, and groundnut, the levels of resistance in most of these varieties released for cultivation are low to moderate. Hence, soybean breeding program in the USA released only four insect-resistant soybean cultivars with limited adoption in their first 26 years (Boethel 1999). Nonetheless, high levels of resistance have been reported in the wild relatives of several pulse crops (Sharma et al. 2003). Classic breeding for resistance to pathogens and pests mostly rely on identifying resistance genes from these wild relatives and then introgressing these into cultivated crops. Wild relatives of pigeon pea such as *Cajanus scarabaeoides*, C. *platycarpus*, C. *acutifolius*, and C. *sericeus* have been identified with high level of resistance to pod borer and pod fly, which can be easily crossed with the cultivated pigeon pea. Larval and pupal developmental periods are all adversely affected when fed on the flowers of wild species such as C. *cajanifolius*, C. *reticulatus*, and C. *sericeus* (Dodia et al. 1996). Hence, mapping population involving C. *cajan* x C. *scarabaeoides* is under development at research institutes like ICRISAT (Bhatnagar-Mathur and Sharma 2016). Similarly in chickpea, accessions belonging to *Cicer bijugum*, C. *judaicum*, C. *cuneatum*, and C. *microphyllum* have also been identified with high levels of resistance to *H. armigera* (Sharma et al. 2002). These wild relatives of chickpea also serve as important source of resistance to the leaf miner and bruchids (Singh and Ocampo 1997). Seven chickpea breeding lines (FLIP 2005-1C, 2C, 3C, 4C, 5C, 6C, 7C) resistant to leaf miner were developed and released through classical breeding in 2006 (Malhotra et al. 2007). In pea, the accessions belonging to the wild relative, *pisum fulvum*, are not preferred for egg laying by the bruchid, *Bruchus pisorum* (Mishra et al. 2017). Similarly, some wild relatives of groundnut (*Arachis cardenasii*, *A. durensis*, A. *kempff-mercadoi*, A. *monticola*, A. *stenosperma*, *A. paraguariensis*, A. *pusilla*, and A. *triseminata* have been shown multiple resistance to *Aproaerema modicella*, *H. armigera*, and *Empoasca kerri* (Sharma et al. 2003). Similarly, *A. cardenasii* (ICG 8216), *A. ipaensis* (ICG 8206), *A. paraguariensis* (ICG 8130), and *A. appressipila* (ICG8946) have shown resistance to leaf feeding and antibiosis to *Spodoptera*. Similarly, *Vigna vexillata* accessions TVNu 72 and TVNu73 have shown high levels of resistance to *M. vitrata* (Jackai and Oghiakhe 1989) and can be used to increase insect resistance in cowpea cultivars. It was found that pod damage by the lepidopteran borers is around 11.72 percent ranging from 7.35 to 28.71 percent out of 24 entries screened for their resistance (Chavan et al. 2009). However, the studies on host plant resistance are restricted to *M. vitrata* and *H. armigera* in case of cowpea, mung bean, and urdbean. Hence, resistance observed in the wild relatives need to be transferred into high-yielding varieties with acceptable agronomic backgrounds. It is necessary to break the linkage between insect resistance and susceptibility to other diseases in some crops. For instance, pod borer-resistant cultivars in chickpea and pigeon pea are often susceptible to wilt. However, it is often difficult to cross crops with wild relatives when they are distantly related. In addition, the so-called linkage drag in which undesirable genes near the desired ones on the same chromosome is brought together in the breeding. However, the precise modification of transferring only resistance from the wild varieties without affecting productivity and other desirable traits is difficult to achieve through conventional breeding methods.

Some of the morphophysiological traits are also responsible for host preference and thus serve as phenotypic markers for breeding insect-resistant crops. Trichomes on the pods of *Vigna vexillata*—a wild relative of cowpea—are partly responsible for resistance to *Clavigralla tomentosicollis* (Chiang and Singh 1988). Likewise, pubescent varieties of soybean are highly resistant to *Empoasca fabae* Harris. Oviposition non-preference is one of the components of resistance to *H. zea* in PI 2227687 soybean (Horber 1978). Pea wild varieties deficient in certain amino acids are resistant to aphid, *Acyrthosiphon pisum*. Interestingly, a wild line of *Pisum sativum* responds to pea weevil eggs by forming callus very similar to *Lathyrus* sp. (Annis and O'Keeffe 1984). Nutritional antibiosis and host preference for feeding are major components of resistance in soybean to *Epilachna varivestis*. However, dearth of accurate evaluation while screening thousands of accessions for resistance to the target pests gives rise to the possibility of missing many potentially good sources of resistance.

7.3.1.2 Molecular Breeding for Insect Resistance in Pulses

In recent years, there is considerable research on exploring the natural variations associated with insect resistance at the molecular level, including both the genomic and transcriptomic levels (Schoonhoven et al. 2005; Rubiales et al. 2015). Although classical breeding is a straightforward approach for insect resistance, the time needed for releasing a resistant variety is the major bottleneck (Sandhu and Kang 2017). Molecular breeding has been evolving with the rapid development of molecular biology techniques since the 1980s, and the past decades has witnessed the acceleration of the selection steps in breeding programs through deploying molecular markers linked to the desired trait—a strategy called marker-assisted selection. With the advent of new technologies and molecular genetic tools that allow genome-wide association studies, it is now possible to dissect the underlying molecular variations in plant defense mechanisms. Such studies not only allow the development of molecular markers but also facilitate marker-assisted breeding in introgressing resistance traits into economically important cultivars (Varshney et al. 2005). Furthermore, the knowledge on the insect-resistant genes may be helpful to develop transgenic crops having greater pest resistance with more precision.

Natural variations in insect resistance can be qualitative or quantitative depending on the number of loci involved. While qualitative traits are controlled by a single (R) gene working based on gene-for-gene principle, quantitative resistance traits are regulated by multiple loci with complex genetics (Yencho et al. 2000). Hence, compared to the resistance based on qualitatively inherited traits (R-genes), quantitative trait loci (QTLs) are more difficult to include in a breeding programs. Besides, only few studies suggests the defense pathways induced by R-genes conferring resistance to insects in plant species (Klingler et al. 2009). Although insect-resistance QTLs have been mapped in few pulse species, the underlying molecular mechanisms of these QTLs remain largely unknown (Wang et al. 2009). Well-known examples include QTLs associated with resistance to various aphids in *M. truncatula* (Kamphuis et al. 2013b), broomrape in pea and in the *D. pinodes*—pea pathosystem (Carrillo et al. 2013). Although more QTLs for insect resistance are being developed, it is important to note that the accuracy and validity of these QTLs may not always be effective in different populations.

Molecular markers can play an important role in accelerating the introgression of genes conferring resistance to target insects into high-yielding cultivars. Preliminary identification of molecular markers for resistance to insects in soybean, chickpea, mung bean, field pea, and cowpea has been reported (Sharma et al. 2016). The most widely used breeding approach is the use of marker-assisted selection where molecular markers linked to resistance loci were used to identify plants containing the resistance genes. However it is pivotal to identify the accurate QTL for molecular marker-assisted selection. Although progress has been made in developing genetic linkage maps of chickpea, cowpea, and soybean, much remains to be done in pigeon pea, beans, lentil, and field pea. A recent study identified a major QTL for bruchid resistance based on recombinant inbred population on chromosome 5 of mung bean genetic map followed by its validation in physical map (Schafleitner et al. 2016). A consensus genetic map was built with 81 QTLs related to insect resistance in soybean based on the public genetic linkage map Soymap2 by BioMercator2.1 (Wang et al. 2009). This study utilized meta-analysis to deduce the locations of QTLs with 95% confidence and also reduced the QTL intervals from 15 cM to 3.67 cM on average. This study provided the basis of QTL fine mapping for resistance to multiple pests including soybean pod borer resistance. A total of eight QTLs contributing bruchid resistance were detected in black gram using the linkage map (Souframanien et al. 2010).

During the last two decades, several studies have reported the identification of DNA markers linked to bruchid resistance in different pulse crops (Mishra et al. 2017). RFLP marker-assisted selection has been used for introgression of the bruchid resistance gene in green gram (Yang et al. 1998), while the random amplified polymorphic DNA (RAPD) markers have developed with linkage to the bruchid resistance in mung bean (Villareal et al. 1998). Efforts were also made in developing molecular markers for resistance to pea weevil in crosses between *P. sativum x P. fulvum* (Byrne et al. 2002). In another study, genetic localization of bruchid resistance in mung bean using RAPD markers revealed that the resistant gene has been located 5.6 cM from pM151 and 11.9 cM from the nearest RAPD marker Q04 sub 900 (Mishra et al. 2017). Two microsatellite markers (BMd8 and BMd26) associated with bruchid resistance and low adult insect emergence have been identified in common bean which could be very effective for MAS (Blair et al. 2010). Similarly, 10 RAPD markers associated with bruchid resistance genes were also identified through BSA in the F12 generation, derived from the cross, *V. radiata Nm92* × *V. sublobata—TC1966* (Chen et al. 2007). Of these, four tightly linked RAPD fragments were cloned and transformed into SCAR (sequence-characterized amplified region) markers.

Previously, a novel defensin encoded by mung bean c-DNA (complementary-DNA) and Br1 locus in LG-09 of SSR marker-based genetic map conferred bruchid resistance in mung bean (Chen et al. 2002a). SSR markers have been deployed to detect polymorphism in breeding populations of pigeon pea as well. However, the number of available markers is a severe limitation to their implication in crop improvement efforts toward insect resistance. For this reason, a major SSR marker development program has been initiated in pigeon pea. QTL associated with insect resistance from PI 229358 and PI 171451 has also been identified using RFLP markers in soybean (Narvel et al. 2001).

With the advancement of genotyping technologies, the identification of desirable traits in legume germplasm has become more accurate and increased the pace of breeding programs. A good number of resistance genes against aphid have been identified from a wild relative of cowpea, TVNu 1158 (Huynh et al. 2015). In a recent study, 105 cowpea cultivars were screened for tolerance to aphids and identified the cultivar IT97K-556-6, from which two resistance loci were mapped against aphids infestation (Huynh et al. 2015). However, the low heritability under field conditions has been the potential factor to decide insect resistance that opens the door for more possibilities to transfer resistance genes from wild relatives or from other resistant species through biotechnological approaches.

The recently built abilities can artificially alter plant genetic material by means of mutagenesis, and genetic engineering has become a significant milestone in developing insect-resistant plants. The concept of induced mutagenesis and mutation breeding has been successfully utilized for the improvement of crops dated back to the beginning of twentieth century. Induced mutation is the viable option to alter the genetics of crop plants that may be difficult to introgress through cross breeding and other traditional breeding procedures. Therefore, during the last several years, different mutagens including ionizing radiations, lower ultraviolet rays, chemical mutagens, and transposons have been used by various researchers to induce genetic variability in various pulse crops such as *Cicer arietinum*, *Vicia faba*, *Vigna mungo*, *Lens culinaris*, *Vigna unguiculata*, *Vigna radiate*, and *Glycine max* (Wani et al. 2014; Usharani and Kumar 2015; Gnanamurthy and Dhanavel 2014; Kozgar 2014; Laskar et al. 2015; Khursheed and Khan 2016; Patil et al. 2007). In tropical bean systems, there have been some success in developing cultivars resistant to single pests, but multiple insect- and disease-resistant varieties are desperately needed to be commercially valuable (Cardona and Kornegay 1999).

7.3.2 *Transgenic Technologies for Engineering Insect-Resistant Plants*

Although many of the insect pests are controlled by insecticides, alternative insect control methods are needed considering the environmental concerns and negative impacts of pesticides on off-target beneficial insects such as pollinators and insectivorous insects (Bakhsh et al. 2009). On the other hand, developing insect-resistant varieties through conventional breeding methods suffers the limitations such as fertility barriers, time-consuming screening procedures, and long gestation periods to release a variety (Rubiales et al. 2015). Nonetheless, the recent advent of recombinant and transformation technologies enabled the gene transfer across species without any taxonomical boundaries (Bakhsh et al. 2015). Besides, genetic engineering techniques also avoid linkage drag by enabling to clone and transfer specific resistance-conferring genes to crops. The recent transgenic technology led to the introduction of many genes of bacterial, plant, or fungal origin encoding insect resistance into many crop species (Bakhsh et al. 2015). However, progress in developing transgenic plants of pulses has been delayed partly due to the lack of established transformation and tissue culture procedures (Popelka et al. 2004). Only few important pulse crops such as chickpea, groundnut, and pigeon pea have standardized tissue culture and regeneration protocols which has been used in the genetic transformation of these crops (Thu et al. 2003; Jayanand et al. 2003; Dayal et al. 2003). Even though the recovery rates are limited to 1% in most cases, most of the pulses can be genetically transformed including pea, chickpea, soybean, alfalfa,

Table 7.1 List of different genes used in genetic engineering to confer insect pest resistance in major pulse crops. *Cry-bacterial crystal proteins*

Pulse crop	Transgene gene	Insect pest	References
Cajanus cajan	*Cry 1E-C* *Cry* 1Ab, *Cry* 1AcF, *Cry* 1Ac	*Spodoptera litura* *Helicoverpa armigera*	Surekha et al. (2005) Ramu et al. (2012)
Cicer arietinum	*Cry* 1Ab *Cry* 1Ac, *Cry* 2Aa, *Cry* 1Aabc Bean α-amylase inhibitor 1 Trypsin inhibitor α-amylase inhibitor gene Allium sativum (*ASAL*)	*Heliothis armigera* *Helicoverpa armigera* *Spodoptera litura* *Callosobruchus spp.* *Helicoverpa armigera* *Callosobruchus maculates* *Aphis craccivora*	Kar et al. (1997) Sanyal et al. (2005); Indurker et al. (2007); Acharjee et al. (2010); Mehrotra et al. (2011); Das et al. (2017) Sarmahet al. (2004) Srinivasan et al. (2005) Ignacimuthu and Prakash (2006); Chakraborti et al. (2009)
Pisum sativum	Bean α-amylase inhibitor, α-amylase inhibitor 1, α-amylase inhibitor Pea Protease Inhibitor Na-PI	*Bruchus pisorum* *Helicoverpa armigera*	Schroeder et al. (1995); Morton et al. (2000) Charity et al. (1999)
Phaseolus vulgaris	Bean α-amylase inhibitor Arcelin-1 (lectin like protein) α-amylase-like α-amylase inhibitor	*Zabrotes subfasciatus* Bruchids *Callosobruchus maculates*, *Callosobruchus* chinensis *Hypothenemus hampei*	Ishimoto et al. (1996) Fabre et al. (1998) Valencia et al. (2000)
Vigna unguiculata	Proteinase inhibitor *Cry* 1Ab Alpha-amylase inhibitor from bean	*Diatraea saccharalis* *Maruca vitrata* *Callosobruchus maculatus*	Mota et al. (2003) Adesoye et al. (2008); Higgins et al. (2012) Solleti et al. (2008)
Glycine max	*Cry* 1Abprotoxin *Cry* 1Ac Proteinase inhibitor *Cry* 1Ac *Cry* 1Ac	*Anticarsia gemmatalis* *Helicoverpa zea* *Pseudoplusia includens* *Heliothis virescens* *Anticarsia gemmatalis* *Diatraea saccharalis* *Helicoverpa zea* *Anticarsia gemmatalis* *Pseudoplusia includes*	Dufourmantel et al. (2005) Stewart et al. (2009) Pompermayer et al. (2001) Walker et al. (2000) Miklos et al. (2007)
Arachis hypogaea	*Cry* 1Ac *Cry* 1E-C	*Elasmopalpus lignosellus* *Spodoptera litura*	Singsit et al. (1997) Beena et al. (2008)

peanuts, common bean, and cowpea (Table 7.1; Somers et al. 2003; Popelka et al. 2004). This low recovery rate is generally ascribed to the recalcitrant nature of pulse tissues to in vitro regeneration and transformation methods used (Somers et al. 2003). Although various transformation methods such as microparticle bombardment have been used for gene transfer to various tissues, the preferred transformation method is *Agrobacterium tumefaciens* in most pulse crops (Somers et al. 2003; Gulati et al. 2002; Tohidfar et al. 2013; Adesoye et al. 2008). In addition, the regulatory issues associated with the transgenic research in individual countries are few limiting factors for growing and exchanging transgenic materials across the globe. For instance, the transgenic pea for resistance to pea weevil through the expression of alpha-amylase inhibitor in Australia is not available to pea breeders in Australia and other countries because of the concerns associated with the use of transgenic food crops (Bakhsh et al. 2015; Morton et al. 2000; Sousa-Majer et al. 2004). Despite these technical and regulatory difficulties, developing insect resistance has been one of the major priority areas for applying genetic engineering technology to pulse crop improvement.

7.3.2.1 Genes with Insecticidal Activities

More than a decade of research demonstrated that *H. armigera* being the most important yield constraint in major pulse crops has no absolute resistance available in the cultivated germplasm. Hence transgenic resistance to *H. armigera* has been the high priority and ideal seed-borne solution to enhance productivity through an integrated pest management in pulse crops. Among the insect-resistant transgenics, one of the main transformation strategies was the introduction of derivatives of insecticidal *Crystal (Cry)* genes from *Bacillus thuringiensis* conferring resistance to many pod borer insects. *Bacillus thuringiensis* is a gram-positive spore-forming soil bacterium, which forms parasporal crystals that consist of one or more or truncated δ-endotoxin (pro-toxin subunits) or Cry proteins of approximately 138 kDa. After ingestion this pro-toxin breaks down to active form inside the gut of the target insect in presence of the high pH and gut enzymes. Thereafter an increased activity of ion channels lead to the leakage of midgut content, paralysis, and eventually death of the insect. In some insects, these toxins cause death due to the loss of appetite (Ravindran 2016).

The usage of *Cry* genes in legumes started with the introduction of Cry 1Ac in alfalfa for *Spodoptera* resistance (Singsit et al. 1997) followed by developing resistance against corn stalk borer in peanut (Kar et al. 1997). Similarly, chickpea cultivars ICCV 1 and ICCV 6 transformed with *cry lAc* gene have been found to inhibit the development of pod borer larvae when fed on transgenic tissues (Kar et al. 1997). Although more than 340 *cry* genes have been identified, only a few have so far found application in plant breeding developing insect-resistant transgenic pulse crops (Tohidfar et al. 2013). Given the high potential of genetic transformation in plant breeding, large efforts have been made to improve the efficiency of transformation in other pulse crops as well. For instance, transgenic pigeon pea plants with *crylAb* and soybean trypsin inhibitor genes are being tested against *H. armigera* at ICRISAT (Rubiales et al. 2015). Transformation of soybean with a synthetic *CrylAc* gene by particle bombardment has resulted in enhanced resistance to multiple pests including com earworm, soybean looper, and velvet bean caterpillar (Stewart et al. 2009). Similarly, transformation of peanut by a synthetic *CrylAc* gene resulted in various levels of resistance to the corn stalk borer ranging from 100% to 66% larval mortality (Singsit et al. 1997). Alfalfa transformation utilizing a synthetic *Cry1Ac* gene accumulated the toxin up to 0.01–0.2% of total soluble protein making them resistant to cotton leaf worm and beet army worm (Tohidfar et al. 2013).

The level of toxin expression has been shown to have a direct correlation with the resistance toward target insects. Hence, efforts have been made to improve the level of toxicity of *cry1Ac* by codon optimization to increase translational efficiency and subsequent protein synthesis (Das et al. 2017). The codon-optimized chimeric *Bt* gene (*cry1Aabc*) using three domains from three different *cry1A* genes has been genetically transformed through *Agrobacterium tumefaciens* into chickpea under constitutive promotor to assess its effect on gram pod borer. The results demonstrated that the chimeric Cry1ac protein is effective against gram pod borer across generations and can serve as good material to utilize in transgenic breeding program (Das et al. 2017).

Transgenic plants expressing Cry genes produce the protein continuously through all the phases of plant growth with varying extents. In a recent study, targeting *cry1Ac* into the chloroplast using homologous ubiquitin promoter and green tissue-specific rbcS promoter restricted the toxin expression only in green tissues and hence reduces unnecessary energy expenditure in expressing the same throughout the plant body (Chakraborty et al. 2016). Nevertheless, various *Cry1Ab* has also been expressed in soybean plastids only (Dufourmantel et al. 2005). In recent years, the problem of narrow specificity has been addressed by pyramiding the insect-resistant genes in combination with other insect-resistant proteins having different specificities (Zhao et al. 2003; Hilder 2003; Beena et al. 2008). In addition, it is also well established to transfer the chimeric *Bt* constructs comprising different elements of other *cry* genes with different modes of action instead of transferring a single *cry1* gene in crop plants to delay the resistance development among insects (Asharani et al. 2011; Mehrotra et al. 2011). Gene pyramiding with two different insecticidal genes and tissue-specific expression expected to reduce the risk of developing insect resistance is viable option to have durable resistance. Expression of a chimeric *cry1AcF* (encoding *cry1Ac* and *cry1F* domains) gene in transgenic pigeon pea has been demonstrated toward resistance to *H. armigera* (Ramu et al. 2012). As an added advantage, combining *Cry1Ac* and *Cry2A* with *Galanthus nivalis* agglutinin (GNA) enhanced resistance against three pests in rice including rice leaf folder, brown hopper, and yellow stem borer (Maqbool et al. 2001). There are also reports on combining QTLs with *Cry1Ac* for better and sustainable lepidopteran insect resistance (Walker et al. 2002). At present many transgenic lines of pigeon pea, chickpea, cowpea, and soybean resistant to the lepidopterans contain only single *cry* gene with the option of stacking two genes by crossing (Sanyal et al. 2005; Adesoye et al. 2008; Ramu et al. 2012). Large number of transgenic events carrying *Bt cry* genes have been developed are currently being evaluated for the larval mortality in pigeon pea and chickpea (Bhatnagar-Mathur and Sharma 2016). These transgenics showed high mortality of the larva and thus able to resist the damage caused by the larvae.

7.3.2.2 Vegetative Insecticidal Protein (VIP) Toxin Protein

Besides *cry* toxins, a number of vegetative insecticidal proteins (VIP) have been isolated from the *B. thuringiensis* at the mid-log phase. Similar to proteins, VIPs are also very specific to the specific class of target insects. For example, Vip3A was found to be effective against Lepidopteran insects such as *Spodoptera exigua*, *Spodoptera litura*, and *Helicoverpa armigera* (Estruch et al. 1996; Chen et al. 2002b).

In spite of great effort involved in developing Bt pulse crops, there are difficulties for consumer acceptance and biosafety of these "anti-natural" products in humans. Hence transgenic technology for insect resistance has found an alternative by exploiting the plant's own defense mechanisms, by introducing the insect-resistant genes derived from plant origin or by manipulating the expression of the endogenous defense proteins. Plant defense compounds include alkaloids, terpenes, cyanogenic glucosides and defense proteins such as chitinase, lectins, cholesterol oxidase, polyphenol oxidase, systemins, and enzyme inhibitors (Mithofer and Boland 2012; Ravindran 2016). At present two major groups of plant-derived genes, namely, enzyme inhibitors acting on key insect gut digestive hydrolases, the α-amylases and proteinases, and lectins have been used to confer insect resistance. Plant protease/proteinase inhibitors are polypeptides or proteins that are part of the plants natural defense system against insect predation and other infections (Gujar et al. 2000). Based on the class of proteases they inhibit, four types of proteases have been identified including serine, cysteine, aspartic, or metalloproteases (Lawrence and Koundal 2002). Some plant serine protease inhibitors also possess trypsin and α-amylase inhibitors activity in addition to protease inhibitor activity proving more potent insecticidal. However, the low levels of these proteins are in their native state may not be sufficient to prevent insect feeding. Hence overexpression of these proteinase inhibitors would offer significant levels of insect protection. The most active inhibitor identified to date is the cowpea trypsin inhibitor, which has been transferred to many crop species. A multi-domain tobacco proteinase inhibitor expressed in transgenic pea under the control of Rubisco small subunit promoter have shown that the mortality of *Helicoverpa armigera* larvae was as compared to controls (Charity et al. 1999). Interestingly, expression of protease inhibitor from *Manduca sexta* (Tobacco hornworm) has resulted in reduced thrips infestation in alfalfa (Thomas et al. 1994).

α-Amylases catalyze the hydrolysis of α-D-(1 → 4)-glucan linkages in starch components and hence play a crucial role in carbohydrate metabolism. Genes for three *α -amylase inhibitors* have been expressed in pea and Adzuki bean. α-amylase inhibitors derived from *Phaseolus vulgaris* seeds showed significant level of protection against cowpea weevil *Callosobruchus maculatus* in transgenic pea plants (Shade et al. 1994). These transgenic plants exhibited complete resistance against bruchids due to high larval mortality at the first or second instars both under lab and field conditions (Schroeder et al. 1995; Morton et al. 2000). Similar results were also obtained when Adzuki bean was transformed with *α-amylase inhibitor* gene and tested for protection against bruchid beetles (Ishimoto et al. 1996). Feeding rats with transgenic peas expressing high levels of α-AI1 showed minimal nutritional differences and can be used as animal feed (Shade et al. 1994; Pusztai et al. 1999).

Lectins are carbohydrate-binding proteins, toxic to various genera of insects. Artificial diet-based insect bioassay revealed that the carbohydrate-binding lectins are highly antinutritional and toxic to various phloem-feeding insect pests. Among these lectins *Allium sativum*, leaf agglutinin and GNA have been proved detrimental to insect pests. The successful efficacy of plant lectins and other non-*Bt* genes against insect pests has been successfully documented in transgenic crop plants. Apart from *Bt* genes for pod borer resistance (Kar et al. 1997; Sanyal et al. 2005; Acharjee et al. 2010; Mehrotra et al. 2011; Ganguly et al. 2014; Chakraborty et al. 2016), transgenic chickpea expressing cowpea trypsin inhibitor (Thu et al. 2003), alpha-amylase inhibitor 1 (*α-ai1*) for bruchid resistance (Sarmah et al. 2004; Ignacimuthu and Prakash 2006), and lectin against aphids have also been developed to improve insect resistance (Chakraborti et al. 2009). Targeted expression of *ASAL* gene in the phloem tissue under the control of *CaMV35S* and phloem tissue-specific *rolC* promoters decreased the survival and fecundity of *A. craccivora* to 11–26% and 22–42%, respectively (Chakraborti et al. 2009). Snowdrop lectin also enhanced the resistance against aphids and rice brown planthopper (Powell 2001).

In addition, transgenic plants expressing chitinase has also shown to be important in controlling the devastating insect pests by dissolution of chitin, which is an insoluble structural polysaccharide that occurs in the exoskeleton and alimentary canal of insects. The previous reports suggest the differential induction of chitinase activity in susceptible and resistant wheat cultivars after Russian wheat aphid infestation (Van der Westhuizen et al. 1998). Exogenous application of chitin synthesis inhibitors also gives 100% mortality in pests supporting the crucial role played by chitinase in insect resistance (Kostyukovsky and Trostanetsky 2006).

Cholesterol oxidase is another enzyme produced by microorganisms of the genera streptomyces, pseudomonas, schyzophyllum, and rhodococcus, which possess potent insecticidal activity against cotton boll weevil (Purcell et al. 1993). The pioneer experiments demonstrated that the filtrate of streptomyces cultures containing cholesterol oxidase killed the boll weevil larvae due to the physical and functional disruption of midgut epithelial membrane (Purcell et al. 1993). This enzyme causes developmental arrest, decrease in the adult female fecundity, and death of the larvae (Purcell et al. 1993). Similarly, polyphenol oxidase gene cloned from poplar expressed in response to wounding and herbivory (Constabel et al. 2000). Upon ingestion, polyphenol oxidase interacts with phenolic substrates to generate highly reactive o-quinones that can covalently modify free amino and sulfhydryl groups within the insect. Hence, polyphenol oxidase significantly reduces protein quality in insect larvae, which has great impact on the larval growth rate (Felton et al. 1992). Oleoresins, phytoalexins (antimicrobial secondary metabolites), etc. are other secondary metabolites shown to have increased effect against herbivore insects and are used in transgenic plants for developing insect resistance. However, some of these enzymes and metabolites are yet to be expressed and validated in pulse crop improvement.

Genetic engineering not just introduces new genes but also offers the possibility to manipulate endogenous expression to create new phenotypes useful to study gene function (Somers et al. 2003). Overexpressing common bean *α-amylase 1 inhibitor* in seeds displays high levels of resistance to different bruchids in pea, chickpea, adzuki bean, and cowpea (Schroeder et al. 1995; Sarmah et al. 2004; Solleti et al. 2008). Overexpressing baculovirus metalloproteases also disrupt peritrophic membrane in the midguts of insects (Lepore et al. 1996). That leads to faster larval mortality due to increased viral infection of the damaged mucin (Cao et al. 2002).

Efforts were also made to shorten the generation time of stable transgenic lines and to improve in vitro regeneration. For instance, the micrografting reduced the generation time by 3 months in pea. Micrografting, as originally described for pea mutants (Murfet 1971), has been continued in pea, lentil, and chickpea (Gulati et al. 2002; Chakraborti et al. 2009). Alternatively, *in-planta* transformation that can bypass the need of tissue culture regeneration steps has been reported for alfalfa, peanuts, pea, pigeon pea, cowpea, lentil, and soybean (Chowrira et al. 1996; Trieu et al. 2000; Rohini and Rao 2000; Ramu et al. 2012; Adesoye et al. 2008).

In general, transgenic technology and its successful utilization in agriculture have contributed significantly to global food security and poverty reduction. However, producing a transgenic line with the proper temporal and spatial expression of the resistance gene is still a challenge. Moreover, despite proper biosafety guidelines and procedures, insect-resistant crops face many challenges in public acceptance before commercialization. Nonetheless, most of the transgenic efforts toward insect resistance have been confined to laboratory except few transgenic lines that are close to commercialization which includes the Bt soybean (Miklos et al. 2007) in Brazil, the BGMV-resistant bean which is in advanced field trials in Brazil, the pod borer-resistant chickpea (Acharjee et al. 2010) which is in early field trials in India, and the pod borer-resistant cowpea (Higgins et al. 2012) which is in multi-location field trials in Nigeria. Although no significant hazard of genetically engineered crops has been evident so far, the discussions on the degradation kinetics of Bt proteins in the soil, horizontal and vertical gene flow, effects on nontarget insects or organisms, other unintended effects, and antibiotic resistance still remain intense. Importantly, these toxins are highly specific due to high-affinity binding sites present in the membrane of target insect midgut that is lacking in human beings or ruminates. However, many researchers as well as common people have also raised concerns about the use of genetically modified organisms, including insect-resistant crops (Godfrey 2000). Hence it is imperative to have a pure scientific and professionally competent regulatory mechanisms for the risk evaluation of insect-resistant transgenic crops for human health and the environment (Baksh 2003).

7.3.3 Management Practices for Insect Resistance

7.3.3.1 Cultural Control

Given the development of resistance to insecticides in several insect species, there is a need to integrate different protecting methods and control tactics against pests in pulses (Oerke 2006) (Fig. 7.2; Table 7.2). Following these cultural practices reduces the pesticide application and contributes to formers economy.

Sowing Time

Sowing time is one of the most important factors affecting crop yield through minimizing infestations (Shinde et al. 2013). It was reported that late sown crops suffered most from the pod borer infestation when compared with the crops sown earlier. The reports also indicate a positive correlation between pod borer larval population and temperature, while the larval population is negatively linked with relative humidity and rainfall (Lomash and Bisht 2013; Shinde et al. 2013). Besides, early termination of flowering and fruiting is also practiced in pulses to reduce the number of generations and to minimize the insect carryover from one season to another. Similarly, the time of harvesting is also important in controlling storage

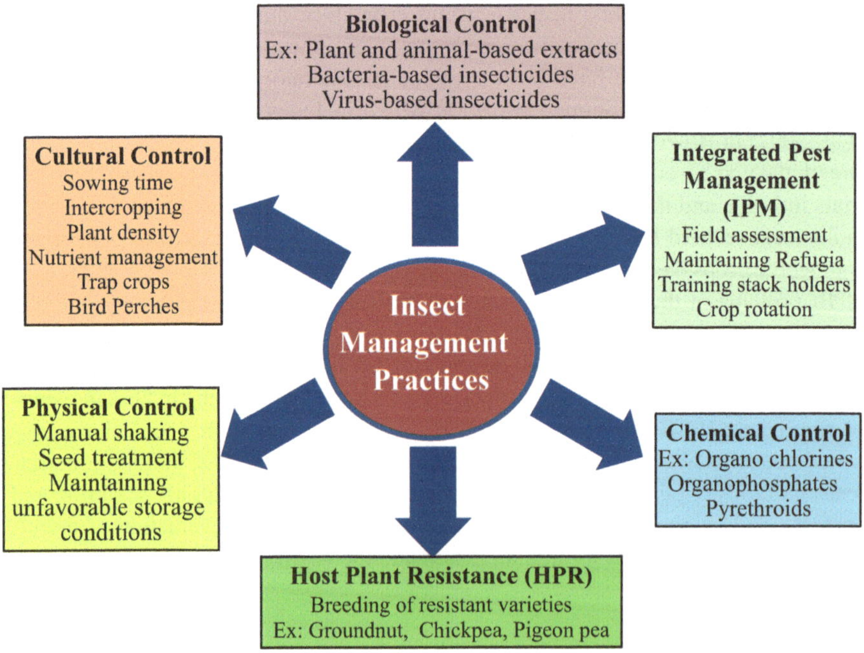

Fig. 7.2 Various agronomic practices used against insect pests in pulses

Table 7.2 Crop management practices for insect pest management in pulses

Operations	Pest	Effect	Reference
Inter Cropping	Leaf miner, jassids, apids, thrips, whiteflies	Reduced larval population	Bhatnagar et al. (1983); Logiswaran and Mohanasundaram (1985); Pattar et al. (2012);
Bird perches	Pod borers	Reduced insect larval population	Visalakshmi (2001)
Manual shaking	Pod borers	Reduced insect population	Ranga Rao and Rameswar Rao (2010)
Spacing	Thrips	Low thrips injury and bud necrosis	Kant et al. (2007)
Irrigation			
Overhead irrigation	Aphids and mites	Reduced insect populations	Rao et al. (1991)
Time of irrigation	Cut worm, army worm	Insect populations reduced due to the activity of birds and other predators during day time irrigation	Ranga Rao and Wightman (1994)
Light traps	Red hairy caterpillar	Reduced insect population and plant damage	Qayum and Sanghi (1994)
Biological control (NPV, Bt)	Pod borers	Insect death	Cherry et al. (2000); Hilder (2003)
Plant derivatives	Leaf and stem miners	Reduced population	Bhushan et al. (2011)
Chemical control (e.g., organochlorines organophosphates pyrethroids)	Pod borers Aphids, Leaf miner Bruchids	Insect death	Shaw et al. (1999); Minja et al. (2000); Sushil et al. (2009).

insect pests of pulses. Timely harvesting which prevents egg laying on mature crops reduces *C. maculatus* infestation by 50–90%.

Intercropping

Pulses have been used in intercropping with a wide variety of crops. In most cases, pulses are intercropped with cereals such as wheat and rice. The intension of intercropping is to diminish the ability of the insects' host seeking behavior and to hinder the development of insect population. As a good example, when compared to chickpea sole cropping, mixed cropping of chickpea with wheat has been shown to reduce pod borer damage by 38.3%. Intercropping of chickpea with mustard, linseed, or safflower has also reduced damage by pod borer (Das 1998; Pattar et al. 2012). Similarly, pigeon pea intercropped with cowpea and sorghum also had better insect resistance. Intercropping groundnut with pearl millet and soybean suppresses thrips, jassids, and leaf miner. Similarly intercropping groundnut with castor suppresses jassids and *Spodoptera*, while pigeon pea intercropping suppresses early leaf spot, late leaf spot, and rust. These plants act as traps or barriers for reducing pest incidence. Sometimes, intercropping is also useful to encourage the activity of natural enemies (Bhatnagar et al. 1983). Hence, growing nonhost crops before growing the insect susceptible pulse crops including pea and faba bean had minimum damage by the red-legged earth mite. Leaf miner larval densities have been reduced when groundnut was intercropped with sorghum or millet in comparison with monoculture groundnut. The efficacy of intercropping against insect damage is possibly by distributing the attaching insect pests among the target pulse crop and intercrop and also due to the conservation of natural enemies.

Besides, trap cropping is also followed by growing pest preferred crops to attract pest populations and limiting the damage to the main crops. Thus, main crops require less amount of insecticides (Pattar et al. 2012). Intercropping pulses with the crops that act as perches for insectivorous birds such as myna—and drengo—increase the predation of target insect. The best example for this is the intercropping sunflower with chickpea reduces the incidence of pod borer damage by attracting predatory birds on the sunflower perch that can reduce pod borer larval number within the shortest time (Pattar et al. 2012). Moreover, predatory wasps carrying a large number of pod borer larvae were also recorded on sunflower plants. Similarly, the two most important defoliators of groundnut (*S. litura*, *H. armigera*) were suppressed by using sunflower as trap crop as the pest prefers sunflower for oviposition and larval feeding than the groundnut crop.

Plant Density

Plant density is another deciding factor for insect pest damage in pulse crops. Crops with high density can harbor more pests and thus lead to more damage (Kant et al. 2007). Consistent with this notion, higher crop density in chickpea accommodated higher population of larvae and pupae of *Helicoverpa* than low plant density planting (Kant et al. 2007). Karungi et al. (2000) demonstrated that aphids are effectively controlled by the cultural practices such as plant density and early planting, while thrips, maruca, and pod-sucking bugs remained less effected.

Sometimes excess application of fertilizers such as NPK results in bushy plant growth making them a better host for insect infestation. For this reason, higher doses of NPK led to higher pod borer damage in chickpea, while moderate NPK doses resulted in lower pod borer infestation. However, high doses of phosphorous substantially reduced the pod borer occurrence and enhanced the seed yield in chickpea. Hence, phosphorus-based fertilizers are preferred for chickpea than nitrogenous fertilizers as the latter enhances larval and pupal development. Organic manures including farm yard manure, neem cake, and vermicompost reduced the pest population in comparison with inorganic fertilizers.

Physical Control

Physical control methods are deployed by treating the seeds and insects using physical agents, including temperature, heat, and moisture content (Sahadia and Aziz 2011), because each insect pest has a specific growth requirement comprising temperature, humidity, and photoperiod depending on the developmental stage (Dahiya et al. 1999). Creating unfavorable environment for their survival is the key to diminish insect population. The physical control practices of insect pests also include removal of egg shells and dead larvae from the plant population. Storage pests such as bruchids can be avoided by physical removal of infested grains before storage and also by keeping the storage environment unfavorable for the survival of storage pests by fumigation and disinfestation of empty store (Casida and Quistad 1998). The lepidopteran insect pest infestations can be minimized by deep ploughing of soil, maintaining spacing between individual plants, shaking of plants to get rid of insects, and weeding to avoid carryover larvae (Dahiya et al. 1999). In pigeon pea, gentle shaking can dislodge 97% of caterpillars followed by removal of borer larvae from their feeding sites found to be very effective and economical (Ranga Rao and Rameswar Rao 2010).

Biological Control

A large number of biocontrol agents have been used successfully for insect and disease control. The biological pest control strategies rely on deploying the natural enemies, predators, parasitoids, and beneficial biological agents applied as formulated products to control insect pests. It is widely incorporated in integrated pest management farming systems although they share only 3% of the overall pesticide market (Hilder 2003). Some predators have been used as biological control agents for pulse resistance. The most common predators of insect pests of pulse crops belong to *Chrysopa* spp., *Geocoris* sp., *Chrysoperla* sp., *Orius* sp., *Polistes* sp., and *Nabis* species (Romeis and Shanower 1996). Given that predators are effective in large numbers, the high cost involved in large-scale production hinders their use as biological control agents for pulse crop insect management. Some of the egg parasitoids including *Trichogramma* species and *Telenomus* species eradicate large number of pod borer eggs. Parasitic flies have been reported to lay eggs in the hemocoel of the *H. armigera* larvae; upon hatching, these parasitoid larvae feed and develop inside the body of *H. armigera*. However, trichome exudates from major pulse crops including chickpea and pigeon pea limit their predatory activity. Among all the biopesticides, nuclear polyhedrosis virus (NPV) and *Bacillus thuringiensis* (Bt) have been very popular. However the time of application is another factor to be considered for better implication of their insect control. The application of Bt formulations proved to be better in insect control when sprayed in the evening than spraying at other times of the day (Cherry et al. 2000). Formulations based on Bt and HaNPV are effective for controlling *H. armigera* on chickpea and pigeon pea. Due to its easy availability and efficacy, various NPV strains have been often used to control *H. armigera* on chickpea (Cherry et al. 2000). Improvements in commercial formulations of NPV including the genetic transformation of insect-selective nerve toxins, hormones, and proteases helped the better perforation of insect cell membrane and specificity of the NPV-based pesticides (Harrison and Bonning 2001). While some of the entomopathogenic fungi including *Beauveria bassiana* and *Metarhizium anisopliae* are also known to be effective in controlling *H. armigera* (Nguyen et al. 2007), their insecticidal activity in pulse crops needs further study.

The use of pheromone traps for gram pod borers, tobacco caterpillars, and leaf miners and the use of plant-derived products against major insect pests are novel strategies that may reduce the cost to former. Plant products derived from neem has been shown to have a broad-spectrum control of more than 200 species of insect pests (Bhushan et al. 2011). Similarly, application of groundnut oil to chickpea seeds significantly reduces the hatching and development of *C. chinensis* eggs. Besides, groundnut oil has been shown to prevent the protoplasmic movement causing coagulation in the eggs and larval death in *C. maculatus* (Ghosh et al. 2007; Bhushan et al. 2011). Some of the essential oils have also been used as insect repellents against *Callosobruchus* species.

Chemical Control

The use of chemical pesticides for controlling most of the insect pests of pulse crops has been well-known strategy from decades. The four major groups of pesticides including organochlorines, organophosphates, carbamates, and pyrethroids have been the most popularly used chemical pesticides for the pest management both in field and storage (Minja et al. 2000). The combined application of insecticide such as spark, a mixture of deltarnethrin 1% + triazophos 35%, and polytron C, mixture of cypermethrin 4% + profenfos 40%, are also effective against pod borer in pulse crops (Shaw et al. 1999). The ability of chemical pesticides to control even advanced stage of infestation has made them the most preferred choice regardless of their intense selection pressure on insects that often argues the resistance against many pesticides (Minja et al. 2000). It was observed that lower pod damage due to *H. armigera* was recorded in endosulfan than *B. thuringiensis* in pigeon pea (Sushil et al. 2009). Similarly, combined application of *B. thuringiensis* (Dipel) and deltamethrin has been shown to be most effective in minimizing the pod borer damage and highest net profit in pigeon pea (Reddy et al. 2001). However, due to the continuous and abundant usage of pesticides, *H. armigera* has been reported to exhibit resistance against a range of pesticides including organochlorides, organophosphates, carbamates, and pyrethroids (Gunning et al. 1998). Besides, chemical pesticides also contaminate the food chain that exerts mutagenic or toxic effects on the natural ecosystems. Alternatively phytochemicals or plant-based extracts are used against many insect pests including bruchids. In contrast to chemical pesticides, these are slow in their action, easily degradable. Nonetheless, they can negatively impact seed germination and also nontarget organisms. On the other hand, inert gases such as nitrogen and carbon dioxide caused a progressive decrease in fertility in *C. maculatus* (Ghosh et al. 2007).

Integrated Pest Management (IPM) Modules

Considering the necessity to exploit the available spectra of natural enemies in order to minimize the dependence on chemical pesticides, various IPM programs have been developed for insect management in pulse crops. IPM begins with understanding the pest situation in the region and field and matching those conditions with pulse varieties that are best suited for managing those stresses. Field assessment, tillage, sowing date, plant density, and weed control can also be an integral part of IPM to minimize pest incursions. The large-scale testing in farmers' field has proved that IPM technology can lower production costs, thus benefiting small-scale farmers (Chaudhary et al. 2008, Nagamani et al. 2013). Detailed monitoring of insect pests invasion on a crop, the variable relationships between trap catches and subsequent egg or larval populations are essential to provide an early warning and to develop predictive model specific to the local ecology of the pest and the particular cropping system (Nagamani et al. 2013). A prediction model for *H. armigera* on chickpea has been developed based on atmospheric parameters including temperature, relative humidity, and rainfall. Similarly, modeling and population dynamics have also been generated for aphids on lupin, pea seed weevil, and pea aphid on field pea (Clement et al. 2000). Phonological models that predict host and parasite development based on temperature have recently been implemented for *O. minor* infecting red clover (Eizenberg et al. 2005) and could be developed for faba bean. Additional supplements of predators or parasitoids may be necessary in further improving insect resistance in pulse crops. At some places, planting is delayed until after conventional crops in the area have emerged to reduce insect infestation. To combat pea seed weevil, border and/or refuge areas may be seeded to a trap crop of Austrian Winter Pea or purple-flowered pea variety a week or two before seeding the organic pea crop. Major successes, for example, have been reported in management of groundnut and chickpea foliar diseases, pod borer in legumes, and groundnut pests. Maintaining non-crop border habitat and/or refuge areas to enhance populations of natural enemies is one of the successful IPM strategy especially used in Bt crops. Maintaining refugia ensures the availability of plenty of susceptible insects nearby the resistant ones to mate with, thereby preventing the creation of a resistant population. IPM strategies were implemented in the pigeon pea- and chickpea-growing areas in both Africa and Asia using a participatory training approach especially to reduce incidences of damage due to *Helicoverpa* (Ranga Rao and Rameswar Rao 2010).

Even though IPM has been campaigned for the past two decades, very less percentage (>10%) of the farmers only adopted IPM practices in various crops. Recently, strict enforcement of the governing regulations for the production and use of pesticides in few countries helped to avoid the unsafe pesticides from the market.

IPM research in the past decade brought out changes in the farmers' attitude in pest management, which resulted 20–100% reduction in pesticide use in different crops. While efforts have been made to use integrated pest management, the number of established IPM programs is scanty due to the unavailability of critical IPM inputs and advisories among rural communities. The recent deployment of farmer participatory approach proved very effective in the exchange of technology. IPM achieved greater yields in pigeon pea and chickpea (8.18–11.97 q/ha vs. 6.78–9.25 q/ha; 9.50–18.83 q/ha vs. 6.0–11.66 q/ha, respectively). Formers got benefited by adopting IPM technologies partly by the cost reduction due to reduced overall application of pesticide and also preserved greater yields when compared with non-IPM practiced fields. Mindful use of insecticides based on regular scouting to assess the need, optimum timing, and selection of insecticides only where natural pest control was inadequate also avoids the fear of pesticide residues in food. Developing new pesticides with minimal nontarget impacts is also a priority (Ganiger 2000). For example, the new insecticide imidacloprid is effective against aphids, wireworms, thrips, and broad bean weevil and hence has considerable potential in faba bean IPM programs (Kaniuczak and Matosz 1998). However, when applied as a foliar spray during podding, it significantly reduced leaf miner *Liriomyza huidobrensis* populations but also suppressed its parasitoid *Diglyphus isaea*. These results indicate that the use of new-generation insecticides coupled with advisories was quite effective against insect pests with less impact on the environment and other nontargets.

Although IPM methods are environment-friendly, cost-effective, and compatible with other strategies, they require skilled personnel. Hence, several management IPM practices have to be developed and tested in farmers' fields. A combination of HPR and weather-based minimal fungicidal protection has led to the rehabilitation of chickpea in BGM-prone areas in Nepal, Bangladesh, and India. Integrated management of groundnut foliar diseases by combining HPR in high-yielding varieties and judicial use of pesticides based on need has been validated in India. However, to achieve breakthrough success, it is necessary to develop integrated pest modules that serve multiple goals including insect forecasting and modeling, field sampling and monitoring, crop diversification, control methods that also increase soil fertility, economic thresholds, and pest management methods.

7.4 Conclusions

The insect pests pose big threat to global food security. It is clear that successful production of insect-resistant crops depends on the integration of genetic resistance, monitoring of the target organisms, safe management, and mindful application of appropriate chemical and biological treatments (Fig. 7.3). HPR can continue to play key role in insect management as it is compatible with other control methods and requires minimum input by the farmer, and there are no negative environmental effects. Additionally, the opportunity for utilizing new insect-resistant genes from

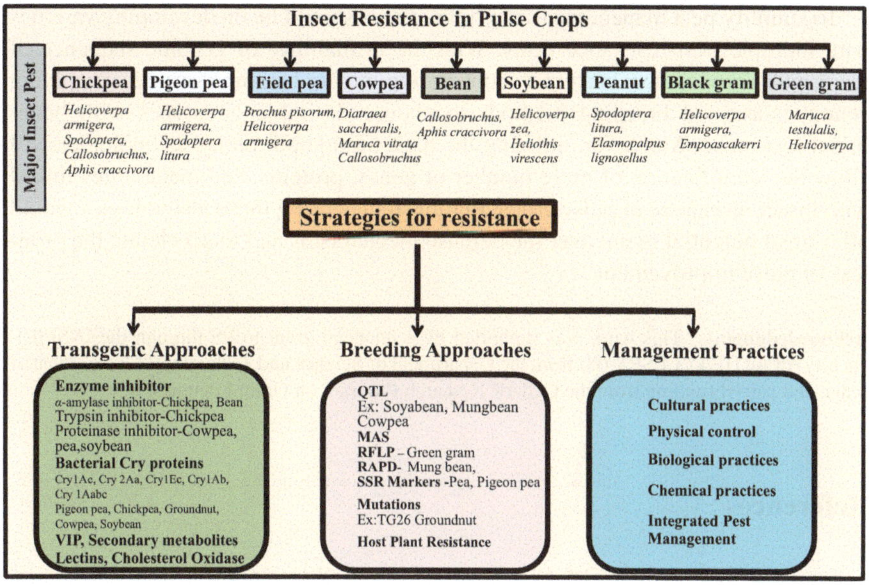

Fig. 7.3 Schematic illustration of the most common insect pests and management strategies conferring resistance in major pulse crops

wild relatives and the ability to move these across plant species through plant biotechnology open new doors to the field of HPR. The present transgenic approaches show certain limitations and are not completely successful in limiting the insect pests. Transgenic plants with reduced expression or efficacy may increase the possibility of developing resistant insect population. However, the upcoming technologies such as sequence-specific gene silencing via RNA interference hold a great promise for effective management of insect pests in pulse crops. Moreover, the possibility of introgressing multiple resistant genes or QTL combinations through gene pyramiding not only increases plant insect tolerance but also delays the development of resistance in insects. These fusion proteins may also confer complete protection against harmful insects by increasing and stabilizing the level of gene expression in target tissues. A viable complementary strategy is to develop transgenic plants with targeted expression of toxin specifically in certain vulnerable tissue of the plant or at particular developmental stage of the plant in the near future. This strategy would minimize the off-target effects on both insect and plant parts. Another important means of resistance management is to apply IPM principles in transgenic crop cultivation. Use of IPM methods coupled with minimal application of chemical insecticides will prolong the life of transgenic crops. However, the complexity of IPM necessitates active involvement of stakeholders, researchers, extension workers, and farmers to lighten apprehensions through participatory research trials. Thus, preference should be given to scale up testing of eco-safe IPM packages to on-farm situations to pulse growers.

To simplify pest management, the future focus should be on developing varieties with multiple pest-resistance traits. With the availability of genome sequence of major pulse crops, there is a great scope to reframe our breeding strategies applying genomics-assisted breeding methods to choose desirable genotypes with better insect resistance. Moreover, utilizing the high-throughput omics technologies will allow the identification of more number of genes, proteins, and metabolites linked with insect resistance in pulse crops. Future studies using these technologies would offer great potential to uncover the defense mechanisms and to accelerate the progress of pulse improvement.

Acknowledgments This work was supported by a financial grant to PS through the INSPIRE Faculty Award (IFA12-LSPA-08) from the Department of Science and Technology, Government of India, and partial funding from the CGIAR Research Program on Grain Legumes.

References

Acharjee S, Sarmah BK, Kumar PA, Olsen K, Mahon R, Moar WJ et al (2010) Transgenic chickpeas (Cicer arietinum L.) expressing a sequence-modified cry2Aa gene. Plant Sci 178(3):333–339

Adesoye A, Machuka J, Togun A (2008) CRY 1AB trangenic cowpea obtained by nodal electroporation. Afr J Biotechnol 7(18):3200–3210

Ali JG, Agrawal AA (2012) Specialist versus generalist insect herbivores and plant defense. Trends Plant Sci 17(5):293–302

Annis B, O'Keeffe LE (1984) Response of two Lathyrus species to infestation by the pea weevil Bruchus pisorum L.(Coleoptera: Bruchidae). Entomol Exp Appl 35(1):83–87

Asharani BM, Ganeshaiah KN, Kumar ARV, Makarla U (2011) Transformation of chickpea lines with Cry1X using in plantatransformation and characterization of putative transformants T1lines for molecular and biochemical characters. J Plant Breed Crop Sci 3:16413–16423

Asif M, Rooney LW, Ali R, Riaz MN (2013) Application and opportunities of pulses in food system – a review. Crit Rev Food Sci Nutr 53:1168–1179

Bakhsh A, Rao AQ, Shahid AA, Husnain T, Riazuddin S (2009) Insect resistance and risk assessment studies in advance lines of Bt cotton harboring Cry1Ac and Cry2A genes. Am Eurasian J Agric Environ Sci 6(1):1–11

Bakhsh A, Khabbazi SD, Baloch FS, Demirel U, Çalişkan ME, Hatipoğlu R et al (2015) Insect-resistant transgenic crops: retrospect and challenges. Turk J Agric For 39(4):531–548

Baksh A (2003) Potential adverse health effects of genetically modified crops. J Toxicol Environ Health B Crit Rev 6:211–225

Bardner R, Fletcher KE (1974) Insect infestations and their effects on the growth and yield of field crops: a review. Bull Entomol Res 64(1):141–160

Beena MR, Tuli R, Gupta AD, Kirti PB (2008) Transgenic peanut (Arachis hypogaea L.) plants expressing cry1EC and rice chitinase cDNA (Chi11) exhibit resistance against insect pest Spodoptera litura and fungal pathogen Phaeoisariopsis personata. Trans Plant J 2:157–164

Bhatnagar VS, Sithanantham S, Pawar CS, Jadhav D, Rao VR, Reed W (1983) Conservation and augmentation of natural enemies with reference to integrated pest management in chickpea (Cicer arietinum L.) and pigeonpea (Cajanus cajan (L.) Millsp). In: Proc International workshop on Integrated Pest Control in Grain legumes. Goiana, Brazil, pp. 157–180

Bhatnagar-Mathur P, Sharma KK (2016) Genetic transformation of pigeonpea: an overview. Legum Perspect 11:35–36

Bhushan S, Singh RP, Shanker R (2011) Bioefficacy of neem and Bt against pod borer, Helicoverpa armigera in chickpea. J Biopest 4(1):87–89

Blair MW, Muñoz C, Buendía HF, Flower J, Bueno JM, Cardona C (2010) Genetic mapping of microsatellite markers around the arcelin bruchid resistance locus in common bean. Theor Appl Genet 121(2):393–402
Boethel DJ (1999) Assessment of soybean germplasm for multiple insect resistance. In: Global plant genetic resources for insect-resistant crops. CRC, Boca Raton, pp 101–129
Bruce TJ (2014) Interplay between insects and plants: dynamic and complex interactions that have coevolved over millions of years but act in milliseconds. J Exp Bot 66(2):455–465
Byrne O, Galwey N, Hardie D (2002) Searching for molecular markers for resistance to pea weevil. In: JA MC (ed) Plant breeding for the 11th milennium: proceedings of the 12th Australasian plant breeding conference. Australian Plant Breeding Association, Perth, pp 62–366
Cao J, Ibrahim H, Garcia J, Mason H, Granados R, Earle E (2002) Transgenic tobacco plants carrying a baculovirus enhancin gene slow the development and increase the mortality of Trichoplusia ni larvae. Plant Cell Rep 21(3):244–250
Cardona C, Kornegay J (1999) Bean germplasm resources for insect resistance. In: Clement SL, Quisenberry SS (eds) Global plant genetic resources for insect-resistant crops. CRC Press LLC, Boca Raton, pp 85–100
Carrillo E, Rubiales D, Pérez-de-Luque A, Fondevilla S (2013) Characterization of mechanisms of resistance against Didymella pinodes in Pisum spp. Eur J Plant Pathol 135(4):761–769
Casida JE, Quistad GB (1998) Golden age of insecticide research: past, present, or future? Annu Rev Entomol 43(1):1–16
Chakraborti D, Sarkar A, Mondal HA, Das S (2009) Tissue specific expression of potent insecticidal, Allium sativum leaf agglutinin (ASAL) in important pulse crop, chickpea (Cicer arietinum L.) to resist the phloem feeding Aphis craccivora. Transgenic Res 18(4):529–544
Chakraborty J, Sen S, Ghosh P, Sengupta A, Basu D, Das S (2016) Homologous promoter derived constitutive and chloroplast targeted expression of synthetic cry1Ac in transgenic chickpea confers resistance against Helicoverpa armigera. Plant Cell Tissue Organ Cult (PCTOC) 125(3):521–535
Charity JA, Anderson MA, Bittisnich DJ, Whitecross M, Higgins TJV (1999) Transgenic tobacco and peas expressing a proteinase inhibitor from Nicotiana alata have increased insect resistance. Mol Breed 5(4):357–365
Chaudhary RG, Saxena H, Dhar V, Prajapati RK (2008) Evaluation and validation of IPM modules against wilt, Phytophthora blight, pod borer and pod fly in pigeonpea. J Food Legum 21:58–60
Chavan AP, Patil SK, Deshmukh GP, Pawar KB, Brahmane RO, Harar PN (2009) Sources of resistance to pigeonpea pod borers. In: International conference on grain legumes: quality improvement, value addition and trade. Indian Institute of Pulses Research, Kanpur, pp 256–257
Chen KC, Lin CY, Kuan CC, Sung HY, Chen CS (2002a) A novel defensin encoded by a mungbean cDNA exhibits insecticidal activity against bruchid. J Agric Food Chem 50(25):7258–7263
Chen JW, Tang LX, Tang MJ, Shi YX, Pang Y (2002b) Cloning and expression product of vip3A gene from Bacillus thuringiensis and analysis of inseceicidal activity. *Sheng wu gong cheng xue bao*. Chin J Biotechnol 18(6):687–692
Chen HM, Liu CA, Kuo CG, Chien CM, Sun HC, Huang CC et al (2007) Development of a molecular marker for a bruchid (Callosobruchus chinensis L.) resistance gene in mungbean. Euphytica 157(1–2):113–122
Cherry AJ, Rabindra RJ, Parnell MA, Geetha N, Kennedy JS, Grzywacz D (2000) Field evaluation of Helicoverpaarmigeranucleopolyhedrovirus formulations for control of the chickpea pod-borer, H. armigera (Hubn.), on chickpea (Cicer arietinum var. Shoba) in southern India. Crop Prot 19(1):51–60
Chiang HS, Norris DM (1983) Morphological and physiological parameters of soybean resistance to agromyzid beanflies. Environ Entomol 12(1):260–265
Chiang HS, Singh SR (1988) Pod hairs as a factor in Vigna vexillata resistance to the pod-sucking bug, Clavigralla tomentosicollis. Entomol Exp Appl 47(2):195–199
Chowrira GM, Akella V, Fuerst PE, Lurquin PF (1996) Transgenic grain legumes obtained byin planta electroporation-mediated gene transfer. Mol Biotechnol 5(2):85–96

Clement SL, Quisenberry SS (eds) (1999) Global plant genetic resources for insect-resistant crops. CRC Press, Boca Raton, p 295

Clement SL, El-Din NEDS, Weigand S, Lateef SS (1993) Research achievements in plant resistance to insect pests of cool season food legumes. Euphytica 73(1–2):41–50

Clement SL, Wightman JA, Hardie DC, Bailey P, Baker G, McDonald G (2000) Opportunities for integrated management of insect pests of grain legumes. In: Linking research and marketing opportunities for pulses in the 21st century. Springer, Dordrecht, pp 467–480

Constabel CP, Yip L, Patton JJ, Christopher ME (2000) Polyphenol oxidase from hybrid poplar. Cloning and expression in response to wounding and herbivory. Plant Physiol 124(1):285–296

Dahiya SS, Chauhan YS, Johansen C, Shanower TG (1999) Adjusting pigeonpea sowing time to manage pod borer infestation. Int Chickpea Newsl 6:44–45

Das SB (1998) Impact of intercropping on Helicoverpa armigera (hub.): incidence and crop yield of chickpea in west Nimar valley of Madhya Pradesh. Insect Environ 4:84–85

Das A, Datta S, Thakur S, Shukla A, Ansari J, Sujayanand GK, Singh NP (2017) Expression of a chimeric gene encoding insecticidal crystal protein Cry1Aabc of Bacillus thuringiensis in Chickpea (Cicer arietinum L.) confers resistance to gram pod borer (Helicoverpa armigera Hubner.). Front Plant Sci 8:1423

Dayal S, Lavanya M, Devi P, Sharma KK (2003) An efficient protocol for shoot regeneration and genetic transformation of pigeonpea [Cajanus cajan (L.) Millsp.] using leaf explants. Plant Cell Rep 21(11):1072–1079

De Moraes CM, Lewis WJ, Pare PW, Alborn HT, Tumlinson JH (1998) Herbivore-infested plants selectively attract parasitoids. Nature 393(6685):570

Dhaliwal GS, Dhaliwal GS (1993) Advances in host plant resistance to insects. Klayan Publishers, New Delhi

Dicke M (1999) Direct and indirect effects of plants on performance of beneficial organisms. In: Handbook of pest management. Marcel Dekker, New York/Basel, pp 105–153

Dodia DA, Patel AJ, Patel IS, Dhulia FK, Tikka SBS (1996) Antibiotic effect of pigeonpea wild relatives on Helicoverpa armigera. Int Chickpea Pigeonpea Newsl 3:100–101

Dufourmantel N, Tissot G, Goutorbe F, Garcon F, Muhr C, Jansens S, Dubald M (2005) Generation and analysis of soybean plastid transformants expressing Bacillus thuringiensis Cry1Ab protoxin. Plant Mol Biol 58(5):659–668

Edwards OR (2001) Interspecific and intraspecific variation in the performance of three pest aphid species on five grain legume hosts. Entomol Exp Appl 100(1):21–30

Edwards OR, Ridsdill-Smith TJ, Berlandier FA (2003) Aphids do not avoid resistance in Australian lupin (Lupinus angustifolius, L. luteus) varieties. Bull Entomol Res 93(5):403–411

Eizenberg H, Colquhoun J, Mallory-Smith C (2005) A predictive degree-days model for small broomrape (Orobanche minor) parasitism in red clover in Oregon. Weed Sci 53(1):37–40

Estruch JJ, Warren GW, Mullins MA, Nye GJ, Craig JA, Koziel MG (1996) Vip3A, a novel Bacillus thuringiensis vegetative insecticidal protein with a wide spectrum of activities against lepidopteran insects. Proc Natl Acad Sci 93(11):5389–5394

Fabre C, Causse H, Mourey L, Koninkx J, Rivière M, Hendriks H (1998) Characterization and sugar-binding properties of arcelin-1, an insecticidal lectin-like protein isolated from kidney bean (Phaseolus vulgaris L. cv. RAZ-2) seeds. Biochem J 329(3):551–560

Felton GW, Donato KK, Broadway RM, Duffey SS (1992) Impact of oxidized plant phenolics on the nutritional quality of dietar protein to a noctuid herbivore, Spodoptera exigua. J Insect Physiol 38(4):277–285

Ganguly M, Molla KA, Karmakar S, Datta K, Datta SK (2014) Development of pod borer-resistant transgenic chickpea using a pod-specific and a constitutive promoter-driven fused cry1Ab/Ac gene. Theor Appl Genet 127(12):2555–2565

Ganiger PC (2000) Bioefficacy of newer insecticides against pod borer complex in pigeonpea Cajanuscajan L. MSc (Agri.) dissertation submitted to Marathwada Agricultural University, Parbhani, 66 pp

Ghosh S, Azhahianambi P, Yadav MP (2007) Upcoming and future strategies of tick control: a review. J Vector Borne Dis 44(2):79

Giri AP, Kachole MS (1998) Amylase inhibitors of pigeonpea (Cajanus cajan) seeds. Phytochemistry 47(2):197–202
Gnanamurthy S, Dhanavel D (2014) Effect of EMS on induced morphological mutants and chromosomal variation in Cowpea (Vigna unguiculata (L.) Walp). Int Let Nat Sci 17 17:33–43
Godfrey J (2000) Do genetically modified foods affect human health? Lancet 355:414
Gopali JB, Sharma OP, Yelshetty S (2013) Effect of insecticides and biorationals against pod bug (Clavigralla gibbosa) in pigeonpea. Indian J Agric Sci 83(5):582–585
Green PWC, Stevenson PC, Simmonds MSJ, Sharma HC (2002) Can larvae of the pod-borer, Helicoverpa armigera (Lepidoptera: Noctuidae), select between wild and cultivated pigeonpea Cajanus sp. (Fabaceae)? Bull Entomol Res 92(1):45–51
Gujar GT, Kumari A, Kalia V, Chandrashekar K (2000) Spatial and temporal variation in susceptibility of the American bollworm, Helicoverpa armigera (Hubner) to Bacillus thuringiensis var. kurstaki in India. Curr Sci:995–1001
Gulati A, Schryer P, McHughen A (2002) Production of fertile transgenic lentil (Lens culinaris Medik) plants using particle bombardment. In Vitro Cell Dev Biol Plant 38(4):316–324
Gunning RV, Moores GD, Devonshire AL (1998) Insensitive acetylcholinesterase and resistance to organophosphates in Australian Helicoverpa armigera. Pestic Biochem Physiol 62(3):147–151
Harrison RL, Bonning BC (2001) Use of proteases to improve the insecticidal activity of baculoviruses. Biol Control 20(3):199–209
Heil M (2004) Induction of two indirect defences benefits Lima bean (Phaseolus lunatus, Fabaceae) in nature. J Ecol 92(3):527–536
Higgins TJV, Gollasch S, Molvig L, Moore A, Popelka C, Armstrong J, Mahon R, Ehlers J, Huesing J, Margam V, Shade R (2012) Insect-protected cowpeas using gene technology. In: Fatokun C (ed) Innovative research along the cowpea value chain proceedings of fifth world cowpea conference on improving livelihood in the cowpea value chain through advancement in science. International Institute of Tropical Agriculture, Ibadan, pp 131–137
Hilder V (2003) GM plants and protection against insects–alternative strategies based on gene technology. Acta Agric Scand (B) 53(S1):34–40
Horber E (1978) Resistance of pests of grain legumes in the U.S.A. In: Singh SR, van HF E, Taylor TA (eds) Pests of grain legumes: ecology and control. Academic Press, London, UK, pp 281–295
Huynh BL, Ehlers JD, Ndeve A, Wanamaker S, Lucas MR, Close TJ, Roberts PA (2015) Genetic mapping and legume synteny of aphid resistance in African cowpea (Vigna unguiculata L. Walp.) grown in California. Mol Breed 35(1):36
Howe GA, Jander G (2008) Plant immunity to insect herbivores. Annu Rev Plant Biol 59:41–66
Ignacimuthu S, Prakash S (2006) Agrobacterium-mediated transformation of chickpea with α-amylase inhibitor gene for insect resistance. J Biosci 31(3):339–345
Indurker S, Misra HS, Eapen S (2007) Genetic transformation of chickpea (Cicer arietinum L.) with insecticidal crystal protein gene using particle gun bombardment. Plant Cell Rep 26(6):755–763
Ishimoto M, Sato T, Chrispeels MJ, Kitamura K (1996) Bruchid resistance of transgenic azuki bean expressing seed α-amylase inhibitor of common bean. Entomol Exp Appl 79(3):309–315
Jackai LEN, Oghiakhe S (1989) Pod wall trichomes and resistance of two wild cowpea, Vigna vexillata, accessions to Maruca testualis (Geyer) (Lepidoptera: Pyralidae) and Clavigralla tomentosicollis Stål (Hemiptera: Coreidae). Bull Entomol Res 79(4):595–605
Jayanand B, Sudarsanam G, Sharma KK (2003) An efficient protocol for the regeneration of whole plants of chickpea (Cicer arietinum L.) by using axillary meristem explants derived from in vitro-germinated seedlings. In Vitro Cell Dev Biol Plant 39(2):171–179
Kamphuis LG, Lichtenzveig J, Peng K, Guo SM, Klingler JP, Siddique KH, Singh KB (2013a) Characterization and genetic dissection of resistance to spotted alfalfa aphid (Therioaphis trifolii) in Medicago truncatula. J Exp Bot 64(16):5157–5172
Kamphuis LG, Zulak K, Gao LL, Anderson J, Singh KB (2013b) Plant–aphid interactions with a focus on legumes. Funct Plant Biol 40(12):1271–1284

Kaniuczak Z, Matosz I (1998) The effect of insecticidal seed dressings upon the broad bean weevil (Bruchus rufimanus Boh.) in the cultivation of the field bean. J Plant Prot Res 38:84–88
Kant K, Kanaujia KR, Kanaujia S (2007) Role of plant density and abiotic factors on population dynamics of Helicoverpa armigera (Hübner) in Chick pea. Ann Plant Prot Sci 15(2):303–306
Kaplan I, Dively GP, Denno RF (2009) The costs of anti-herbivore defense traits in agricultural crop plants: a case study involving leafhoppers and trichomes. Ecol Appl 19(4):864–872
Kar S, Basu D, Das S, Ramkrishnan NA, Mukherjee P, Nayak P, Sen SK (1997) Expression of cryIA (c) gene of Bacillus thuringiensis in transgenic chickpea plants inhibits development of pod-borer (Heliothis armigera) larvae. Transgenic Res 6(2):177–185
Kareiva P, Sahakian R (1990) Tritrophic effects of a simple architectural mutation in pea plants. Nature 345(6274):433–434
Karungi J, Adipala E, Ogenga-Latigo MW, Kyamanywa S, Oyobo N (2000) Pest management in cowpea. Part 1. Influence of planting time and plant density on cowpea field pests infestation in eastern Uganda. Crop Prot 19(4):231–236
Keneni G, Bekele E, Getu E, Imtiaz M, Damte T, Mulatu B, Dagne K (2011) Breeding food legumes for resistance to storage insect pests: potential and limitations. Sustainability 3(9):1399–1415
Kessler A, Baldwin IT (2001) Defensive function of herbivore-induced plant volatile emissions in nature. Science 291(5511):2141–2144
Khursheed S, Khan S (2016) Genetic improvement of two cultivars of Vicia faba L. using gamma-irradiation and ethyl methanesulphonate mutagenesis. Legum Res Int J 40(2):338–344
Klingler JP, Nair RM, Edwards OR (2009) A single gene, AIN, in *Medicago truncatula* mediates a hypersensitive response to both bluegreen aphid and pea aphid, but confers resistance only to bluegreen aphid. J Exp Bot 60:4115–4127
Koona P, Osisanya EO, Jackai LEN, Tamo M, Reeves J (2002) Pod surface characteristics in wild and cultivated Vigna species and resistance to the coreid bug Clavigralla tomentosicollis Stal. (Hemiptera: Coreidae). Int J Trop Insect Sci 22(1):1–7
Kostyukovsky M, Trostanetsky A (2006) The effect of a new chitin synthesis inhibitor, novaluron, on various developmental stages of Tribolium castaneum (Herbst). J Stored Prod Res 42(2):136–148
Kozgar I (2014) Mutation breeding in chickpea: perspectives and prospects for food security. Walter de Gruyter GmbH & Co KG, Warsaw
Kumar J, Choudhury AK, Solanki RK, Pratap A (2011) Towards MAS in pulses: a review. Plant Breed 130:297–313
Kusnierczyk A, Winge P, Midelfart H, Armbruster WS, Rossiter JT, Bones AM (2007) Transcriptional responses of Arabidopsis thaliana ecotypes with different glucosinolate profiles after attack by polyphagous Myzus persicae and oligophagous Brevicoryne brassicae. J Exp Bot 58(10):2537–2552
Lam WKF, Pedigo LP (2001) Effect of trichome density on soybean pod feeding by adult bean leaf beetles (Coleoptera: Chrysomelidae). J Econ Entomol 94(6):1459–1463
Laskar RA, Khan S, Khursheed S, Raina A, Amin R (2015) Quantitative analysis of induced phenotypic diversity in chickpea using physical and chemical mutagenesis. J Agron 14(3):102
Lawrence PK, Koundal KR (2002) Plant protease inhibitors in control of phytophagous insects. Electron J Biotechnol 5(1):5–6
Lepore LS, Roelvink PR, Granados RR (1996) Enhancin, the granulosis virus protein that facilitates nucleopolyhedrovirus (NPV) infections, is a metalloprotease. J Invertebr Pathol 68(2):131–140
Logiswaran G, Mohanasundaram M (1985) Effect of inter-cropping, spacing and mulching in the control of groundnut leaf miner, *Aproaerema modicella* (Deventer) (Gelechiidae: Lepidoptera). Madras Agric J 72:695–700
Lomash K, Bisht RS (2013) Population dynamics of Helicoverpa armigera (Hubner) on chickpea crop. Pantnagar J Res 11(1):35–38
Malhotra RS, El-Bouhssini M, Joubi A (2007) Registration of seven improved chickpea breeding lines resistant to leaf miner. J Plant Regist 1(2):145–146
Maqbool SB, Riazuddin S, Loc NT, Gatehouse AM, Gatehouse JA, Christou P (2001) Expression of multiple insecticidal genes confers broad resistance against a range of different rice pests. Mol Breed 7(1):85–93

Mehrotra M, Singh AK, Sanyal I, Altosaar I, Amla DV (2011) Pyramiding of modified cry1Ab and cry1Ac genes of Bacillus thuringiensis in transgenic chickpea (Cicer arietinum L.) for improved resistance to pod borer insect Helicoverpa armigera. Euphytica 182(1):87–102

Michler, J. D., & Josephson, A. L. (2017). To specialize or diversify: agricultural diversity and poverty dynamics in Ethiopia. *World Development, 89*, 214–226.

Miklos JA, Alibhai MF, Bledig SA, Connor-Ward DC, Gao AG, Holmes BA, Kolacz KH, Kabuye VT, Macrae TC, Paradise MS, Toedebusch AS, Harrison LA (2007) Characterization of soybeanexhibiting high expression of a synthetic transgene that confers a high degreeof resistance to Lepidopteran pests. Crop Sci 47:148–157

Minja EM, Shanower TG, Silim SN, Karuru O (2000) Efficacy of different insecticides for pigeonpea pest management in Kenya. Int Chickpea Pigeonpea Newsl 7:53–55

Mishra SK, Macedo MLR, Panda SK, Panigrahi J (2017) Bruchid pest management in pulses: past practices, present status and use of modern breeding tools for development of resistant varieties. Ann Appl Biol. https://doi.org/10.1111/aab.12401

Mithofer A, Boland W (2012) Plant defense against herbivores: chemical aspects. Annu Rev Plant Biol 63:431–450

Morton RL, Schroeder HE, Bateman KS, Chrispeels MJ, Armstrong E, Higgins TJ (2000) Bean α-amylase inhibitor 1 in transgenic peas (Pisum sativum) provides complete protection from pea weevil (Bruchus pisorum) under field conditions. Proc Natl Acad Sci 97(8):3820–3825

Mota AC, DaMatta RA, Lima Filho M, Silva CP, Xavier-Filho J (2003) Cowpea (Vigna unguiculata) vicilins bind to the peritrophic membrane of larval sugarcane stalk borer (Diatraea saccharalis). J Insect Physiol 49(9):873–880

Murfet IC (1971) Flowering in Pisum: reciprocal grafts between known genotypes. Aust J Biol Sci 24(4):1089–1102

Nagamani P, Viswanath K, Sharma OP, Bhagat S, Reddy PL (2013) Demonstration of IPM module for management of Helicoverpa armigera at village level. Ann Plant Prot Sci 21(2):432–434

Nanda UK, Sasmal A, Mohanty SK (1996) Varietal reaction of pigeonpea to pod borer Helicoverpa armigera (Hubner) and modalities of resistance. Curr Agric Res 9:107–111

Narvel JM, Walker DR, Rector BG, All JN, Parrott WA, Boerma HR (2001) A retrospective DNA marker assessment of the development of insect resistant soybean. Crop Sci 41(6):1931–1939

Nguyen NT, Borgemeister C, Poehling HM, Zimmermann G (2007) Laboratory investigations on the potential of entomopathogenic fungi for biocontrol of Helicoverpaarmigera (Lepidoptera: Noctuidae) larvae and pupae. Biocontrol Sci Tech 17(8):853–864

Oerke EC (2006) Crop losses to pests. J Agric Sci 144:31–43

Pandey SP, Srivastava S, Goel R, Lakhwani D, Singh P, Asif MH, Sane AP (2017) Simulated herbivory in chickpea causes rapid changes in defense pathways and hormonal transcription networks of JA/ethylene/GA/auxin within minutes of wounding. Sci Rep 7:44729

Parde VD, Sharma HC, Kachole MS (2012) Protease inhibitors in wild relatives of pigeonpea against the cotton bollworm/legume pod borer, Helicoverpa armigera. Am J Plant Sci 3:627–635

Patil A, Taware SP, Oak MD, Tamhankar SA, Rao VS (2007) Improvement of oil quality in soybean [Glycine max (L.) Merrill] by mutation breeding. J Am Oil Chem Soc 84(12):1117–1124

Pattar PS, Mansur CP, Alagundagi SC, Karbantanal SS (2012) Effect of intercropping systems on gram pod borer Helicoverpa armigera hubner and its natural enemies in chickpea. Indian J Entomol 74(2):136–141

Petitt FL, Wietlisbach DO (1992) Intraspecific competition among same-aged larvae of Liriomyza sativae (Diptera: Agromyzidae) in lima bean primary leaves. Environ Entomol 21(1):136–140

Pompermayer P, Lopes AR, Terra WR, Parra JRP, Falco MC, Silva-Filho MC (2001) Effects of soybean proteinase inhibitor on development, survival and reproductive potential of the sugarcane borer, Diatraea saccharalis. Entomolo Exp Appl 99(1):79 85

Popelka JC, Terryn N, Higgins TJV (2004) Gene technology for grain legumes: can it contribute to the food challenge in developing countries? Plant Sci 167(2):195–206

Powell KS (2001) Antimetabolic effects of plant lectins towards nymphal stages of the planthoppers Tarophagous proserpina and Nilaparvata lugens. Entomol Exp Appl 99(1):71–78

Purcell JP, Greenplate JT, Jennings MG, Ryerse JS, Pershing JC, Sims SR et al (1993) Cholesterol oxidase: a potent insecticidal protein active against boll weevil larvae. Biochem Biophys Res Commun 196(3):1406–1413

Pusztai A, Bardocz GG, Alonso R, Chrispeels MJ, Schroeder HE, Tabe LM, Higgins TJ (1999) Expression of the insecticidal bean alpha-amylase inhibitor transgene has minimal detrimental effect on the nutritional value of peas fed to rats at 30% of the diet. J Nutr 129:1597–1603

Qayum MA, Sanghi NK (1994). Red hairy caterpillar management through group action and non-pesticidal methods. Programme coordinated by ASW & Oxfarm (India) Trust

Ramu SV, Rohini S, Keshavareddy G, Gowri Neelima M, Shanmugam NB, Kumar ARV, Udayakumar M (2012) Expression of a synthetic cry1AcF gene in transgenic Pigeon pea confers resistance to Helicoverpa armigera. J Appl Entomol 136(9):675–687

Ranga Rao GV, Rao VR (2010) Status of IPM in Indian agriculture: a need for better adoption. Indian J Plant Prot 38(2):115–121

Ranga Rao GV, Wightman JA (1994) First annual Rabi/summer groundnut research workers. In: Group meeting held at Indian institute of Technology, Kharagpur, pp 12–15

Ranjekar PK, Patankar A, Gupta V, Bhatnagar R, Bentur J, Kumar PA (2003) Genetic engineering of crop plants for insect resistance. Curr Sci 84(3):321–329

Rao GR, Wightman JA, Rao DR (1991) The development of a standard pheromone trapping procedure for Spodoptera Iitura (F) (Lepidoptera: Noctuidae) population in groundnut (Arachis hyopogaea L) crops. Int J Pest Manag 37(1):37–40

Ravindran BM (2016) Transgenic Pest resistance. Devagiri J Sci 2(1):1–31

Reddy AA (2009) Pulses production technology: Status and way forward. Econ Polit Wkly 44:73–80

Reddy NC, Singh Y, Dureja P, Singh SV (2001) Bioefficacy of insecticides, biopesticides and their combinations against podborers in pigeonpea. Indian J Entomol 63(2):137–143

Rohini VK, Rao KS (2000) Transformation of peanut (Arachis hypogaea L.): a non-tissue culture based approach for generating transgenic plants. Plant Sci 150(1):41–49

Romeis J, Shanower TG (1996) Arthropod natural enemies of Helicoverpa armigera (Hübner) (Lepidoptera: Noctuidae) in India. Biocontrol Sci Tech 6(4):481–508

Rubiales D, Fondevilla S, Chen W, Gentzbittel L, Higgins TJ, Castillejo MA et al (2015) Achievements and challenges in legume breeding for pest and disease resistance. Crit Rev Plant Sci 34(1–3):195–236

Sahadia E, Aziz AE (2011) Control strategies of stored product pests. J Entomol 8:101–122

Sandhu S, Kang MS (2017) Advances in Breeding for Resistance to Insects. In: Arora R, Sandhu S (eds) Breeding Insect Resistant Crops for Sustainable Agriculture. Springer, Singapore, pp 67–99. https://doi.org/10.1007/978-981-10-6056-4_3

Sanyal I, Singh AK, Kaushik M, Amla DV (2005) Agrobacterium-mediated transformation of chickpea (Cicer arietinum L.) with Bacillus thuringiensis cry1Ac gene for resistance against pod borer insect Helicoverpa armigera. Plant Sci 168(4):1135–1146

Sarmah BK, Moore A, Tate W, Molvig L, Morton RL, Rees DP, Higgins TJV (2004) Transgenic chickpea seeds expressing high levels of a bean α-amylase inhibitor. Mol Breed 14(1):73–82

Schafleitner R, Huang SM, Chu SH, Yen JY, Lin CY, Yan MR, Krishnan B, Liu MS, Lo HF, Chen CY, Long-fang OC (2016) Identification of single nucleotide polymorphism markers associated with resistance to bruchids (Callosobruchus spp.) in wild mungbean (Vigna radiata var. sublobata) and cultivated V. radiata through genotyping by sequencing and quantitative trait locus analysis. BMC Plant Biol 16(1):159

Schoonhoven LM, Van Loon JJ, Dicke M (2005) Insect-plant biology. Oxford University Press, Oxford

Schroeder HE, Gollasch S, Moore A, Tabe LM, Craig S, Hardie DC, Chrispeels MJ, Spencer D, Higgins TJ (1995) Bean [alpha]-amylase inhibitor confers resistance to the pea weevil (Bruchus pisorum) in transgenic peas (Pisum sativum L.). Plant Physiol 107(4):1233–1239

Shade RE, Schroeder HE, Pueyo JJ, Tabe LM, Murdock LL, Higgins TJV, Chrispeels MJ (1994) Transgenic pea seeds expressing the α-amylase inhibitor of the common bean are resistant to bruchid beetles. Bio/Technology 12:793–796

Shanower TG, Romeis JMEM, Minja EM (1999) Insect pests of pigeonpea and their management. Annu Rev Entomol 44(1):77–96

Sharma BC, Mann K, Kashyap SL, Pampapathy G, Ridsdill-Smith J (2002) Identification of *Helicoverpa* resistance in wild species of chickpeas. In: McComb JA (ed) Plant breeding for the 11th millennium: proceedings of the 12th Australian plant breeding conference - 15–20 Sept 2002. Australian Plant Breeding Inc., Perth, pp 277–280

Sharma HC, Pampapathy G, Dwivedi SL, Reddy LJ (2003) Mechanisms and diversity of resistance to insect pests in wild relatives of groundnut. J Econ Entomol 96(6):1886–1897

Sharma HC, Varshney RK, Gaur PM, Gowda CLL (2008) Potential for using morphological, biochemical, and molecular markers for resistance to insect pests in grain legumes. J Food Legum 21(4):211–217

Sharma HC, Dhillon MK, Bhatnagar-Mathur P, Sharma KK, Butterfield M (2010) Potential of transgenic grain legumes for pest management and sustainable crop production. In: Pests and pathogens: management strategies. BS Publications, Hyderabad, pp 135–158 ISBN 978-81-7800-227-9

Sharma KK, Ortiz R (2000) Program for the application of genetic transformation for crop improvement in the semi-arid tropics. In Vitro Cell Dev Biol Plant 36(2):83–92

Sharma OP, Patange NR, Rachappa V, Venilla S (2016) Integrated disease and insect pest management for enhancing production of pulse crops. Indian J Genet Plant Breed 76(4):451–458

Shaw SS, Choudhary RK, Verma RS, Badaya AK, Mandloi KC (1999) Efficacy of some insecticidal mixture against bollworm complex of cotton under rainfed conditions. Shashya 6(1):71–74

Shimoda T, Takabayashi J, Ashihara W, Takafuji A (1997) Response of predatory insect Scolothrips takahashii toward herbivore-induced plant volatiles under laboratory and field conditions. J Chem Ecol 23(8):2033–2048

Shinde YA, Patel BR, Mulekar VG (2013) Seasonal incidence of gram caterpillar, Helicoverpaarmigera (Hub.) in chickpea. Curr Biotica 7(1):2

Singh SR, Emden HV (1979) Insect pests of grain legumes. Annu Rev Entomol 24(1):255–278

Singh KB, Ocampo B (1997) Exploitation of wild Cicer species for yield improvement in chickpea. Theor Appl Genet 95(3):418–423

Singsit C, Adang MJ, Lynch RE, Anderson WF, Wang A, Cardineau G, Ozias-Akins P (1997) Expression of a Bacillus thuringiensis cryIA (c) gene in transgenic peanut plants and its efficacy against lesser cornstalk borer. Transgenic Res 6(2):169–176

Solleti SK, Bakshi S, Purkayastha J, Panda SK, Sahoo L (2008) Transgenic cowpea (Vigna unguiculata) seeds expressing a bean α-amylase inhibitor 1 confer resistance to storage pests, bruchid beetles. Plant Cell Rep 27(12):1841–1850

Somers DA, Samac DA, Olhoft PM (2003) Recent advances in legume transformation. Plant Physiol 131(3):892–899

Souframanien J, Gupta SK, Gopalakrishna T (2010) Identification of quantitative trait loci for bruchid (Callosobruchus maculatus) resistance in black gram [Vigna mungo (L.) Hepper]. Euphytica 176(3):349–356

Sousa-Majer MJD, Turner NC, Hardie DC, Morton RL, Lamont B, Higgins TJ (2004) Response to water deficit and high temperature of transgenic peas (Pisum sativum L.) containing a seed-specific α-amylase inhibitor and the subsequent effects on pea weevil (Bruchus pisorum L.) survival. J Exp Bot 55(396):497–505

Srinivasan A, Giri AP, Harsulkar AM, Gatehouse JA, Gupta VS (2005) Kunitz trypsin inhibitor from chickpea (Cicer arietinum L.) that exerts anti-metabolic effect on podborer (Helicoverpaarmigera) larvae. Plant Mol Biol 57(3):359–374

Stam JM, Kroes A, Li Y, Gols R, van Loon JJ, Poelman EH, Dicke M (2014) Plant interactions with multiple insect herbivores: from community to genes. Annu Rev Plant Biol 65:689–713

Stewart SA, Hodge S, Ismail N, Mansfield JW, Feys BJ, Prospéri JM, Powell G (2009) The RAP1 gene confers effective, race-specific resistance to the pea aphid in Medicago truncatula independent of the hypersensitive reaction. Mol Plant-Microbe Interact 22(12):1645–1655

Surekha C, Beena MR, Arundhati A, Singh PK, Tuli R, Dutta-Gupta A, Kirti PB (2005) Agrobacterium-mediated genetic transformation of pigeon pea (Cajanus cajan (L.) Millsp.) using embryonal segments and development of transgenic plants for resistance against Spodoptera. Plant Sci 169(6):1074–1080

Sushil K, Chauhan R, Roshan L (2009) Evaluation of native strains of Bacillus thuringiensis var. kurstaki against Helicoverpaarmigera (Hübner) on pigeonpea. J Insect Sci (Ludhiana) 22(2):139–143

Thomas JC, Wasmann CC, Echt C, Dunn RL, Bohnert HJ, McCoy TJ (1994) Introduction and expression of an insect proteinase inhibitor in alfalfa Medicago sativa L. Plant Cell Rep 14(1):31–36

Thu TT, Mai TTX, Dewaele E, Farsi S, Tadesse Y, Angenon G, Jacobs M (2003) In vitro regeneration and transformation of pigeonpea [Cajanus cajan (L.) Millsp]. Mol Breed 11(2):159–168

Tohidfar M, Zare N, Jouzani GS, Eftekhari SM (2013) Agrobacterium-mediated transformation of alfalfa (Medicago sativa) using a synthetic cry3a gene to enhance resistance against alfalfa weevil. Plant Cell Tissue Org Cult 113(2):227–235

Trieu AT, Burleigh SH, Kardailsky IV, Maldonado-Mendoza IE, Versaw WK, Blaylock LA, Shin H, Chiou TJ, Katagi H, Dewbre GR, Weigel D (2000) Transformation of Medicago truncatula via infiltration of seedlings or flowering plants with Agrobacterium. Plant J 22(6):531–541

Usharani KS, Kumar CA (2015) Mutagenic effects of gamma rays and EMS on frequency and spectrum of chlorophyll mutations in urdbean (Vigna mungo (L.) Hepper). Indian J Sci Technol 8(10):927–933

Valencia A, Bustillo AE, Ossa GE, Chrispeels MJ (2000) α-Amylases of the coffee berry borer (Hypothenemus hampei) and their inhibition by two plant amylase inhibitors. Insect Biochem Mol Biol 30(3):207–213

Van der Westhuizen AJ, Qian XM, Botha AM (1998) Differential induction of apoplastic peroxidase and chitinase activities in susceptible and resistant wheat cultivars by Russian wheat aphid infestation. Plant Cell Rep 18(1–2):132–137

Varshney RK, Graner A, Sorrells ME (2005) Genomics-assisted breeding for crop improvement. Trends Plant Sci 10(12):621–630

Villareal JM, Hautea DM, Carpena AL (1998) Molecular mapping of the bruchid resistance gene in mungbean Vigna radiata L. Philipp J Crop Sci 23(Suppl. 1):1–9

Visalakshmi V (2001) Effect of different IPM components on Helicoverpa armigera (Hubner) and their impact on natural enemies in chickpea. Doctoral dissertation, Acharya NG Ranga Agricultural University

Walker DR, All JN, McPherson RM, Boerma HR, Parrott WA (2000) Field evaluation of soybean engineered with a synthetic cry1Ac transgene for resistance to corn earworm, soybean looper, velvetbean caterpillar (Lepidoptera: Noctuidae), and lesser cornstalk borer (Lepidoptera: Pyralidae). J Econ Entomol 93(3):613–622

Walker D, Boerma HR, All J, Parrott W (2002) Combining cry1Ac with QTL alleles from PI 229358 to improve soybean resistance to lepidopteran pests. Mol Breed 9(1):43–51

Wang SF, Liu AY, Ridsdill-Smith TJ, Ghisalberti EL (2000) Role of alkaloids in resistance of yellow lupin to red-legged earth mite Halotydeus destructor. J Chem Ecol 26(2):429–441

Wang J, Song W, Zhang W, Liu C, Hu G, Chen Q (2009) Meta-analysis of insect-resistance QTLs in soybean. Hereditas 31:953–961

Wani MR, Kozgar MI, Khan S, Ahanger MA, Ahmad P (2014) Induced mutagenesis for the improvement of pulse crops with special reference to mung bean: a review update. In: Improvement of crops in the era of climatic changes. Springer, New York, pp 247–288

War AR, Paulraj MG, Ahmad T, Buhroo AA, Hussain B, Ignacimuthu S, Sharma HC (2012) Mechanisms of plant defense against insect herbivores. Plant Signal Behav 7(10):1306–1320

War AR, Paulraj MG, Hussain B, Buhroo AA, Ignacimuthu S, Sharma HC (2013) Effect of plant secondary metabolites on legume pod borer, Helicoverpa armigera. J Pest Sci 86(3):399–408

Yang G, Espelie KE, Todd JW, Culbreath AK, Pittman RN, Demski JW (1993) Cuticular lipids from wild and cultivated peanuts and the relative resistance of these peanut species to fall armyworm and thrips. J Agric Food Chem 41(5):814–818

Yang TJ, Kim DH, Kuo GC, Kumar L, Yong ND, Park HG (1998) RFLP marker-assisted selection in backcross breeding for introgression of the bruchid resistance gene in mungbean. Korean J Breed (Korea Republic) 30:8–15

Yencho GC, Cohen MB, Byrne PF (2000) Applications of tagging and mapping insect resistance loci in plants. Annu Rev Entomol 45(1):393–422

Yoshida M, Cowgill SE, Wightman JA (1997) Roles of oxalic and malic acids in chickpea trichome exudate in host-plant resistance to Helicoverpa armigera. J Chem Ecol 23(4):1195–1210

Zhao JZ, Cao J, Li Y, Collins HL, Roush RT, Earle ED, Shelton AM (2003) Transgenic plants expressing two Bacillus thuringiensis toxins delay insect resistance evolution. Nat Biotechnol 21(12):1493–1497

Zhu-Salzman K, Luthe DS, Felton GW (2008) Arthropod-inducible proteins: broad spectrum defenses against multiple herbivores. Plant Physiol 146(3):852–858

Chapter 8
Genetic and Genomic Approaches for Improvement in Mungbean (*Vigna radiata* L.)

Alok Das, Prateek Singh, Neetu Singh Kushwah, Shallu Thakur, Meenal Rathore, Aditya Pratap, and N. P. Singh

8.1 Origin and Distribution

Mungbean is believed to be originated in India ~3500 years ago, based on genetic diversity studies and archaeological evidence. India has a wide range of genetic diversity of cultivated as well as wild types of mungbean. Archaeological studies of carbonized grains of wild types of mungbean were reported from Daimabad – a chalcolithic site in Ahmednagar district of Western Maharashtra, the site tentatively dated as circa 2200 to 1000 BC (Kajale 1977). However, Vedic text such as Kautilya's 'Arthasashtra' and in Charaka Samhita point to their origin further beyond the pre-Christian era (Jain and Mehra 1980). During the domestication process, its cultivation might have migrated to other Asian countries and Africa. The putative wild progenitor of the species, *V. radiata* var. *sublobata*, found in abundance as weeds in waste land areas in different parts of India and other countries as well (Singh and Singh 1974; Lawn and Cottrell 1988).

8.2 *Vigna* Germplasm and Resources

Grouping of plant species to different *gene pools* (Harlan and de Wet 1971) is a very important aspect of breeding programme that helps to decide parents in hybridization programme. *Vigna species* have been grouped into primary, secondary, and tertiary gene pools on the basis of crossability, cytogenetic, phylogenetic and molecular data (Table 8.1). Useful traits are mostly available in wild species in secondary and tertiary gene pools (Mallikarjuna et al. 2006; Tullu et al. 2006), which was ignored

A. Das (✉) · P. Singh · N. S. Kushwah · S. Thakur · M. Rathore · A. Pratap · N. P. Singh
ICAR-Indian Institute of Pulses Research, Kanpur, India
e-mail: adas@icar.org.in

S. H. Wani, M. Jain (eds.), *Pulse Improvement*,
https://doi.org/10.1007/978-3-030-01743-9_8

Table 8.1 Gene Pool of *Vigna* species

Primary gene pool	Secondary gene pool	Tertiary gene pool	References
Vigna radiata var. *radiata*, *V. radiata* var. *sublobata*, *V. radiata* var. *setulosa*	*V. mungo* var. *mungo*, *V. mungo* var. *silvestris*, *V. aconitifolia*, *V. trilobata*	*V. angularis*, *V. dalzelliana*, *V. glabrescens*, *V. grandis*, *V. umbellata*, *V. vexillata*	Chandel and Lester (1991), Dana and Karmakar (1990), Smartt (1981, 1985), Kumar et al. (2004)

Source: Kumar et al. (2011)

during the process of domestication. However, successful hybridization between the cultivated *Vignas* and their wild relatives in secondary and tertiary gene pools is constrained by pre- or postfertilization barriers, and their successful utilization in *Vigna* improvement programmes requires interventions such as embryo rescue, polyploidization, reciprocal crossing, hormonal manipulations and use of bridge species.

Ex situ germplasm collections are important aspects of conservation of biodiversity, and mungbean genetic resources are maintained at different centres throughout the globe including the ICAR-National Bureau of Plant Genetic Resources, New Delhi, and ICAR-Indian Institute of Pulses Research, Kanpur, India; AVRDC-The World Vegetable Centre, Taiwan; University of the Philippines; Institute of Crop Germplasm Resources, Chinese Academy of Agricultural Sciences and Plant Genetic Resources Conservation Unit, University of Georgia, USA.

Screening of large mungbean germplasm collections for traits of interest is an arduous task. Establishing subsets of collections (core or mini core) that represent the diversity of the whole collection makes screening easier and more practical. Mungbean core/mini core collections were reported from various countries of the world (Bisht et al. 1998; Liu et al. 2008; Barkley et al. 2008; Moe et al. 2011; Schafleitner et al. 2015). Molecular analyses of such representative collection highlight the genetic diversity that can be used for broadening the genetic base of mungbean cultivars (Sangiri et al. 2007).

8.3 Biotic and Abiotic Stresses

A plethora of diseases (powdery mildews, leaf spots, blights, rusts, mosaics) caused by fungi, bacteria, viruses and nematodes adversely affect the yield potential of mungbean. Notably, the yellow mosaic disease in mungbean is devastating and recurrent throughout India. Similarly, storage grain pests, bruchids (*Callosobruchus* spp.), cause damage to the grains during storage. Mungbean also suffers from a number of abiotic stresses, notably, temperature and drought stress. It is sensitive to temperature stress especially at full-bloom stage, and exposure to high temperature and moisture stresses often leads to heavy yield reductions. Low temperature affects germination in *spring* season in North India, and heat and drought stresses cause adverse effect at reproductive stage. High sensitivity to photoperiod and

Table 8.2 Important biotic and abiotic stresses in mungbean

Season/Niches	Biotic stress	Abiotic stress
Kharif (North, West, East and Central India)	Mungbean yellow mosaic virus (MYMV), *Cercospora* leaf spot (CLS), web blight, defoliators, sucking insect pests.	Preharvest sprouting, terminal drought
Spring/summer (North and Central India)	MYMV, root and stem rot, powdery mildew, rust, *Cercospora* leaf spot	Preharvest sprouting, temperature stress and drought stress, sensitivity to photoperiod and temperature, terminal drought

Modified from Singh and Singh (2016)

temperature is another major bottleneck in realizing the yield potential and predicting desired harvest index in mungbean (Table 8.2).

Development of mungbean genotypes with drought and salinity tolerance, which can retain large number of flowers with pods during high temperature (> 40 °C), are crucial factors to increase mungbean productivity in India. Under the Indian Council of Agricultural Research-National Initiative on Climate Resilient Agriculture (NICRA), a large number of mungbean genotypes have been identified based upon multilocation trials and controlled environments exhibiting enhanced tolerance to drought, heat, waterlogging and other parameters.

8.4 Selection of Useful Genes from the Germplasm

Selection of elite genotypes from germplasm (indigenous and exotic) and landraces has been important aspect of mungbean breeding. In the initial phases of the mungbean breeding programme, virtually all the varieties were developed by selection of superior genotypes from the samples of local cultivars or exotic materials. Desirable genotypes were selected, and after their progeny testing, the superior pure lines were evaluated for yield, yield traits and reaction to diseases, and the best pure line was released for cultivation.

Breeding for resistance to diseases and insect pests has been one of the major objectives in mungbean breeding. Two most important stresses, mungbean yellow mosaic virus and storage grain pest (*Callosobruchus*), have been the centre of major research activities. The physical, biochemical and genetic basis of bruchid resistance in mungbean has been reviewed extensively (War et al. 2017). Bruchids (*Callosobruchus maculatus* Fab. and *C. chinensis* L.) are the most important insect causing both field and storage infestation. Wild sources of bruchid resistance were reported and used to develop bruchid-resistant lines; however, undesirable genetic linkages (linkage drag) threaten exploitation of wild germplasm into cultivars. Further, evolution of new biotype of bruchids has rendered mungbean lines susceptible that were earlier reported to be resistant. Host-plant resistance is an effective alternative to control bruchids in mungbean and is associated with morphological, biochemical and molecular traits. These traits affect various aspects of insect growth and development, thereby, alleviate loss due to insect attack. Bruchid resistance gene has been mapped using molecular markers in different mapping populations of mungbean (War et al. 2017). Genetic engineering using bean alpha-amylase inhibitor holds promise, details of which were discussed later.

Identification of disease resistance sources has been at the forefront of mungbean breeding activities. Resistance to mungbean yellow mosaic virus (MYMV) is an important component, and more than 50 MYMV-resistant varieties have been released for cultivation in India. Resistance is often compromised as the virus mutates fast to generate novel pathotypes. However for durable resistance, genetic engineering using RNAi or genome editing appears promising. Reports of commercially released varieties with resistant to powdery mildew, *Macrophomina* blight and leaf crinkle virus are plenty (Kaur et al. 2008). Promising AVRDC accessions V 4281, V 2396 and V 3495 were reported resistant to agromyzids, whereas, accessions V 2709 and V 2802 were resistant to bruchids. Reports of introgression of MYMV and bruchids resistance through wide crosses, *V. radiata* x *V. radiata* var. *sublobata*, were also attempted. Disease-resistant (*R*) genes were also identified from amphidiploids of mungbean x rice bean crosses (Dar et al. 1991). Mutagenic treatment has also been successfully employed for generating variability for resistance against *Cercospora* leaf spot (CLS) and MYMV. Details of genomic resources in mungbean for future breeding programmes were reviewed (Kim et al. 2015).

8.5 Genetic Studies in Mungbean

Studies on genetics of quantitative and qualitative traits in mungbean and their inheritance have been of utmost importance, since last few decades (Table 8.3). The first cross in mungbean was made to study inheritance of colour of ripe pods and seed coat surface in mungbean. Subsequently, attempts were made to understand the inheritance of various morphological traits including plant type, plant colour, leaf type, flower colour, inflorescence type, pod pubescence, pod shape, pod colour, shattering habit, seed coat colour, seed coat surface, and hard-seededness in these crops (Singh 2014). In general, most of the traits are monogenic, except seed colours which were controlled by two independent genes. Contrasting results were obtained for twining habit trait reported to be governed by a single recessive gene (Pathak and Singh 1963) or single dominant gene (*T*) (Khattak et al. 1999). Similarly, semi-spreading habit was reported to be dominant over erect habit and governed by a single dominant gene (Pathak and Singh 1963). Trifoliate leaf trait is dominant over the entire leaf and is monogenic (Singh 1980; Chhabra 1990; Talukdar and Talukdar 2003). However, two dominant genes, 'Tlb1' and 'Tlb2', with duplicate gene action for trilobed leaves were also reported (Sareen 1985). Narrow lanceolate leaf is reported to be controlled by two recessive genes, 'nl1' and 'nl2' (Dwivedi and Singh 1985). Purple hypocotyls are dominant over purple spotted and green hypocotyls and purple spotted over green hypocotyls. The purple pigmentation on stem, petiole and veins of the leaves is reported to be conditioned by a single dominant gene 'Ppp1' with pleiotropic effect (Dwiwedi and Singh, 1986). Similarly, stem fascination is controlled by a single recessive gene ('fs1') having a pleiotropic effect

Table 8.3 Inheritance and gene action of economically important traits in mungbean

Trait	Inheritance	Reference
Plant type and growth habit	Single dominant/recessive gene, semi-spreading is dominant over erect habit	Sen and Ghosh (1959), Pathak and Singh (1963), Khattak et al. (1999)
Pubescence	Single dominant gene	Murty and Patel (1973), Sen and Ghosh (1959)
Nodulation	Additive and nonadditive gene action	Singh et al. (1985)
Pigmentation	Single dominant/recessive gene, anthocyanin in hypocotyl governed by two supplementary genes	Pathak and Singh (1963), Mishra et al. (1970), Mukherjee and Pradhan (2002)
Leaf traits	Single dominant gene Large leaflet is dominant over small leaflet; lobbed is dominant over entire type	Singh and Singh (1995), Singh and Mehta (1953), Talukdar and Talukdar (2003)
Stem fasciation	Single recessive gene	Dwiwedi and Singh (1990)
Inflorescence type	Simple types controlled by two dominant genes and compound types are double recessive homozygous; number of clusters controlled by single gene	Sen and Ghosh (1959), Singh and Singh (1970)
Flower colour	Single dominant gene	Bose (1939)
Yield components	Additive and nonadditive gene action	Singh and Singh (1972), Yohe and Poehlman (1975), Dasgupta et al. (1998), Khattak et al. (2002)
Pod colour	Single dominant gene	Bose (1939), Sen and Ghosh (1959), Murty and Patel (1973)
Pod shattering	Single dominant gene	Verma and Krishi (1969)
Seed coat colour	One or few genes; mottling governed by single gene	Khattak et al. (1999), Chen and Liu (2001), Lambrides et al. (2004)
Seed coat surface	Two complementary genes	Bose (1939), Sen and Ghosh (1959), Murty and Patel (1973)
Cotyledon colour	Single recessive gene control green colour	Thakare et al. (1988)
Hard-seededness	One or few dominant genes involved	Lambrides (1996), Singh (1983), Humphry et al. (2005)
Preharvest sprouting	Additive and nonadditive gene action; high G x E interaction	Durga and Kumar (1997)
Crop duration	Additive, nonadditive and epistatic gene action	Khattak et al. (2001)
Seed weight	Small is dominant over larger size	Sen and Murty (1960), Fatokun et al. (1992), Humphry et al. (2005)
Protein content	Additive and nonadditive gene action	Chandra and Tickoo (1998)

Source: Singh and Singh et al. (2016)

on the number of floral organs. The pubescence of pods is reported to be dominant over non-pubescence and is governed by independent duplicate genes (Khadilkar 1963). Inheritance of seed coat colour was extensively studied (Khattak et al. 1999). Black, black-spotted and dull-green seed coat colours were found to be dominant over green, non-spotted and shiny green colour, respectively. Inheritance studies on disease resistance genes were conducted on viral diseases like MYMV and bacterial pustule. Resistance to MYMV is reported to be governed by two recessive genes and in few cases reported due to a single dominant/recessive gene. Resistance to bacterial pustule in mungbean is governed by dominant gene. The discordance in the nature of inheritance of disease resistance could be ascribed to racial differences in these studies. Recent approaches for cloning and tagging of genes for disease resistance shall augment understanding genetics of disease resistance.

8.6 Genomic Resources

Genomic resources of mungbean: chloroplast genome sequence (Accession KPS1; NCBI Genome Database GQ893027.1) (Tangphatsornruang et al. 2010), mitochondrial genome sequence (cv. Berken; NCBI Genome Database HM367685.1) (Alverson et al. 2011) and whole genome sequence (cv. VC1973A; NCBI Genome Database PRJNA243847) (Kang et al. 2014) are invaluable resource for mungbean community.

Chloroplast Genome Chloroplasts (cp) are organelles pivotal for photosynthesis and harbour highly conserved circular double-stranded DNA molecule (72–217 kb) necessary for functional plant cell. Characteristic features of chloroplast genome are a pair of large inverted repeats (IRa and IRb) that are usually 10–28 kb in length that delineates the genome into one large single-copy (LSC) region and a small single-copy (SSC) region.

Chloroplast genome of Fabaceae family is known to have undergone more rearrangements than other angiosperms. Based on 454 pyrosequencing technology, gene content and structural organization of mungbean cp genome were reported to be similar to common bean, *Phaseolus vulgaris*. With an average AT content of 64.82%, the mungbean chloroplast genome is 151.27 kb in length including a pair of IRs of 26.474 kb separated by small single-copy region of 17.427 kb and large single-copy region of 80.896 kb. Notably, chloroplasts of *Cicer* and *Medicago* have lost one copy of IR and are grouped together in IR-lacking clade (IRLC). Mungbean cp genome contains 108 unique genes including 29 tRNA genes (representing 20 amino acids identified in the genome), 4 rRNA genes and 75 predicted protein coding genes and 19 unique genes duplicated in IR region making a total of 127 genes (Table 8.4). Non-coding sequences, including intergenic spaces and introns, comprise about 41.45% of the mungbean cp genome. Two distinct rearrangements, a 50 kb inversion and 78 kb rearrangement, were observed. The first inversion is common in papilionoid legumes indicating an early split in the diversification of papilionoid members. The second inversion of 78 kb is a distinct rearrangement

Table 8.4 Genes present in mungbean chloroplast genome

Sl. No.	Process/unit	Genes
1	Photosystem I	psaA, psaB, psaC, psaI, psaJ, ycf3, ycf4
2	Photosystem II	psbA, psbB, psbC, psbD, psbE, psbF, psbH, psbI, psbJ, psbJ, psbK, psbL, psbM, psbN, psbT, psbZ
3	Cytochrome b6/f	petA, petB, petD, petG, petL, petN
4	ATP synthase	atpA, atpB, atpE, atpF, atpH, atpI
5	RuBisCO	rbcL
6	NADH oxidoreductase	ndhA, ndhB, ndhC, ndhD, ndhE, ndhF, ndhG, ndhH, ndhI, ndhJ, ndhK
7	Large subunit ribosomal proteins	rpl2, rpl14, rpl16, rpl20, rpl23, rpl32, rpl36
8	Small subunit ribosomal proteins	rps2, rps3, rps4, rps7, rps8, rps11, rps12, rps14, rps15, rps18, rps19
9	RNAP	rpoA, rpoB, C1, C2
10	Other proteins	accD, ccsA, cemA, clpP, matK
11	Proteins of unknown function	ycf1, ycf2
12	Ribosomal RNAs	rrn16, rrn23, rrn4.5, rrn5
13	Transfer RNAs	trnA (UGC), C (GCA), D (GUC), E (UUC), F (GAA), G (UCC), H (GUG), I (CAU), I (GAU), K (UUU), L (CAA), L (UAA), L (UAG), fM (CAU), M (CAU), N (GUU), P (UGG), Q (UUG), R (ACG), R (UCU), S (GCU), S (GGA), S (UGA), T (GGU), T(UGU), V(GAC), V(UAC), W(CCA), Y(GUA)

Adapted from Tangphatsornruang et al. (2010)

which is found in subtribe Phaseolinae and encompasses the entire fragment of LSC, which also supports expansion-contraction mechanism of rearrangement.

Variations in cp genome sequence are useful for evolutionary studies from population level processes to more distant phylogenetic relationships. Cp-derived markers (*matK* gene and the *trnL-trnF* intergenic spacer) have been used to study the evolutionary studies of plants. Repetitive sequences like microsatellites within the cp genome can also be potentially useful for ecological and evolutionary studies because they are nonrecombinant, haploid and uniparentally inherited. Besides phylogenetic studies, sequence information will also facilitate cp transformation to express foreign proteins.

8.7 Mitochondrial Genome

Mitochondria are rod-shaped organelles considered power generator (ATP) of cells and generally harbour circular double-stranded DNA molecules of variable sizes, structure and sequence complexity. Characteristic features of mitochondrial genome are size and number of repetitive sequences (sites of intramolecular recombination) and variation in gene content (reflecting ongoing gene loss and functional gene transfer to nucleus). Recombination frequency is proportional to the size of the repeats and recombination across inverted repeats inverts the intervening sequences, whereas recombination across directly oriented repeats separates the genome into pairs of sub-genomic molecules. Such processes create a structurally dynamic assemblage of genomic molecules in vivo and have led to a virtual scrambling in the gene orders, sequence duplications and deletions, resulting in shifts in genome size.

Based on shotgun Sanger sequencing, mungbean mitochondrial genome is protein-gene-poor 401.262 kb in length with that lacks large, recombinationally active repeats and *promiscuous* sequences from chloroplast and nuclear genomes. The coding region constitutes 37.9% of the genome including protein exons, *cis*- and *trans*-spliced introns, 3 rRNA and 16 tRNA and conserved syntenic regions, while non-coding (mitochondria-like, chloroplast-like and nuclear-like) and uncharacterized constitute the rest. Unlike other seed plant mitochondrial genome, a very meagre 0.5% and 1.6% of chloroplast and nuclear (mostly transposable elements)-derived sequences were identified in the genome. Duplication of *atp*9 and loss of *cox*2 gene exemplify dynamic assemblage of the mitochondrial genome. Although respiratory genes were never lost during evolution, yet some of the genes are either absent or are present as pseudogenes in various stages of attrition.

Repeat content in *Vigna* genome is very low (ca. 2.7%) as compared to 8–62% coverage in other genomes and is skewed towards fewer and shorter repeats (<100 nt in length), which are recombinationally active. However, very few recombination products were identified in vivo in very low abundance. For better understanding, deep sequencing of 1000–10,000 X coverage shall uncover variable in genome size, sequence content, mutation rate and RNA editing frequency.

8.8 Nuclear Genome

Nuclear genome of three *Vigna* species (*Vigna radiata* var. *radiata*, $2n = 2x = 22$, accession *VC1973A;* wild relative *Vigna radiata* var. *sublobata*, $2n = 2x = 22$, accession TC1966 and polyploid relative *Vigna reflexo-pilosa* var. *glabra*, $2n = 4x = 44$, accession V1160) was sequenced using next-generation sequencing platforms, viz. IlluminaHiSeq2000 and Roche/454 GS FLX+. Approximately, 80–84% genome corresponding to 11 linkage groups of the three species was uncovered by the sequencing effort. Comparative studies indicate well-conserved macro-synteny blocks between other members of *Fabaceae* family, although blocks were highly dispersed.

Repetitive sequence and transposable elements (LTR retrotransposons, LTR/Gypsy, LTR/Copia, class II transposons like CACTA, Mutator, PIF-Harbinger, hAT, etc.) constitute approx. 50.1% of the mungbean genome. A total of 22,427 genes were identified in the genome including 160 *Vigna* gene clusters and 1850 genes encoding transcription factors (TFs). Another 2310 non-coding genes were predicted, including 629 transfer RNAs, 280 ribosomal RNAs, 537 microRNAs, 717 small nucleolar RNAs, 110 small nuclear RNAs and 37 regulatory RNAs. In contrast, the allotetraploid V1160 has a total of 41,844 genes, almost twice the number of mungbean genes. The estimated genome size of polyploid *Vigna* genome (968 Mb) is almost twice that of mungbean genome (579 Mb). The genome of VC1973A and its wild relative TC1966 can serve as model to understand mungbean domestication. Repetitive elements are the major driving force behind plant evolution. Higher proportions of repetitive elements were found in domesticated compared to wild. A total of 2,922,833 SNPs were identified between cultivated and wild mungbean with SNP frequency of 6.78 per 1 kb. Members of the Papilionoideae family shared a whole genome duplication (WGD) event approx. 58 million years ago (MYA), before the family split into major groups including warm-season and cool-season clades. Mungbean has only experienced only one ancient WGD, in contrast to soybean that underwent another round of WDG (approx. 5–13 MYA). Bayesian studies of allotetraploid species indicated that allopolyploidy event occurred around 0.09 MYA, with one diploid genome and chloroplast emanating from *Vigna trinervia* and the second possible from related species including *Vigna minima*.

8.9 Genetic Engineering Studies in Mungbean

Germplasm resources are plenty in mungbean, and understanding genetic basis of the traits to mitigate various biotic and abiotic stresses is currently underway. Genetic engineering is an alternative approach to improve these traits. However, success of this technique depends upon rapid, efficient and reproducible regeneration and transformation protocol as well as well-characterized and cloned genes for target traits. In grain legumes, direct organogenesis pathway from cotyledonary nodes, embryonic axis, shoot apices and primary leaf explants has been reported the most common pathway for regeneration and transformation (Chandra and Pental 2003). Axillary meristem at the junction of cotyledon and terminal meristem at the apices of embryonic axis explants contain cells that are highly competent for regeneration and *Agrobacterium* accessibility. The advantage of this pathway is that it does not require callus phase and hence chances of getting somaclonal variation is less. Also, time required for regeneration is less compared to indirect organogenesis via callus phase. Further, as mungbean is recalcitrant to regeneration and the frequency of transformation is low, additional efforts are required to overcome the host tissue recalcitrance and enhance the *Agrobacterium* virulence. Significant progress has been made to improve the transformation frequency of cotyledonary node explants-based transformation system in mungbean by optimizing several factors which influences *Agrobacterium*-mediated transformation. Traits targeted for genetic improvement in mungbean include oxidative stress and salinity tolerance, bruchid resistance, seedling rot pathogen resistance, etc. (Table 8.5).

Mungbean transformation has been attempted since the last two decades employing both direct and indirect organogenesis mode; however, first successful recovery of mungbean transgenic plants has been reported from cotyledonary node explants

Table 8.5 Reports of genetic transformation of mungbean

Sl. No.	Cultivar	Explants	Gene	Selection agent	Transformation frequency	Reference
1	K-851, Pusa-105, ML-5, ML-323, ML-337, PDM-11, PS-7	Hypocotyl, 4-day-old cotyledonary node with both cotyledons, primary leaf	*npt*II, *uid*A	Kanamycin (75 mg/L)	0.9%	Jaiwal et al. (2001)
2	K-851	Primary leaf explant from 10-day-old seedlings	*hpt*	Hygromycin (2.5 mg/L)	2.0%	Mahalaxmi et al. (2006)
3	PUSA-105	Cotyledonary node explant without cotyledon from 16-h-old seedlings	*bar, α-ai*	Phosphinothricin (1.0 mg/L)	1.51%	Sonia et al. (2007)
4	ML-267	Cotyledonary node explants from 2-day-old seedlings	*nptII, BjNPR1*	Kanamycin (100 mg/L)	–	Vijayan and Kirti (2011)
5	ML-267	Double cotyledonary node explant from 3-day-old seedlings	*nptII, uidA, Annexin1 bj*	Kanamycin (100 mg/L)	–	Yadav et al. (2012)
5	K-851	Four-day-old cotyledonary node explant derived from seeds treated with 5 μM TDZ	*AtNHX1*	Kanamycin (100 mg/L)	–	Sahoo et al. (2016)
6	–	Cotyledonary node explant without cotyledon from 16-h-old seedlings	*nptII, codA*	Kanamycin (100 mg/L)	–	Baloda et al. (2017)
7	K-851	Four-day-old cotyledonary node explant with both cotyledons	*AtNHX1 and bar*	Phosphinothricin (1.5 mg/L)	–	Kumar et al. (2017)

uidA β-glucuronidase gene, *nptII* neomycin phospho transferase II, *bar* basta (bialaphos) resistance, *α-ai Phaseolus vulgaris* alpha-amylase inhibitor, BjNPR1 *Brassica juncea* non-expressor of pathogenesis-related gene-1, Annexin 1bj *Brassica juncea* annexin, At*NHX1 Arabidopsis thaliana* tonoplast Na^+/H^+ antiporter gene, *codA* choline oxidase A in the mungbean cv. K-851 using *A. tumefaciens* strain LBA4404 harbouring pTOK233 vector carrying β-glucuronidase (*gusA*) and neomycin phosphotransferase II (*npt*II) marker gene at an overall efficiency of 0.9% (Jaiwal et al. 2001). Histochemical assay indicated stable *β*-glucuronidase (*GUS*) expression and molecular analyses like PCR and Southern hybridization confirmed the presence and integration of transgene in T_0 plants. However, transmission of transgenes (*GUS and nptII*) to the progeny has not been presented. Authors also attempted indirect shoot organogenesis using primary leaves and hypocotyl explants which produced transgenic calli after cocultivation with *Agrobacterium*. Transgenic nature of calli was confirmed by selection on kanamycin containing medium and *GUS* assay. However, the callus failed to regenerate shoots.

In another report, mungbean transgenic plants were shown to regenerate from primary leaf explants (cv. K-851), cocultivated with disarmed *Agrobacterium tumefaciens* strain, C-58 harbouring pCAMBIA–1301 plasmid comprising hygromycin and phosphotransferase genes (Mahalaksmi et al. 2006). Hygromycin-selected shoots were rooted and transferred to glasshouse to produce seeds. Presence and stable inheritance of *gus* gene were confirmed by PCR, and Southern hybridization and histochemical *GUS* assay confirmed the stable gene expression. However, lower regeneration efficiency of primary leaf explants and lot of escapes on hygromycin-containing medium may limit the use of these explants for routine introduction of desirable genes to mungbean.

An improved protocol of genetic transformation of mungbean (*cv*. Pusa 105) was reported using phosphinothricin as selective agent and *Phaseolus vulgaris α*-amylase inhibitor-1 (*α*AI-1) gene for resistance to bruchids (Sonia et al. 2007). The protocol is rapid and takes around 8–10 weeks for recovery of transgenic plants with the transformation efficiency of 1.51%. Transgenic plants showed stable integration, inheritance and expression of transgene as evidenced by PCR analysis, Southern hybridization and PPT leaf paint assay, respectively.

For enhancing resistance against seedling rot pathogen, *Rhizoctonia solani*, transgenic mungbean (*cv*. ML-267) plants were regenerated from cotyledonary node explants using kanamycin selection (Vijayan and Kirti 2011). Gene integration in primary transformants was confirmed by Southern hybridization and the expression of *Bj*NPR1 gene was confirmed by RT-PCR. Resistance against *R. solani* was demonstrated by seedling bioassay of transgenic T_2 plants. Further, parameters enhancing genetic transformation such as *A. tumefaciens* strains, pH, L-cysteine, tobacco leaf extract and kanamycin concentrations were optimized.

For enhanced salt tolerance, *Arabidopsis thaliana* tonoplast Na^+/H^+ antiporter (AtNHX1) gene was employed in transgenic mungbean (*cv*. K-851) with the transformation frequency of 2.07% based on kanamycin monosulphate selection (Sahoo et al. 2016). These transgenic lines exhibited enhanced tolerance to salt as confirmed by physiological and biochemical studies. However, these lines were dwarf and low yielder compared to wild type which may be due to metabolic pay-offs and pleiotropic effects caused by constitutive expression of transgene in transgenic lines. To avoid the unintended effects on growth and yield due to constitutive expression of transgene, the use of AtNHX1 gene under the control of stress-inducible promoter *rd29A* was reported (Kumar et al. 2017). As a result of stress-inducible expression, transgenic lines are phenotypically normal and exhibited enhanced salt tolerance under salt stress-inducible condition. Another strategy demonstrated for enhanced salt tolerance is the use of choline oxidase (*codA*) for synthesis of osmoprotectants and compatible solute, glycine betaine (GB). Transgenic mungbean lines exhibited enhanced salt tolerance as confirmed by leaf disc senescence test. However, inheritance study of transgene in progeny was not presented in the study (Baloda et al. 2017). For oxidative stress tolerance, annexin1 gene from *Brassica juncea* (*annexin1 bj*) was mobilized in mungbean (Yadav et al. 2012).

Although, these transgenics are in experimental state, yet a unifying theme demonstrates the suitability of *Agrobacterium*-mediated transformation using cotyledonary node-based system in mungbean. This system could be utilized for development of transgenic in mungbean for targeted trait.

8.10 Prospects

Mungbean has emerged as one of the important grain legumes in Asia and particularly in India, owing to short life cycle and that the seeds are rich source of folate and iron, besides protein. Research initiatives to generate genetic and genomic resources have taken centre stage, realizing the vast potential of the crop (Das et al. 2016). It is a well-suited model organism among grain legumes, particularly, *Vignas*, because of small genome size, short life cycle and self-pollinating nature. Once considered *orphan* crop, mungbean currently has vast repositories of genomic resources, and worldwide efforts are on to enrich understanding genetic basis of agriculturally important traits.

References

Alverson AJ, Zhuo S, Rice DW, Sloan DB, Palmer JD (2011) The mitochondrial genome of the legume *Vigna radiata* and the analysis of recombination across short mitochondrial repeats. PLoS One 6(1):e16404

Baloda A, Modanpotra S et al (2017) Transformation of mungbean plants for salt and drought tolerance by introducing a gene for an osmoprotectants glycine betaine. J Plant Stress Physiol 22:678–680

Barkley NA, Wang ML, Gillaspie AG, Dean RE, Pederson GA, Jenkins TM (2008) Discovering and verifying DNA polymorphism in a mung bean [V. *radiata* (L). R. Wilczek] collection by Ecotilling and sequencing. BMC Res Notes 1:28

Bisht IS, Mahajan RK, Kawalkar TG (1998) Diversity in greengram (*Vigna radiata* (L.) Wilczek) germplasm collection and its potential use in crop improvement. Ann Appl Biol 132:301–312

Bose RD (1939) Studies in Indian pulses. IX contribution to te genetics of mung (Phaseolus radiates Linn. Syn. Ph. Aureus Roxb.). Indian J Agric Sci 9:607–624

Chandel KPS, Lester RN (1991) Origin and evolution of Asiatic Vigna species. In: Proc Golden Jublee celebrations Symp grain legumes. Indian Soc Genet Plant Breed, New Delhi, pp 25–45

Chandra N, Tickoo JL (1998) Genetic analyses of protein content in mungbean (Vigna radiata L. Wilczek). Indian J Genet Plant Breed 58(4):475–478
Chandra A, Pental D (2003) Regeneration and genetic transformation of grain legumes: an overview. Curr Sci 84:381–387
Chhabra AK (1990) Inheritance of lobed and pentafoliate leaf in mungbean. Indian J Pulses Res 3:69–72
Chen H, Liu X (2001) Inheritance of seed color and lustre in mungbean (Vigna radiata). Agric Sci Technol Hunan 2:8–12
Dana S, Karmakar PG (1990) Species relation in Vigna subgenus Ceratotropis and its implications in breeding. Plant Breed Rev 8:19–42
Dar GM, Verma MM, Gosal SS, Brar JS (1991) Characterization of some inter-specific hybrids and amphidiploids in Vigna. In: Sharma B, Mehra BB (eds) Golden Jubilee celebrations symposium in grain legumes. Indian Society of Genetics and Plant Breeding, New Delhi, pp 73–78
Das A, Poornima KN, Thakur S, Singh NP (2016) The mungbean genome sequence: a blueprint for *Vigna* improvement. Curr Sci 111(7):1144–1145
Dasgupta T, Banik A, Das S (1998) Combining ability in mungbean. Indian J Pulses Res 11:28–32
Durga KK, Kumar SS (1997) Screening for preharvest sprouting in pulses. Legum Res 20:193–197
Dwivedi S, Singh DP (1985) Inheritance of narrow trifoliate leaf in mungbean (*Vigna radiata* (L.) Wilczek). SABRAO J 17:177–180
Dwiwedi S, Singh DP (1986) Inheritance of purple pigmentation and seed colour in green gram. Crop Improv 13:63–66
Dwiwedi S, Singh DP (1990) Inheritance of fascination in mungbean (Vigna radiata L. Welczek). Indian J Genet 50:80–82
Fatokun CA, Menancio-Hautea D, Danesh D, Young ND (1992) Evidence for orthologous seed weight genes in cowpea and mungbean based on RFLP mapping. Genetics 132:841–846
Harlan JR, de Wet JMJ (1971) Towards a rational classification of cultivated plants. Taxon 20:509–517
Humphry ME, Lambrides CJ, Chapman SC, Aitken EAB, Imrie BC, Lawn RJ, McIntyre CL, Liu CJ (2005) Relationships between hard-seededness and seed weight in mungbean (Vigna radiata) assessed by QTL analysis. Plant Breed 124:292–298
Jain HK, Mehra KL (1980) Evolution, adaptation, relationships and uses of the species of vigna cultivated in Asia. In: Summerfield RJ, Bunting AH, Sherrey R (eds) Advances in legume science, Royal Botanic Gardens, Kew, vol 1, pp 459–468
Jaiwal PK, Kumari R, Ignacimuthu S, Potrykus I, Sautter C (2001) *Agrobacterium tumefaciens*-mediated transformation of mungbean (*Vigna radiata* L. Wilczek)-a recalcitrant grain legume. Plant Sci 161:239–247
Kajale MP (1977) Ancient grains from excavations at Nevasa. Maharashtra Geophytol 7:98–106
Kang YJ, Kim SK, Kim MY, Lestari P, Kim KH, Ha BK, Jun TH, Hwang WJ, Lee T, Lee J, Shim S, Yoon MY, Jang YE, Han KS, Taeprayoon P, Yoon N, Somta P, Tanya P, Kim KS, Gwag JG, Moon JK, Lee YH, Park BS, Bombarely A, Doyle JJ, Jackson SA, Schafleitner R, Srinives P, Varshney RK, Lee SH (2014) Genome sequence of mungbean and insights into evolution within Vigna species. Nat Commun 11(5):5443
Kaur L, Tripathi HS, Vishwa D, Reddy MV, Singh G and Kharkwal MC (2008) Management of diseases in food legumes. Ed: Kharkwal MC. Proceedings of the 4th international food legumes research conference, Vol 1, 18–22 October, 2005, New Delhi, pp 608–637
Khadilkar BT (1963) Inheritance of two characters in greengram. Akola Agril College Magazine 3:29–32
Khattak GSS, Haq MA, Rana SA, Srinives P, Ashraf M (1999) Inheritance of resistance to Mungbean yellow mosaic virus (MYMV) in mungbean (Vigna radiata (L.) Wilczek). Thai J Agric Sci 32:49–54
Khattak GSS, Maq MA, Ashraf M, Tahir GR (2001) Genetics basis of synchrony in pod in Mungbean [Vigna radiata (L.) Welczek]. Kasetsart J Natl Sci 35:1–7

Khattak GSS, Maq MA, Ashraf M, Tahir GR (2002) Triple test cross for some morphological traits in Mungbean [Vigna radiata (L.) Welczek]. Euthopia 126:413–420

Kim SK, Nair RM, Lee J, Lee SH (2015) Genomic resources in mungbean for future breeding programs. Front Plant Sci 6:626

Kumar S, Gupta S, Chandra S, Singh BB (2004) How wide is the genetic base of pulse crops? In: Ali M, Singh BB, Kumar S, Dhar V (eds) Pulses in new perspective. Indian Society of Pulses Research and Development, Kanpur, pp 211–221

Kumar S, Imtiaz M, Gupta S, Pratap A (2011) Distant hybridization and alien gene introgression. In: Pratap A, Kumar J (eds) Biology and breeding of food legumes. CABI, Oxfordshire, pp 81–110

Kumar S, Kalita A, Srivastava R, Sahoo L (2017) Co-expression of *Arabidopsis* NHX1 and bar improves the tolerance to salinity, oxidative stress, and herbicide in transgenic mungbean. Front Plant Sci 8:1896

Lambrides CJ (1996) Breeding for improved seed quality traits in mungbean {Vigna radiata (L,) Wilczek} using DNA markers. PhD thesis. University of Queensland, Australia

Lambrides CJ, Godwin ID, Lawn RJ, Imrie BC (2004) Segregation distortion for seed testa colour in mungbean (Vigna radiata L. Wilcek). J Hered 95:532–535

Lawn RJ, Cottrell A (1988) Wild mungbean and its relatives in Australia. Biologist (Columbus, Ohio) 35:267–273

Liu C, Wang S, Wang L, Sun L, Mei L, Xu N et al (2008) Establishment of candidate core collection in Chinese mungbean germplasm resources. Acta Agron Sin 34:700

Mahalaksmi LS, Leela B, Kumar K, Naresh B, Devi P (2006) *In vitro* regeneration from the petioles of primary leaves of mungbean (*Vigna radiata*). Plant Biotechnol 23(4):409–411

Mallikarjuna RC, Koteswara RY, Reddy MV (2006) Evaluation of mungbean germplasm for yield and yield components. Legum Res 29:73–79

Mishra RC, Sahu RC, Tripathy D (1970) A note on heterosis in greengram (Phaseolus aureus Roxb.). Curr Sci 39:190–191

Moe KT, Chung JW, Cho YI, Moon JK, Ku JH, Jung JK (2011) Sequence information on simple sequence repeats and single nucleotide polymorphisms through transcriptome analysis of Mungbean. J Integr Plant Biol 53:63–73

Mukherjee A, Pradhan K (2002) Genetics of lobation in trifoliate leaf of mungbean. Abstracted in: Perspect Cytol Genet 11:88

Murty BK, Patel GJ (1973) Inheritance of some morphological characters in mungbean. Bansilal Amritalal Coll Agric Mag 25:1–9

Pathak GN, Singh B (1963) Inheritance studies in greengram. Indian J Genet Pl Breed 23:215–218

Sahoo DP, Kumar S, Mishra S, Kobayashi Y, Panda SK, Sahoo L (2016) Enhanced salinity tolerance in transgenic mungbean overexpressing Arabidopsis antiporter (NHX1) gene. Mol Breed 36:144

Sangiri C, Kaga A, Tomooka N, Vaughan D, Srinives P (2007) Genetic diversity of the mungbean (*Vigna radiata*, Leguminosae) genepool on the basis of microsatellite analysis. Aust J Bot 55:837–847

Sareen PK (1985) Genetic analysis of the trilobate leaf mutants in mungbean (*Vigna radiata*, var. aureus (L.) Wilczek). Curr Sci 54:930–931

Schafleitner R, Nair RM, Rathore A, Wang Y, Lin C, Chu S et al (2015) The AVRDC–the world vegetable center mungbean (*Vigna radiata*) core and mini core collections. BMC Genomics 16:344

Sen NK, Ghosh AK (1959) Genetic studies in green gram. Indian J Genet 19:210–227

Sen NK, Murty ASN (1960) Inheritance of seed weight in green gram (Phaseolus aurcus Roxb). Genetics 45:1559–1562

Singh DP (1980) Inheritance of resistance to yellow mosaic virus in blackgram (*Vigna mungo* (L.) Hepper). Theor Appl Genet 57:233–235

Singh DP (1983) A high yielding mungbean variety through mutation breeding. Mutat Breed Newsl 22:1–2

Singh DP (2014) Genetics and breeding of pulse crops, 3rd edn. Kalyani Publishers, Ludhiana, p 473

Singh D, Mehta TR (1953) Inheritance of lobed leaf margin in mung (Phaseolus aureus L.). Curr Sci 22:348

Singh TP, Singh KB (1970) Inheritance of clusters per node in mungbean (Phaseolus aureus Roxb.). Curr Sci 39:265

Singh TP, Singh KB (1972) Combining ability in mungbean. Indian J Genet 32:67–72

Singh TP, Singh KB (1974) Components of genetic variance and dominance pattern for some quantitative traits in mungbean (*Phaseolus aureus* Roxb.). Z Pflanzenzucht 71:233–242

Singh BB, Singh DP (1995) Inheritance of a small leaf mutant in mungbean. Indian J Genet 55:69–70

Singh DP, Singh BB (2016) Varietal improvement of pulse crops in India: introspection and prospectives. Indian J Agric Sci 86:3–24

Singh BB, Singh SR, Adjadi O (1985) Bruchid resistance in cowpea. Crop Sci 25:736–739

Smartt J (1981) Gene pools in Phaseolus and Vigna cultigens. Euphytica 30:445–459

Smartt J (1985) Evolution of grain legumes. III. Pulses in the genus Vigna. Exp Agric 21:87–100

Sonia SR, Saini R, Singh RP, Jaiwal PK (2007) *Agrobacterium tumefaciens* mediated transfer of *Phaseolus vulgaris* á-amylase inhibitor-1 gene into mung bean (*Vigna radiata* (L.) Wilczek) using bar as selectable marker. Plant Cell Rep 26:187–198

Talukdar D, Talukdar T (2003) Inheritance of growth habit and leaf-shape in mungbean [*Vigna radiata* (L.) Wilczek.]. J Indian J Genet 63(2):165–166

Tangphatsornruang S, Sangsrakru D, Chanprasert J, Uthaipaisanwong P, Yoocha T, Jomchai N, Tragoonrung S (2010) The chloroplast genome sequence of mungbean (*Vigna radiata*) determined by high-throughput pyrosequencing: structural organization and phylogenetic relationships. DNA Res 17(1):11–22

Thakare RG, Gadgil JD, Mitra R (1988) Origin and evolution of seed protein genes in Vigna mungoand Vigna radiata. In: Mungbean: proceedings of the second international symposium. Asian vegetable research and development centre, Taipei, pp 47–52

Tullu A, Taran B, Breitkreutz C, Buchwalt L, Banniza S, Warkentin TD (2006) A quantitative-trait locus for resistance to *Ascochyta blight* (Ascochyta lentis) maps close to a gene resistance to anthracnose (*Colletotrichum truncatum*) in lentil. Can J Plant Pathol 28:588–595

Verma SNP, Krishi JN (1969) Inheritance of some qualitative characters in greengram (Phaseolus aureus Roxb.). Indian J Hered 1:105–106

Vijayan SB, Kirti P (2011) Mungbean plants expressing BjNPR1 exhibit enhanced resistance against the seedling rot pathogen, *Rhizoctonia solani*. Transgenic Res 21:193–200

War AR, Murugesen S, Beddepalli VN, Srinivasan R, Nair RM (2017) Mechanism of resistance in mungbean (*Vigna radiata* (L.) R. Wilczek var radiata) to bruchids, *Callosobruchus* spp (Coleoptera: Bruchidae). Front Plant Sci 8:1031

Yadav SK, Katikala S, Yellisetty V, Kannepalle A, Narayana JL, Maddi V, Mandapaka M, Shanker AK, Bandi V, Bharadwaja KP (2012) Optimization of *Agrobacterium* mediated genetic transformation of cotyledonary node explants of *Vigna radiata*. Springerplus 1:59

Yohe JM, Poehlman JM (1975) Regression, correlations, and combining ability in mungbean (Vigna radiata (L.) Wilczek). Trop Agric (Trinidad) 52:343–352

Chapter 9
Phosphate Homeostasis: Links with Seed Quality and Stress Tolerance in Chickpea

Poonam Mehra, Ajit Pal Singh, Jyoti Bhadouria, Lokesh Verma, Poonam Panchal, and Jitender Giri

9.1 Introduction

Chickpea (*Cicer arietinum* L., $2n = 16$) is the second most important grain legume crop of the world (Garg et al. 2016). India holds first place in terms of area (66%) and production (70%) of chickpea worldwide (Galav et al. 2018). It is a primary crop in many developing countries that contribute significantly to human food as well as animal feed (Pang et al. 2018). Chickpea is a cheap and rich source of dietary proteins, vitamins and minerals (phosphorus, calcium, magnesium, iron, zinc) (Tan et al. 2017). On an average, chickpea seeds contain 23% protein, 64% carbohydrates, 5% fat, 6%fibre and 3% minerals (Rasool et al. 2015). Furthermore, chickpea contains several compounds of nutritional and medicinal importance (Jukanti et al. 2012). On the basis of origin and seed type, cultivated chickpea is grouped as *Desi* (Indian origin; white- or brown-coloured seed) and *Kabuli* (Mediterranean and Middle Eastern origin; reddish brown-coloured seed) (Bampidisa and Christodoulou 2011; Tripathi et al. 2013). *Desi* types (microsperma) are small seeded with thinner seed coat, while *Kabuli* types (macrosperma) are large seeded with thick seed coat (Tripathi et al. 2013).

Besides humans' food, chickpea crop also has significant role in improving soil health and fertility. The deep tap root system of chickpea improves soil aeration, and heavy leaf drop adds to soil organic manure (Singh et al. 2018). It can fix 25–30 kg nitrogen/ha annually through symbiotic nitrogen fixation (Reddy and Reddy 2005) and thus, has reduced demand for nitrogenous fertilizers. Further, chickpea is majorly grown on marginal soils with poor soil fertility. Although chickpea can fix N, soil phosphorus (P) deficiency in crop growing areas is a major limiting factor for chickpea production (Srinivasarao et al. 2006). Remarkably, P deficiencies in chickpea cause an annual yield loss of 653,000 tonnes globally (Rasool et al. 2015).

P. Mehra · A. P. Singh · J. Bhadouria · L. Verma · P. Panchal · J. Giri (✉)
National Institute of Plant Genome Research, New Delhi, India
e-mail: jitender@nipgr.ac.in

S. H. Wani, M. Jain (eds.), *Pulse Improvement*,
https://doi.org/10.1007/978-3-030-01743-9_9

Adequate P nutrition is required for proper root growth, seed development, optimum yield and also symbiotic nitrogen fixation (Tripathi et al. 2013). Chickpea seeds contain high amount of phytate which is stored in the form of P and considered as anti-nutritive. Therefore, reducing seed phytate is target of many studies. Apart from this, some studies indicate broader roles of P nutrition on other abiotic/biotic stresses such as drought, cold, heat, salinity, pathogen, insect pests etc. This chapter briefly describes the different mechanisms of maintaining P homeostasis in chickpea and further explores the impact of P nutrition on chickpea response to other stresses.

9.2 Phosphate Acquisition and Homeostasis in Chickpea

Bioavailable form of P, inorganic phosphate (Pi) is an essential macronutrient for proper growth and development of plants. Inside cell, it acts as building block of nucleic acids that contain genetic information. It is a component of ATP which is used as energy currency in the cell. Cellular membranes which make different compartment within the cell as well as separate cell interior from outside environment also require P in the form of phospholipids as structural component. Phosphate is also involved in many enzymatic reactions of different anabolic and catabolic pathways (Theodorou and Plaxton 1993). After N, P is the most limiting macronutrients that affect plants productivity. In soil, most of the phosphate is present in the form of insoluble inorganic or organic complexes that are not readily absorbed by plant roots (Mengel and Kirkby 1978). Therefore, most of the world's agricultural lands are either inherently low in P or now facing scarcity of Pi. To get optimum crop yields, these lands require high amount of phosphate fertilizers. Ironically, P-reserves (P-rocks) are limited in world which makes situation more arduous. To cope with P deficiency, plants have evolved series of molecular, biochemical, physiological and morphological adaptations. We summarize here these set of adaptations for legume crops with emphasis on chickpea.

9.2.1 P Deficiency and Adaptations in Chickpea

Among different crops, legumes are being increasingly used in rotation with other crops as a part of conservation agriculture. Legumes are ideal crops for conservation agriculture as they provide soil cover and also increase soil N content as these crops are able to fulfil their N need by symbiotic nitrogen fixation (Siddique et al. 2012). In addition to N fixation, legumes increase N rich organic matter of soil as these crops have high N to C ratio (Siddique et al. 2012). Besides, legume crops are also effective in breaking disease cycle of particular cereal crop because of different genetic make-up from cereals. However, P deficiency is a major constrain that limits legume cultivation and affects their productivity (Lemma et al. 2013).

Legumes are heavy feeder of P, therefore, require optimum P fertilization (Singh et al. 2018). In legumes, P affects N fixation as it is a high energy demanding process that needs ATP which is P rich compound (Lemma et al. 2013). Cultivation of legume crops in P-deficient soils leads to poor nodulation and poor plant vigour (Amijee and Giller 1998). However, P deficiency is difficult to diagnose in many legume crops which show only stunted growth at early stages of seedling growth (Martins et al. 2017). In some species, abnormal leaf coloration has been reported under P deficiency as a result of anthocyanin accumulation. In mung bean, mash bean and soybean, biomass, N fixation and photosynthetic rate have been found to be reduced under P deficiency. In lentil P deficiency leads to reduced accumulation of reducing sugar in leaves but higher accumulation in roots (Sarker and Karmoker 2011).

In chickpea, P deficiency is more pronounced in acidic and alkaline soils with pH less than 6.0 and more than 7.5, respectively. P-deficient chickpea plants show stunting, restricted root development and dark green leaves (Kumar and Sharma 2013). Stem and top edges of older leaves show reddish purple pigmentation. Discoloration of older leaves starts from upper surface of bottom leaflets and proceeds towards upper leaflets (Kumar and Sharma 2013). Later on, with prolonged deficiency, older leaves turn yellow followed by bronzing. P deficiency in chickpea also induces release of protons and carboxylic acids from roots into the rhizosphere (Neumann and Römheld 1999). Moreover, crop maturity gets delayed and yield is severely affected.

9.2.2 Morphological Adaptations: Root System Architecture (RSA) Plays Big Roles

Phosphate is maximally present in the upper soil layers, and its concentration goes down with depth in soil horizon (Tanner et al. 2016; Hansel et al. 2017). Being immobile, plant modulates their plastic root system architecture (RSA) in order to optimize nutrient uptake like P from the soil. RSA denotes the spatial configuration of roots of all subtypes including root hairs (Lynch 1995). Nutrient availability can modulate almost every aspect of RSA, i.e. root length, diameter, number, angle and density (Lynch 1995; Gruber et al. 2013). Low Pi availability is known to inhibit primary root length by disturbing meristem in dicots like *Arabidopsis* (Sánchez-Calderón et al. 2005). It was proposed that root tip perceives the low Pi signals and activates LPR (low phosphate root) protein in the root tip which in turn modulates hormonal fluxes resulting in the determinate root apical meristem (Svistoonoff et al. 2007). Primary root length inhibition is also well-documented in legumes like white lupin (Péret et al. 2014) and chickpea (Singh et al. 2015). Pi deficiency inhibits primary root length in most of the plants and simultaneously promotes lateral root elongation in *Arabidopsis*, rice, chickpea and white lupin (Péret et al. 2014; Gruber et al. 2013; Alloush 2003). Along with the lateral root length, lateral density was also found to be increased (Gruber et al. 2013; Alloush 2003). This modulated RSA is more efficient for top soil foraging and improves plants' ability to survive under low Pi conditions (Lynch and Brown 2011; Rubio et al. 2003; Zhu et al. 2005). As observed in common bean, the secondary root growth was more in P-efficient genotypes than P-inefficient genotypes, while the shoot growth remains stalled (Nielsen et al. 2001). Experiments conducted by He et al. (2003) signify that Pi deficiency signals are produced in the Pi-deficient zone of rhizosphere, and these signals induce secondary root growth in the region where Pi is actually available. Similar results were obtained in barley; however, the response was not unique to Pi as nitrate; ammonium and potassium deficiency also produced similar secondary growth in the region of nutrient availability (Drew 1975). This implies that plant induces secondary root growth in the region of high nutrient availability to optimize nutrient uptake, and this response is more pronounced when plant sense nutrient deficiency. These newly produced root branching also contain non-respiring xylem and sclerenchyma tissue, therefore minimizing the energy cost while maintaining maximum absorption of nutrients (Nielsen et al. 2001).

A special type of root modification is observed in legumes like white lupin in the form of cluster roots in response to Pi deficiency (Neumann et al. 2000). Cluster root formation is regulated by the internal Pi concentration of the plant (Dinkelaker et al. 1995). The rhizosphere around cluster roots is full of exudative burst and therefore facilitates P solubilization and uptake (Shane and Lambers 2005). However, there is no report of cluster root formation in many Leguminosae members like chickpea.

P deficiency is also known to induce root hair formation and elongation in number of plants (López-Bucio et al. 2003; Bates and Lynch 1996; Giri et al. 2018), while this elongation of root hairs seems to be Pi deficiency specific as neither of boron, manganese and sulphur deficiency produced any change in the root hair length (Bates and Lynch 1996). Root hairs are the epidermal cell elongations which enhance the root surface area many folds. Plant can produce root hairs at the minimum metabolic cost; therefore, this morphological phene seems to be most economic strategy used by the plant in order to improve phosphate acquisition (Lynch 2011). It has also been well established that root hairs participate in phosphate acquisition (Gahoonia and Nielsen 1998), and plants acquire more phosphate when plants have longer root hairs (Bates and Lynch 2001). Information on chickpea root architecture under Pi deficiency is limited. Experiments conducted by Gahoonia et al. (2007) concluded that chickpea genotypes with longer and denser root hairs can absorb nutrients more effectively (including Pi) than the genotypes with smaller root hairs. On contrary, recent study found no clear correlation between P uptake and root hair length of the 100 chickpea genotypes (Pang et al. 2018). The possible explanation of their observation may be because of small root hair population whereas root hairs can only help in P acquisition if they are of some significant length (Clarkson 1991). It is also noteworthy that root hairs are the active surface area of the root and can participate in the acidification of the soil. Chickpea genotypes with longer and denser root hairs like B-chhola-5 were able to better acidify the soil (Gahoonia et al. 2007). This acidification of the soil could be the result of root exudates including secondary metabolites like oxalic acid, citric acids which have roles in the release of inorganic phosphate from the bound forms. Roots also secrete acid phosphatases (APases) which actively break down phosphate from various organic phosphate complexes and make it available for the roots (Bhadouria et al. 2017).

Leguminous plants including chickpea contain special root structures called 'root nodules'. This symbiotic association of rhizobia and roots of leguminous plants play great role in plant nutrition and fertility of the soil. Pi deficiency adversely affects nodule development and therefore, affects optimal nitrogen fixation as well. The threshold phosphate demand is also high for nodules than shoot as nodule phosphate content is quite high than other tissues of the plant (Jakobson 1985; Israel 1987; Hoch-Jensen et al. 2002; Schulze et al. 2006). Reports have suggested that milder P deficiency results in smaller root nodules, while the number of root nodules increased (Ribet and Drevon 1995). It is not still clear if the P partitioning is regulated by the bacterial P demand or the plant divert P to the nodule as some adaptive measures.

Legume genotypes have vast genotypic variability in terms of root system architecture; therefore in order to develop genotypes for better phosphate acquisition efficiency (PAE), there is need to target root phenes which can improve phosphate uptake like cluster roots, lateral roots, root angle and root hair. Crop breeding strategy targeting these useful root phene provides an opportunity to meet the global food demand. For such kind of initiative, screening of germplasms for such beneficial root phenes is the most initial step. Few of the phenes like root hair length and density have already been targeted for the improvement of bean varieties. The newly developed bean lines with improved roots (IIAM 2010) have high PAE and yield under the Pi deficient conditions.

9.2.3 Biochemical and Physiological Adaptations: For Increasing Pi Availability

Plants respond to Pi deficiency by exhibiting series of biochemical adaptations. These include membrane lipid remodelling where P containing lipids are replaced by non-P lipids (Nakamura 2013). Besides, several organophosphate hydrolysing enzymes such as phosphatases and phytases are biosynthesized and/or secreted in to rhizosphere. In many plant species pigments such as anthocyanins are also synthesized which also visually mark Pi deficiency. In some plants alternate Pi- and adenylate-dependent steps are bypassed through alternative reactions in key metabolic processes. Such bypasses conserve cellular Pi for essential processes (Plaxton and Tran 2011; Mehra et al. 2016). However, most of these adaptations have been studied in details in *Arabidopsis*, cereals and few of legumes such as common bean and white lupin. In chickpea, there are very few studies which uncover Pi-induced biochemical adaptations.

9.2.3.1 Membrane Lipid Remodelling Under P Deficiency

In addition to acquire more P from soil, plants initiate different mechanisms to use the acquired P more efficiently under its deficiency (Denton et al. 2007). One of these mechanisms is membrane lipid remodelling by which Pi-starved plants utilize phospholipids present in their membranes as source for internal Pi supply (Nakamura 2013). Phospholipids are the most abundant class of lipids in cell membranes. Cellular membranes harbour up to 30% of organically bound P in the form of phospholipids. Under P deficiency, plants activate phospholipid degradation pathway to utilize Pi present in these membranes (Pant et al. 2015). Degradation of phospholipids is coupled with the synthesis of non-P containing lipids like MGDG, DGDG and SQDG. These non-P-containing lipids replace phospholipids in order to maintain membrane integrity.

In plants, three pathways of phospholipid degradation have been proposed (Pant et al. 2015). One pathway is mediated by Phospholipase C (PLC) in which phospholipids are degraded by non-specific PLC (NPC4 and NPC5) into DAG and Pi containing head group that could be utilized as valuable source by Pi-starved plants (Nakamura et al. 2005). Second pathway is initiated by Phospholipase D (PLD) in which phospholipids are first converted into Phosphatidic acid (PA), which is then converted into diacylglycerol (DAG) and Pi by Phosphatidic acid phosphatase (PAP) (Gaude et al. 2008). Third pathway is mediated by lipid acyl hydrolase (LAH) and glycerophosphodiester phosphodiesterase (GDPD) in which phospholipids are catalysed into glycerophosphodiesters (GPD) and fatty acids via deacylation at their *sn-1* and *sn-2* positions by LAH, and then GPD is hydrolysed into *sn*-glycerol-3-phosphate (G-3-P) and corresponding alcohol (e.g. choline if the phospholipid is PC, ethanolamine if the phospholipid is phosphatidylethanolamine, etc.) by GDPD. In legumes such as white lupin, glycerophosphodiester phosphodiesterase gene *GPX-PDE1* and *GPX-PDE2* have been reported to be induced under P deficiency and play important role in root hair development (Cheng et al. 2011). In chickpea, 6 GDPD encoding genes have been identified. *CaGDPD1* was found to be highly induced by Pi deficiency and also encodes an active glycerophosphodiester phosphodiesterase (Mehra and Giri 2016). However, it remains to see if *CaGDPD1* functions similar to its white lupin counterpart.

In soybean, Gaude et al. (2004) reported accumulation of MGDG and DGDG in roots and leaves under P deficiency. In legumes, DGDG synthase encoding genes are also induced during late stages of nodule development that leads to accumulation of DGDG in peribacteroid membrane (PBM) which is important to fulfil increased demand of phosphate for N fixation (Gaude et al. 2004). In Chinese milk vetch, knockdown of *As246* gene (nodule specific lipid transfer protein) resulted in reduction of DGDG content in the symbiosomes that affects nodule development. This may be due to the scarcity of Pi as nodulation requires large amount of P, and the absence of DGDG restricts utilization of membrane P (Lei et al. 2014). In soybean, root hairs which are important for nutrients uptake, also have been found to be undergoing lipidomic changes in which galactolipids are increased and phospholipids are decreased under Pi deficiency, and such changes are more responsive than the overall root growth (Wei et al. 2016). These results indicate crucial roles of lipid remodelling in maintaining Pi homeostasis for nodule and root hair development in legumes. However, such studies are so far lacking in important legume crop, chickpea and needs in-depth investigations.

9.2.3.2 Organic Acids (OAs) Secretion for P Solubilization

Orthophosphate (Pi) form of phosphorus is considered as ‘plant available’ form. Highly reactive nature of orthophosphate renders it immobility in soil (Herrera-Estrella and López-Arredondo 2016). About 70% of world’s total arable land is Pi deficient due to high pH fluctuations. Agricultural land with acidic pH consists of cationic species like Al^{3+} and Fe^{3+} that readily react with orthophosphate leading to formation of inorganic phosphate compounds. Same way, presence of Ca^{2+} and Mg^{2+} in alkaline soil is the prime reason for complex formation of orthophosphate (Fink et al. 2016). Low molecular weight organic acids (LMWOAs) like citrate, malate can dissociate these phosphate complexes owing to their high chelating properties (Raghothama 1999). Plants, then easily take up the free orthophosphate from soil at the root soil interface. Organic acids (OAs) are also involved in enhancing various P-solublizing micro-organisms. OA exudation in response to Pi deficiency in chickpea was first reported by Neumann and Römheld (1999). These include citric acid, malic acid and malonic acid. The citrate exudation was in proportion to amount of citrate accumulated in root tissue. Accumulation in root tissue in turn depends upon increased activity of OAs biosynthesis enzymes like phosphoenolpyruvate carboxylase (PEPC) and decreased activity of catabolic enzyme like aconitase (Anoop et al. 2003). However, in another study, the exudation correlated with content of OAs in leaves rather than root (Ohwaki and Hirata 1992). OAs exudation in chickpea helps in Pi dissociation from phosphates of calcium that plant can take up for its growth and development in calcareous soil (Hocking et al. 2000). Experiments by Veneklaas et al. (2003) also showed enhanced malonate, malate and citrate concentration in soil in response to low Pi irrespective to the varieties of chickpea used. However, few other studies indicated lack of correlation between Pi deficiency and organic acid exudation from plant roots in chickpea (Chen et al. 2013). The probable reason for contrasting results in different experiments may vary from the use of different cultivars to microbial degradation or varying properties of soil (Wouterlood et al. 2004). Legumes like white lupin naturally secrete OAs under low Pi stress. They undergo formation of specialized proteoid roots (cluster roots) under Pi deficiency for secretion of carboxylates. However, chickpea secretes OAs from the whole root (Wouterlood et al. 2005). Few symbiotic associations also help in plant’s adaptation to low Pi stress by changing metabolic flux including enhanced OAs secretion owing to their high Pi requirement during the process of nitrogen fixation (Esfahani et al. 2016). Overall genetic manipulations or breeding techniques that target organic acid metabolism may help improve low Pi-tolerant varieties in chickpea.

9.2.3.3 Acid Phosphatases Play Dual Roles in P Solubilization and Signalling

Besides inorganic complexes, P is also present in the form of organic complexes which include inositol phosphates, sugar phosphates, nucleotides, phosphoprotein, phosphonates and phospholipids in both cell and decaying biomass in soil. These organic complexes are broken down with the help of enzymes: phosphatases and phytases (Bhadouria et al. 2017; Mehra et al. 2017; Pandey et al. 2017). Phosphatases are non-specific in nature and are divided into two categories: acid or alkaline phosphatases (Vance et al. 2003). Acid phosphatases are widely distributed in all organisms and are principally involved in mineralization of organic P. Purple acid phosphatases (PAPs) belong to class of acid phosphatases. PAPs are non-specific enzymes exhibiting high activity against different phosphomonoesters which include ATP, phosphoenolpyruvate and phosphoproteins (Bozzo et al. 2004). PAPs have been identified from a wide range of crops/plants, i.e. *Arabidopsis*, tomato, wheat, barley, maize and rice (Bozzo et al. 2002; Li et al. 2002; Dionisio et al. 2011; Liang et al. 2010; Mehra et al. 2017). The five conserved domains DXG, GDXXY, GNH (D/E), VXXH and GHXH are present in all the PAPs across different kingdoms (Cashikar et al. 1997). Most of the PAPs are low Pi inducible and involved in Pi acquisition and utilization. According to the various studies, the overexpression of PAPs provides low Pi tolerance. However, some studies have indicated much more diverse role of PAPs such as carbon metabolism (*AtPAP2*), nodule formation, nitrogen fixation (*AsPPD1*), salt stress tolerance (*GmPAP3*), cell wall synthesis (*NtPAP12*) and extracellular dNTPs utilization (*SgPAPs*) (Liao et al. 2003; Kaida et al. 2010; Sun et al. 2012; Wang et al. 2015; Liu et al. 2016). *PvPAP1* and *PvPAP3* are reported to enhance Pi mobilization from extracellular dNTPs (Liang et al. 2012). *PgPAP18* from Pennisetum glaucum is involved in providing tolerance against environment stress (Reddy et al. 2017). PAPs are also reported to be involved in defence response as AtPAP5 provide resistance against *Pseudomonas syringae* (Ravichandran et al. 2015). Some PAPs (*NtPAP*, *AtPAP15*, *AtPAP23*, *GmPhy*, *HvPaPhyOsPHY1* and *CaPAP7*) have also shown phytase activity (Hegeman and Grabau 2001; Kaida et al. 2003; Zhu et al. 2005; Kuang et al. 2009; Dionisio et al. 2011; Li et al. 2012a; Bhadouria et al. 2017). The overexpression of phytase, a purple acid phosphatase with phytase activity (*HvPAPhy_a*) led to increased phytase activity in grains, green leaves and dead mature straw (Holme et al. 2017). Phytase from wheat, barley and rice belonging to PAPs family has also been identified (Rasmussen et al. 2003). PAPs have also been well studied in legumes such as *Lupinus albus* (white lupin), *Phaseolus vulgaris*, *Glycine max*, *Medicago truncatula* and *Vigna radiata* (Gilbert et al. 1999; Li et al. 2012b; Liang et al. 2010; Xiao et al. 2006; Wongkaew et al. 2013). In legumes, proteoid roots are well known adaptation to overcome Pi deficiency. These proteoid roots also secrete acid phosphatase along with organic acids to adapt in Pi-deficient condition (Gilbert et al. 1999).

In chickpea, 25 PAPs have been recently identified, and most of them are low Pi inducible (Bhadouria et al. 2017). Association analysis of SNPs underlying these PAPs revealed significant association of *CaPAP7* with chickpea seed weight and phytate content. CaPAP7 can hydrolyse different organic P compounds abundant in cell cytoplasm. Interestingly, CaPAP7 was also found to be a functional PAP with phytase activity. Phytases catalyse breakdown of phytate into inorganic phosphates. Phytases are enzymes of agricultural importance as 80% of phosphate is bound in form of phytate. Phytases are also reported to increase phosphate uptake from organic sources (Xiao et al. 2005). Therefore, engineering of chickpea phytases can play dual role in combatting P deficiency as well as improving seed quality by reducing phytate content.

9.3 Effect of Phosphate Nutrition on Abiotic and Biotic Stress Response in Chickpea

Chickpea production in India and other parts of world experiences huge losses due to several abiotic and biotic stresses (reviewed in Jha et al. 2014). Among the abiotic stresses, drought, cold, terminal heat and edaphic factors (such as soil salinity, acidity, nutrient deficiencies and metal toxicities) are the major challenges that limit chickpea production worldwide. Among biotic stresses, *Ascochyta* blight, *Fusarium* wilt and infestation by pod borer are the chief constraints for chickpea production. Abiotic and biotic stresses in chickpea cause an annual yield loss of approximately 6.4 and 4.8 million tonnes, respectively (Ryan 1997). Emerging studies suggest extensive cross talks between Pi starvation and other adverse environmental conditions. Several genes that are differentially expressed under Pi deficiency also respond to other abiotic and biotic stresses (Hammond et al. 2003). This indicates existence of redundant signalling cascades regulating Pi deficiency and other stresses. Here, we summarize the leads, evidences and knowledge gap involving roles and mechanisms of P nutrition in multiple stress tolerance in chickpea.

9.3.1 Drought Stress and Phosphorus

Chickpea cultivation is largely rain fed; therefore, drought stress is the primary constraint that limits chickpea yield up to 40–50% (Ahmad et al. 2005). Most of the chickpea genotypes encounter terminal drought stress at pod setting stages owing to significant rise in temperatures towards crop maturity (Ramamoorthy et al. 2016). In an interesting study conducted by Kahraman et al. (2015), nutritional variations in chickpea seeds were determined under drought stress at vegetative, flowering and pod setting stages. Notably, significant reduction in P content was observed when no irrigation was applied at flowering and pod setting stages. Similarly, other study revealed that post-anthesis drought causes larger depletion in P content in drought susceptible chickpea cultivars as compared to drought tolerant genotypes (Gunes et al. 2006). These studies emphasize that drought stress, especially terminal drought (at post-anthesis stages) has significant influence on P accumulation in chickpea.

Drought stress has negative effect on plant nutrients especially N and P concentrations (He and Dijkstra 2014). Drought decreases uptake and transport of nutrients by reducing 'transpiration driven mass flow of nutrients' and decreasing rate of nutrient diffusion in soils (Hu and Schmidhalter 2005). Crops like *Zea mays* are known to accumulate up to 5% of the total P by transpiration driven mass flow and rest 95% by diffusion or root uptake (Barber 1995). In a recent study with 266 chickpea genotypes, positive correlation was observed between leaf transpiration driven mass flow and P acquisition (Pang et al. 2018). This indicates important role of water availability in P acquisition in chickpea. Selection of drought tolerant chickpea genotypes may also improve low P tolerance as in the case of other legumes such as common bean (Beebe et al. 2008). Conversely, P fertilization was found to mitigate drought stress in many plant species (Rodriguez et al. 1996; Jones et al. 2005; Kuwahara et al. 2016), especially legumes such as white clover (Singh et al. 2000) and soybean (Jin et al. 2005, 2006). Increased hydraulic conductivity of roots, more root proliferation and increased synthesis of osmoprotectants for maintaining leaf water potential have been suggested as possible reasons (Al-Karaki et al. 1996; Jin et al. 2015; Tariq et al. 2018). Such studies could be useful in enhancing combined drought and low P tolerance of chickpea genotypes.

9.3.2 Salinity and Phosphorus

Soil salinity adversely affects chickpea growth and development resulting in an annual yield loss of 8–10% (Flowers et al. 2010). Salinity affects seed germination, shoot biomass, pod setting and pod filling in chickpea. Reproductive stages of chickpea are more sensitive to salt stress (Vadez et al. 2012). Salt stress reduces the P use efficiency of chickpea, whereas P fertilization significantly increased plant biomass and P use efficiency under saline conditions (Sadji-Ait Kaci et al. 2018). Salt stress also affects chickpea rhizobia interaction leading to poor rhizobial colonization, decreased nodule formation, leghaemoglobin content and premature senescence of nodules (Sadji-Ait Kaci et al. 2016). However, application of low doses of P fertilizer significantly improved nodulation and increased leghaemoglobin content (up to 92%) under saline conditions in chickpea (Sadji-Ait Kaci et al. 2016). Interestingly, P fertilization under salt stress was found to increase chickpea yield by 30% (Sadji-Ait Kaci et al. 2016). These results indicate multifaceted role of phosphorus in alleviating salt stress in chickpea.

Some common molecular regulators have been reported to improve both salt and low P tolerance. For instance, overexpression of vacuolar H^+-pyrophosphatase AVP1 improved drought and salt tolerance in *Arabidopsis* (Gaxiola et al. 2001). Later on, direct correlation was revealed between salt tolerance and *AVP1* expression in *Arabidopsis* ecotypes with differential salinity tolerance (Jha et al. 2010). Interestingly, induction of *AVP1* mRNA was also observed under limiting Pi conditions in *Arabidopsis* (Yang et al. 2007). Overexpression of *AVP1* in *Arabidopsis*, rice and tomato significantly improved root architecture, increased rhizosphere acidification and organic acid exudation leading to improved low P tolerance (Yang et al. 2007). Another PSR (P starvation response) gene, Pi transporter *PHT4;6* was reported as a determinant of salt stress in *Arabidopsis*. *Arabidopsis* mutants of *pht4;6* showed increased salt sensitivity as compared to wild type (Cubero et al. 2009). These kinds of biotechnological interventions can also be targeted to engineer elite chickpea genotypes for low P and salinity tolerance as homologs of these genes do exist in chickpea.

9.3.3 Heat Stress and Phosphorus

In chickpea, high temperature adversely affects pollen germination, fertility and pollen tube growth at reproductive stages. Elevated temperature also influences seed germination, photosynthesis as well as transpiration rate (reviewed by Devasirvatham et al. 2012). The impact of heat is very prominent in roots. Heat stress reduces root growth, number and biomass which directly decrease uptake of nutrients such as N and P (Giri et al. 2017). Moreover, heat stress also disrupts nutrient metabolism by disrupting function of many enzymes. Pacak et al. (2016) showed that heat stress affects expression of phosphate starvation-responsive high-affinity P transporters. Promoter regions of many P transporters also contain HSE elements (heat shock elements) (Baek et al. 2017) which may probably drive their expression under heat stress. However, elaborate studies to refine the cross talks between P and heat stress are significantly lacking in crops plants including chickpea.

9.3.4 Cold Stress and Phosphorus

Low soil temperatures induce P deficiency in crop plants by reducing P uptake and translocation in plants (Barber 1986; Ercoli et al. 1996). Cold stress is a serious threat to vegetative growth of chickpea plants in the temperate regions of the world. Cold stress creates chlorosis, necrosis and curling of chickpea leaves. At reproductive stages, cold stress is more deleterious and leads to juvenile bud drop, reduced pollen viability, pollen tube growth inhibition and reduced stigma receptivity (reviewed by Jha et al. 2014). Overall, both seed quality and seed yield are adversely affected. Mantri et al. (2007) reported induction of low phosphate induced proteins in flower/pods of cold tolerant chickpea genotype under cold stress. Phosphorus nutrition has been considered essential for flower and seed formation as well as early flowering. Therefore, it was suggested that tolerant chickpea genotype tends to maintain P homeostasis under cold stress in order to sustain flower and pods.

Some PSR genes have also been reported to mediate cross talks between cold stress and P homeostasis (Hammond et al. 2003). In rice, silencing of one of the key regulators of P homeostasis, *OsSPX1*, was found to increase sensitivity to cold stress (Wang et al. 2013). On the other hand, overexpression of *OsSPX1* in *Arabidopsis* and tobacco enhanced cold tolerance (Zhao et al. 2009) suggesting SPX proteins as important candidates in cold and low P stress tolerance. Similarly, some transcription factors such as *WRKY74* are responsive to both cold stress and Pi deficiency (Dai et al. 2016). Recently, in silico analysis of many Pi-responsive genes revealed presence of LTR (low-temperature-responsive elements) in promoter regions (Baek et al. 2017). Moreover, cold and low P stresses have been reported to share somewhat similar biochemical adaptations in terms of membrane lipid alterations (Zheng et al. 2014; Murakami et al. 2018). Such studies can also be extrapolated to chickpea for expanding our understanding about common mechanisms of improved low P and cold stress tolerance.

9.3.5 Heavy Metal Stress and Phosphorus

Heavy metal contamination of cultivable soils and its toxicities to crop plants is emerging as threat to modern agriculture. Heavy metals severely affect growth and biomass of chickpea plants. Different heavy metals such as Cd, Ni, Pb and Cr suppress root nodulation capacity of chickpea plants causing appreciable decrease in number of nodules (Naz et al. 2015). Among the various heavy metals, contamination of soils with arsenic is rising at alarming rate in India (De 2005). Chickpea plants grown in arsenic contaminated soils show severe root and shoot growth inhibition, membrane damage, reduction in chlorophyll, leaf water and sucrose content (Malik et al. 2011). Also, the number of pods and seed yield shows marked reduction. Arsenate treatment also declines chickpea seed quality (Malik et al. 2011). Accumulation of starch, sugars, storage proteins (albumins, globulins, glutelins and prolamins), many amino acids (lysine, methionine, cystine, phenylalanine, tyrosine, proline, threonine, tryptophan, and valine) and minerals (Ca, P and Fe) is drastically reduced in arsenate treated chickpea plants (Malik et al. 2011).

Arsenate is chemically analogous to Pi; therefore, both arsenate [As(V)] and inorganic phosphate are taken up by roots through same uptake system which involves Pi transporters (Kamiya et al. 2013; Shukla et al. 2015). Arsenate replaces phosphorus in ATP and forms unstable ADP-As which causes disruption of energy flow (Meharg et al. 1994). Arsenate competes with Pi, therefore, disturbs Pi metabolism in plants (Tripathi et al. 2007; Zhao et al. 2010). Arsenic application induces oxidative stress and increases concentrations of H_2O_2 and activity of catalase and ascorbate peroxidase in chickpea (Gunes et al. 2009). Application of P increases availability of arsenic, thereby further increasing its accumulation. However, P supply reduces activity of catalase and ascorbate peroxidase which helps in alleviating arsenic-induced oxidative stress in chickpea (Gunes et al. 2009). Given the effect that P fertilization increases accumulation of arsenic in plants, optimum doses of P fertilizers should be carefully determined in arsenic contaminated soils. Additionally, P fertilization could also be useful in phytoextraction for cleaning arsenic contamination form soils (Fayiga and Ma 2006). Indeed, some studies have proved utility of chickpea as a hyper accumulator in phytoremediation of arsenic as well as other heavy metals such as copper and chromium (Gupta et al. 2008; Kambhampati and Vu 2013; Velez et al. 2016).

9.3.6 *Other Nutrient Deficiencies and Phosphorus*

Micronutrient deficiencies are widely prevalent in Indian soils with zinc and iron being the most deficient ones (Singh 2008). In a study with 20 chickpea genotypes, increasing P levels decreased concentration of micronutrients such iron, zinc, copper and manganese (Srinivasarao et al. 2006). Moreover, chickpea grown on calcium carbonate-Fe oxide systems showed Fe chlorosis under high P supply (Sánchez-Rodriguez et al. 2014). Artificial application of organic acids (citrate, oxalate and malate) solubilizes Fe in rhizosphere and alleviates Fe chlorosis under high P supply in chickpea (Sánchez-Rodriguez et al. 2014). Several studies have recognized molecular and physiological interplay between Fe and P supply and revealed P supply-dependent availability of iron (Rai et al. 2015; Balzergue et al. 2017; Mora-Macías et al. 2017). Similarly, P and Zn uptake are also negatively correlated with each other in chickpea (Hamilton et al. 1993). Zn deficiency is proposed to increase membrane permeability of root cells leading to enhanced P uptake (Marschner and Cakmak 1986). Uptake of P is highly regulated by high-affinity P transporters in both dicots and monocots. These transporters stay in repressed state under high P supply and are specifically induced by P deficiency (Dong et al. 1999; Liu et al. 1998; Muchhal et al. 1996; Muchhal and Raghothama 1999). However, studies reveal that this tight regulation is lost under Zn deficiency as high-affinity Pi transporters are upregulated under Zn deficiency irrespective of levels of exogenous P supply (Huang et al. 2000). Some other candidate genes such as *PHR1* transcription factor and *PHO1* have been implicated in coregulation of Zn and P homeostasis in *Arabidopsis* (Khan et al. 2014; reviewed by Bouain et al. 2014). Therefore, doses of both the nutrients should be optimally balanced for nutrient deficient chickpea soils (Krishna and George 2017). Balanced doses of P and Zn increased nodulation, leghaemoglobin content, activity of N fixing organism and, thus, overall N fixation in chickpea (Shukla and Yadav 1982). Additionally, plant traits that co-ordinately enhance both Zn and P acquisition efficiency can be simultaneously targeted in crop improvement breeding programmes (Rose et al. 2013).

Among the macronutrients, P affects nitrogen nutrition of plants by influencing symbiotic nitrogen fixation (SNF) capacity. Pi deficiency leads to reduced concentration of ATP and adenylate energy charge in nodules leading to decreased nitrogenase activity and SNF capacity in legumes (Valentine et al. 2011). Application of Pi fertilizers in chickpea increased nodulation and symbiotic nitrogen fixation (Yahiya et al. 1995). Due to symbiotic relationship with effective and compatible *Rhizobium* strains, chickpea possess superior ability to fix atmospheric nitrogen in soils (Esfahani et al. 2016). This can improve soil fertility and reduce demand of chemical fertilizers. However, Pi deficiency in chickpea growing areas is a major constraint for chickpea production (Srinivasarao et al. 2006). Thus, efforts have been made for selection of *Rhizobium*-chickpea symbiotic associations that can efficiently fix atmospheric nitrogen under low Pi conditions (Esfahani et al. 2016). It was observed that *Mesorhizobium*-chickpea associations that are able to reprogramme whole plant metabolism (in order to maintain nodular Pi concentrations under Pi deficiency) possess high symbiotic efficiency under Pi deficiency (Esfahani et al. 2016). This reprogramming includes alterations in expression of myriad of genes leading to activation of alternative carbon metabolism pathways, enhanced biosynthesis and secretion of organic acids in rhizosphere and protection of nodule metabolism from low Pi-induced oxidative stress (Esfahani et al. 2014, 2016). Such understanding will help in formulating efficient breeding and genetic engineering strategies for improving symbiotic nitrogen fixation in chickpea genotypes under Pi deficiency.

9.3.7 Disease and Phosphorus

Ascochyta blight of chickpea, caused by *Ascochyta rabiei*, is the most prevalent biotic constraint which can cause complete yield loss and decline in seed quality (Pande et al. 2005). In case of lentils, it was observed that higher plant P concentrations increase plant susceptibility to *Ascochyta* blight caused by *Ascochyta lentis* (Sahi et al. 2007). However, in chickpea, resistant genotypes showed more P accumulation than susceptible genotypes upon inoculation with pathogen (Khirbat and Jalali 1999). Apart from *Ascochyta* blight, fusarium wilt is another most devastating disease that affects chickpea production worldwide. Fusarium wilt caused by *Fusarium oxysporum* f. sp. *ciceri.* causes an annual yield loss of 10–15% in India (Jiménez-Díaz et al. 2015); however, 90–100% yield loss can occur under favourable environmental conditions (Halila and Strange 1996; Jendoubi et al. 2017). P fertilization has significant influence on composition of *Fusarium* population in Asparagus (Yergeau et al. 2006). Disease reduction was observed with P fertilization in case of root rot of peas caused by *Fusarium oxysporum* (Maheshwari et al. 1981). However, no extensive case studies were carried out to explore the effect of P fertilization on either *Ascochyta* blight or fusarium wilt of chickpea. In other crops, P fertilization proved effective in reducing diseases incidence. For example, in wheat, P fertilization eliminated economic losses due to pythium root rot (Huber 1980). Similarly, P application reduces downy mildew, blue mould, leaf curl virus disease in tobacco, brown stripe disease in sugarcane, blast and bacterial leaf blight disease in rice, blight in soybean, yellow dwarf virus disease in barley and powdery mildew in cucumber and mango (reviewed by Dordas 2008). Given the significant impact of P in disease resistance, further research is required to establish the role of P nutrition in disease management in chickpea.

In recent past, some studies elucidated mechanistic explanations for close coordination between phosphate starvation response and plant defence. In citrus, P application was found to reduce disease symptoms of devastating disease, Huanglongbing (HLB) (Zhao et al. 2013). Interestingly, miRNA profiling of diseased plants showed induction of PSR miRNA, miR399 in diseased plants. Nutrient quantitation results revealed 65% depletion in P contents in diseased plants as compared to control (Zhao et al. 2013). Recently in a breakthrough study, Castrillo et al. (2017) showed that MYB transcription factor, PHR1 directly regulates and integrates PSR and plant immune system genes under P deficiency.

9.3.8 Insect Pests and Phosphorus

Chickpea is infested by approximately 60 insect species (Sharma et al. 2007). Among the major pests of chickpea, pod borer and leaf miner are the major insect pests worldwide (Rasool et al. 2015). Plant nutrition status affects the susceptibility of plants to the attack of insect pests. Fast-growing plants with plenty of N and P are preferred food for herbivores (Grutters et al. 2016). Khan et al. (2016) showed that plants experiencing P deficiency possess enhanced resistance to insect herbivores. P-starved plants induce biosynthesis of jasmonic acids (JA) and its bioactive conjugate JA-isoleucine. Also, various genes of JA signalling pathway are differentially expressed by P deficiency which collectively provides resistance to generalist herbivore *Spodoptera littoralis* in *Arabidopsis*, tobacco and tomato. These evidences suggest role of P supply in plant defence against herbivores. Contradictory to this, in some crops such as mustard, increased P supply reduced incidence of damage by major pests such as sawfly and aphids (Ram and Gupta 1988). Similarly, P supply enhanced plant nutrition status to withstand damage by pests (*Aphis craccivora, Megalurothrips sjostedti and Maruca vitrata*) in cowpea genotypes (Asiwe 2009). Such results encourage more research in understanding chickpea defence responses against herbivores from perspective of phosphate nutrition.

9.4 Regulating Seed P Content for Improving Seed Quality

Legumes seed size is comparatively larger than cereals. Almost 1–5% of the seed weight corresponds to phytate, a stored form of P (Vats and Banerjee 2004). Phytate stores maximum amount of P and various micronutrient like iron, calcium and magnesium as complexes. About 60% of P absorbed by plants is finally allocated to seeds during its development (Yamaji et al. 2016). Major form of this P is phytate, and its content varies between 7.48 and 8.00 mg g^{-1} in chickpea seeds (Duhan et al. 1989). Further, monogastric animals cannot digest phytate; therefore, they are not able to absorb these micronutrients. This results in high amount of P in their excreta which ultimately ends up in water bodes leading to pollution. Milling and soaking, fermentation and germination are some of the procedure which reduces the phytate content to some extent (Gupta et al. 2015). Therefore, reducing phytate accumulation in seeds can lead to more economical and environment-friendly agriculture.

Molecular approaches have been used to reduce seed phytate content in plants. The *lpa* (low phytic acid) mutants in maize, barley and rice (Larson et al. 1998, 2000; Raboy et al. 2000) show reduced seed phytate content. The seeds produced from plants with *lpa* characteristics have normal total P content whereas reduced phytate (Raboy 2002) and almost 70% of the total P were available to plant. Though seeds with strong *lpa-1* allele show reduced germination percentage and seed weight, whereas *lpa241* mutation caused impaired vegetative growth and seed development. All these drawbacks limit the utilization of these mutants for generating low seed phytate varieties (Raboy et al. 2000; Raboy 2001; Pilu et al. 2005). One of the ways to reduce phytate content is blocking the activity of enzymes involved in phytate biosynthesis. One such enzyme, MIPS, Ins (3) P synthase converts glucose-6-phosphate to myo-inositol 3-phosphate (Ins (3) P). When the expression of MIPS is blocked, myo-inositol content reduces, but seed germination gets impaired. The knockout mutants of maize Ins (1, 3, 4) P3 5/6 kinase and *Arabidopsis* IPK2β has lead to development of viable seeds (Shi et al. 2003; Stevenson-Paulik et al. 2005). In *Arabidopsis ipk1* mutants, most of the Pi is still bound in inositol complexes, and there is no increase in Pi accumulation. However, the double mutant *ipk1ipk2* produced low phytic acid seed, but inositol complexes InsP4 and InsP5 were still present (Stevenson-Paulik et al. 2005). Given these limitations, this approach has met with limited success. It was found that multidrug resistance-associated protein (MRP) ATP-binding cassette (ABC) transporter gene does not express in maize *lpa-1* mutants. Given its role as transporter, the silencing of this gene in embryo-specific manner (as highly expressed in embryos) produced viable seeds with low phytate content and high Pi and showed no germination problems (Shi et al. 2007). Recently, in another approach, node-based transporters have been introduced by Yamaji et al. (2016). These transporters are localized in different cells in nodes of plants and are responsible for mineral distribution within leaf blades and panicles. For instance, a node-based transporter, SULTR-like phosphorus distribution transporter (SPDT), is responsible for allocation of phosphorus in rice grains (Yamaji et al. 2016). As it is responsible for P accumulation in grains, its knockout mutant had reduced P and therefore less phytate accumulation in grains. Moreover, reduction in P content did not affect other agronomic traits of plants. Similar genes or strategies could be a possible target for producing high-quality seeds in chickpea without compromising other agronomic traits.

Alternatively, phytases could also be targeted for reducing phytate content. One of the chickpea PAPs, CaPAP7, having phytase activity shows association with seed weight and phytate content (Bhadouria et al. 2017). This gene also expresses in flower parts. Therefore, there is a possibility that PAPs with phytase activity could provide a way to engineer plants with reduce seed phytate content. Even phytase expressing in germinating seedling could also be targeted for producing seeds with low phytate content. Combination of more than one candidate genes could also be studied to get the varieties with best agronomic traits while reducing seed phytate.

9.5 Conclusions and Future Perspectives

Legumes possess the ability to fix atmospheric nitrogen and therefore, are usually grown on marginal soils of the world with poor soil fertility. Although, the demand for nitrogenous fertilizers is significantly low for legume crops, but the requirement of phosphatic fertilizers is still very high. Notably, legumes relying on SNF for meeting their demands for nitrogen require more P as compared to non-symbiotic legumes (Leidi and Rodríguez-Navarro 2000). Root nodules in leguminous crops like chickpea act as a sink for P and possess high Pi demand for optimal metabolism and energetic transformations (Sulieman and Tran 2015). Therefore, balanced P fertilization is a key component of chickpea cultivation. However, application of P fertilizers is not a judicious alternative from both economic and ecological perspectives. Despite of being an important legume crop, vigorous efforts for improving low P tolerance of chickpea are still lagging behind. Improving chickpea varieties for enhanced low P tolerance and increased P use efficiency can support sustainable agriculture by preventing environmental deterioration and minimizing expenditure on phosphatic fertilizers. Hence, studying low P-associated morphological, physiological, biochemical and molecular responses in chickpea will provide new insights and leads for designing strategies for low P-tolerant chickpea genotypes.

Seeds of chickpea contain high P reserves in the form of phytate. However, high phytate in seeds acts as a potential anti-nutrient that decreases the bioavailability of micronutrients, otherwise abundant in chickpea. Investigations in crops have revealed strategies to manipulate seed phytate content. These attempts have both positive and negative consequences on seed quality especially germination. Recent research work in cereal crops like rice has revealed existence of node-specific P transporters which regulate P distribution to seeds. Efficient engineering of these transporters can successfully reduce seed phytate content without declining seed quality. Such studies can be smartly extrapolated to chickpea for developing low phytate chickpea. Additionally, the consequence of such manipulations on seed quality, germination and yield needs to be thoroughly assessed.

Apart from the pivotal roles of P in plant's growth and development, there are series of evidences suggesting its roles in various biotic and abiotic stresses. Of these, the role of P in disease management is particularly intriguing. P fertilization helps to reduce disease incidence in myriad of crops. Therefore, it would be worth investigating the role of P in management of prevalent and devastating diseases of chickpea such as *Ascochyta* blight. Understanding broader roles of phosphate nutrition in multiple stress tolerance would pave wave for tailoring crop management practices and engineering chickpea for better seed yield and quality in environmentally sustainable manner.

Acknowledgements We acknowledge research fellowships from DBT, UGC and CSIR, India. Our research is funded by core grant from NIPGR, DBT and DBT-IYBA grant to JG.

References

Ahmad F, Gaur P, Croser J (2005) Chickpea (*Cicer ariatinum* L.). In: Singh R, Jauhar P (eds) Genetic resources, chromosome engineering and crop improvement, Grains legumes. CRC Press, Boca Raton, pp 185–214

Al-Karaki GN, Al-Karaki RB, Al-Karaki CY (1996) Phosphorus nutrition and water stress effects on proline accumulation in Sorghum and bean. J Plant Physiol 148:745–751

Alloush G (2003) Responses of hydroponically-grown chickpea to low phosphorus: pH changes, nutrient uptake rates, and root morphological changes. Agronomie 23:123–133

Amijee F, Giller KE (1998) Environmental constraints to nodulation and nitrogen fixation of Phaseolus vulgaris L. in Tanzania I. A survey of soil fertility and root nodulation. Afr Crop Sci J 6:159–169

Anoop VM, Basu U, McCammon MT, McAlister-Henn L, Taylor GJ (2003) Modulation of citrate metabolism alters aluminum tolerance in yeast and transgenic canola overexpressing a mitochondrial citrate synthase. Plant Physiol 132:2205–2217

Asiwe J (2009) The impact of phosphate fertilizer as a pest management tactic in four cowpea varieties. Afr J Biotechnol 8:7182–7188

Baek D, Chun HJ, Yun DJ, Kim MC (2017) Cross-talk between phosphate starvation and other environmental stress signaling pathways in plants. Mol Cells 40:697–705

Balzergue C, Dartevelle T, Godon C, Laugier E, Meisrimler C, Teulon JM et al (2017) Low phosphate activates STOP1-ALMT1 to rapidly inhibit root cell elongation. Nat Commun 8:15300

Bampidisa VA, Christodoulou V (2011) Chickpeas (*Cicer arietinum* L.) in animal nutrition: a review. Anim Feed Sci Technol 168:1–20

Barber SA (1986) Soil-plant interactions in the phosphorus nutrition of plants. In: Khasawneh FE, Sample EC, Kamprath EJ (eds) The role of phosphorus in agriculture. ASA, CSSA, and SSSA, Madison

Barber SA (1995) Soil nutrient bioavailability: a mechanistic approach, 2nd edn. Wiley, New York

Bates TR, Lynch JP (1996) Stimulation of root hair elongation in *Arabidopsis thaiiana* by low phosphorus availability. Plant Cell and Environment 19:529–538

Beebe SE, Rao IM, Cajiao C, Grajales M (2008) Selection for drought resistance in common bean also improves yield in phosphorus limited and favorable environments. Crop Sci 48:582–592

Bhadouria J, Singh AP, Mehra P, Verma L, Srivastawa R, Parida SK, Giri J (2017) Identification of purple acid phosphatases in chickpea and potential roles of CaPAP7 in seed Phytate accumulation. Sci Rep 7:11012

Bouain N, Shahzad Z, Rouached A, Khan GA, Berthomieu P, Abdelly C et al (2014) Phosphate and zinc transport and signalling in plants: toward a better understanding of their homeostasis interaction. J Exp Bot 65:5725–5741

Bozzo GG, Raghothama KG, Plaxton WC (2002) Purification and characterization of two secreted purple acid phosphatase isozymes from phosphate-starved tomato (*Lycopersicon esculentum*) cell cultures. FEBS J 269:6278–6286

Bozzo GG, Raghothama KG, Plaxton WC (2004) Structural and kinetic properties of a novel purple acid phosphatase from phosphate-starved tomato (Lycopersicon esculentum) cell cultures. Biochem J 377:419–428

Bates TR, Lynch JP (2001) Root hairs confer a competitive advantage under low phosphorus availability. Plant Soil 236:243–250

Cashikar AG, Kumaresan R, Rao NM (1997) Biochemical characterization and subcellular localization of the red kidney bean purple acid phosphatase. Plant Physiol 114:907–915

Castrillo G, Teixeira PJPL, Paredes SH, Law TF, De Lorenzo L, Feltcher ME, Finkel OM, Breakfield NW, Mieczkowski P, Jones CD, Paz-Ares J (2017) Root microbiota drive direct integration of phosphate stress and immunity. Nature 543(7646):513

Chen YL, Dunbabin VM, Diggle AJ, Siddique KH, Rengel Z (2013) Phosphorus starvation boosts carboxylate secretion in P-deficient genotypes of Lupinus angustifolius with contrasting root structure. Crop Pasture Sci 64:588–599

Cheng Y, Zhou W, Peters C, Li M, Wang X, Huang J (2011) Characterization of the Arabidopsis glycerophosphodiester phosphodiesterase (GDPD) family reveals a role of the plastid-localized AtGDPD1 in maintaining cellular phosphate homeostasis under phosphate starvation. Plant J 66:781–795

Clarkson TD (1991) Root structure and site of ion uptake. In: Waisel Y, Eshel A, Kafkafi U (eds) Plant roots: the hidden half. Marcel Dekker, New York

Cubero B, Nakagawa Y, Jiang XY, Miura KJ, Li F, Raghothama KG, Bressan RA, Hasegawa PM, Pardo JM (2009) The phosphate transporter PHT4;6 is a determinant of salt tolerance that is localized to the golgi apparatus of arabidopsis. Mol Plant 2:535–552

Dai X, Wang Y, Zhang WH (2016) OsWRKY74, a WRKY transcription factor, modulates tolerance to phosphate starvation in rice. J Exp Bot 67:947–960

De M (2005) Arsenic-India's health crisis attracting global attention. Curr Sci 88:683–684

Denton MD, Veneklaas EJ, Freimoser FM, Lambers H (2007) Banksia species (Proteaceae) from severely phosphorus-impoverished soils exhibit extreme efficiency in the use and remobilization of phosphorus. Plant Cell Environ 30:1557–1565

Devasirvatham V, Tan DKY, Gaur PM, Raju TN, Trethowan RM (2012) High temperature tolerance in chickpea and its implications for plant improvement. Crop Pasture Sci 63:419–428

Dinkelaker B, Hengeler C, Marschner H (1995) Distribution and function of proteoid roots and other root clusters. Botanica Acta 108:183–200

Dionisio G, Madsen CK, Holm PB, Welinder KG, Jørgensen M, Stoger E, Arcalis E, Brinch-Pedersen H (2011) Cloning and characterization of purple acid phosphatase phytases from wheat, barley, maize, and rice. Plant Physiol 156:1087–1100

Dong B, Ryan PR, Rengel Z, Delhaize E (1999) Phosphate uptake in Arabidopsis thaliana: dependence of uptake on the expression of transporter genes and internal phosphate concentrations. Plant Cell Environ 22:1455–1461

Dordas C (2008) Role of nutrients in controlling plant diseases in sustainable agriculture: a review. In: Lichtfouse E, Navarrete M, Debaeke P, Véronique S, Alberola C (eds) Sustainable agriculture. Springer, Dordrecht

Drew MC (1975) Comparison of the effects of a localized supply of phosphate, nitrate, ammonium and potassium on the growth of the seminal root system, and the shoot, in barley. New Phytol 75:479–490

Duhan A, Chauhan BM, Punia D, Kapoor AC (1989) Phytic acid content of chickpea (*Cicer arietinum*) and black gram (*Vigna mungo*): varietal differences and effect of domestic processing and cooking methods. J Sci Food Agric 49:449–455

Ercoli L, Mariotti M, Masoni A, Massantini F (1996) Effect of temperature and phosphorus fertilization on phosphorus and nitrogen uptake by sorghum. Crop Sci 36:348–354

Esfahani MN, Sulieman S, Schulze J, Yamaguchi-Shinozaki K, Shinozaki K, Tran LS (2014) Mechanisms of physiological adjustment of N2 fixation in Cicer arietinum L. (chickpea) during early stages of water deficit: single or multi-factor controls. Plant J 79:964–980

Esfahani MN, Kusano M, Nguyen KH, Watanabe Y, Ha CV, Saito K, Sulieman S, Herrera-Estrella L, Phan TLS (2016) Adaptation of the symbiotic *Mesorhizobium*-chickpea relationship to phosphate deficiency relies on reprogramming of whole-plant metabolism. Proc Natl Acad Sci U S A 113:E4610–E4619

Fayiga AO, Ma LQ (2006) Using phosphate rock to immobilize metals in soil and increase arsenic uptake by hyperaccumulator Pteris vittata. Sci Total Environ 359:17–25

Fink JR, Inda AV, Bavaresco J, Barrón V, Torrent J, Bayer C (2016) Adsorption and desorption of phosphorus in subtropical soils as affected by management system and mineralogy. Soil Tillage Res 155:62–68

Flowers TJ, Gaur PM, Gowda CL, Krishnamurthy L, Samineni S, Siddique KH, Turner NC, Vadez V, Varshney RK, Colmer TD (2010) Salt sensitivity in chickpea. Plant Cell Environ 33:490–509

Gahoonia TS, Nielsen NE (1998) Direct evidence on participation of root hairs in phosphorus (P-32) uptake from soil. Plant Soil 198:147–152

Gahoonia TS, Ali R, Malhotra RS, Jahoor A, Rahman MM (2007) Variation in root morphological and physiological traits and nutrient uptake of chickpea genotypes. J Plant Nutr 30:829–841

Galav A, Bhowmick AK, Joshi N, Singh KK, Mehta V, Sharma S (2018) Impact of weather parameters on population fluctuation of *Helicoverpa armigera* (H) on chickpea. Int J Adv Sci Res Manage 1:121–123

Garg R, Shankar R, Thakkar B, Kudapa H, Krishnamurthy L, Mantri N, Varshney RK, Bhatia S, Jain M (2016) Transcriptome analyses reveal genotype- and developmental stage-specific molecular responses to drought and salinity stresses in chickpea. Sci Rep 6:19228

Gaude N, Tippmann H, Flemetakis E, Katinakis P, Udvardi M, Dörmann P (2004) The galactolipid digalactosyldiacylglycerol accumulates in the peribacteroid membrane of nitrogen-fixing nodules of soybean and lotus. J Biol Chem 279:34624–34630

Gaude N, Nakamura Y, Scheible WR, Ohta H, Dormann P (2008) Phospholipase C5 (NPC5) is involved in galactolipid accumulation during phosphate limitation in leaves of Arabidopsis. Plant J 56:28–39

Gaxiola RA, Li J, Undurraga S, Dang LM, Allen GJ, Alper SL, Fink GR (2001) Drought- and salt-tolerant plants result from overexpression of the AVP1 H^+-pump. Proc Natl Acad Sci U S A 98:11444–11449

Gilbert GA, Knight JD, Vance CP, Allan DL (1999) Acid phosphatase activity in phosphorus-deficient white lupin roots. Plant Cell Environ 22:801–810

Giri A, Heckathorn S, Mishra S, Krause C (2017) Heat stress decreases levels of nutrient-uptake and -assimilation proteins in tomato roots. Plants 6:6

Giri J, Bhosale R, Huang G, Pandey BK, Parker H, Zappala S, Yang J, Dievart A, Bureau C, Ljung K, Price A, Rose T, Larrieu A, Mairhofer S, Sturrock CJ, White P, Dupuy L, Hawkesford M, Perin C, Liang W, Peret B, Hodgman CT, Lynch J, Wissuwa M, Zhang D, Pridmore T, Mooney SJ, Guiderdoni E, Swarup R, Bennett MJ (2018) Rice auxin influx carrier OsAUX1 facilitates root hair elongation in response to low external phosphate. Nat Commun 9:1408

Gruber BD, Giehl RF, Friedel S, von Wirén N (2013) Plasticity of the Arabidopsis root system under nutrient deficiencies. Plant Physiol 163:161–179

Grutters BMC, Gross EM, Bakker ES (2016) Insect herbivory on native and exotic aquatic plants: phosphorus and nitrogen drive insect growth and nutrient release. Hydrobiologia 778:209–220

Gunes A, Cicek N, Inal A, Alpaslan M, Eraslan F, Guneri E, Guzelordu T (2006) Genotypic response of chickpea (*Cicer arietinum* L.) cultivars to drought stress implemented at pre- and post-anthesis stages and its relations with nutrient uptake and efficiency. Plant Soil Environ 52:368–376

Gunes A, Pilbeam DJ, Inal A (2009) Effect of arsenic-phosphorous interaction on arsenic-induced oxidative stress in chickpea plants. Plant Soil 314:211–220

Gupta DK, Tripathi RD, Mishra S, Srivastava S, Dwivedi S, Rai UN, Yang XE, Huanji H, Inouhe M (2008) Arsenic accumulation in roots and shoot Vis-a-Vis its effects on growth and level of phytochelatins in seedlings of *Cicer arietinum* L. J Environ Biol 3:281–286

Gupta RK, Gangoliya SS, Singh NK (2015) Reduction of phytic acid and enhancement of bioavailable micronutrients in food grains. J Food Sci Technol 52:676–684

Halila M, Strange R (1996) Identification of the causal agent of wilt of chickpea in Tunisia as Fusarium oxysporum f. sp ciceri race 0. Phytopathologia mediterranea 35:67–74

Hamilton MA, Westermann DT, James DW (1993) Factors affecting zinc uptake in cropping systems. Soil Sci Soc Am J 57:1310–1315

Hammond JP, Bennett MJ, Bowen HC, Broadley MR, Eastwood DC, May ST, Rahn C, Swarup R, Woolaway KE, White PJ (2003) Changes in gene expression in Arabidopsis shoots during phosphate starvation and the potential for developing smart plants. Plant Physiol 132:578–596

Hansel FD, Diaz DAR, Amado TJC, Rosso LHM (2017) Deep banding increases phosphorus removal by soybean grown under no-tillage production systems. Agron J 109:1091–1098

He M, Dijkstra FA (2014) Drought effect on plant nitrogen and phosphorus: a meta-analysis. New Phytol 204:924–931

He Y, Liao H, Yan X (2003) Localized supply of phosphorus induces root morphological and architectural changes of rice in split and stratified soil cultures. Plant Soil 248:247–256

Hegeman CE, Grabau EA (2001) A novel phytase with sequence similarity to purple acid phosphatases is expressed in cotyledons of germinating soybean seedlings. Plant Physiol 126:1598–1608

Herrera-Estrella L, López-Arredondo D (2016) Phosphorus: the underrated element for feeding the world. Trends Plant Sci 21:461–463
Hoch-Jensen H, Schjoerring JK, Soussana JF (2002) The influence of phosphorus deficiency on growth and nitrogen fixation of white clover plants. Ann Bot 90:745–753
Hocking P, Randall P, Delhaize E, Keerthisinghe G (2000) The role of organic acids exuded from roots in phosphorus nutrition and aluminium tolerance in acidic soils. Manage Conserv Trop Acid Soils Sustainable Crop Prod 33:61–73
Holme IB, Dionisio G, Madsen CK, Brinch-Pedersen H (2017) Barley *HvPAPhy_a* as transgene provides high and stable phytase activities in mature barley straw and in grains. Plant Biotechnol J 15:415–422
Hu Y, Schmidhalter U (2005) Drought and salinity: a comparison of their effects on the mineral nutrition of plants. J Plant Nutr Soil Sci 168:541–549
Huang C, Barker SJ, Langridge P, Smith FW, Graham RD (2000) Zinc deficiency up-regulates expression of high-affinity phosphate transporter genes in both phosphate-sufficient and -deficient barley roots. Plant Physiol 124:415–422
Huber DM (1980) The role of mineral nutrition in defense. In: Horsfall JG, Cowling EB (eds) Plant disease. An advanced treatise. Academic, London
IIAM (2010) Relatório de Actividades do Centro Zonal Sul. Centro Zonal Sul, Instituto de Investigação Agrária de Moçambique, Ministério da Agricultura, Chokwe, Gaza
Israel DW (1987) Investigations of the role of phosphorus in symbiotic dinitrogen fixation. Plant Physiol 84:835–840
Jakobson I (1985) The role of phosphorus in nitrogen fixation by young pea plants (Pisum sativum). Physiol Plant 88:190–196
Jendoubi W, Bouhadida M, Boukteb A, Béji M, Kharrat M (2017) Fusarium wilt affecting chickpea crop. Agriculture 7:23
Jha D, Shirley N, Tester M, Roy SJ (2010) Variation in salinity tolerance and shoot sodium accumulation in Arabidopsis ecotypes linked to differences in the natural expression levels of transporters involved in sodium transport. Plant Cell Environ 33:793–804
Jha UC, Chaturvedi SK, Bohra A, Basu PS, Khan MS, Barh D (2014) Abiotic stresses, constraints and improvement strategies in chickpea. Plant Breed 133:163–178
Jiménez-Díaz RM, Castillo P, del Mar Jiménez-Gasco M, Landa BB, Navas-Cortés JA (2015) Fusarium wilt of chickpeas: biology, ecology and management. Crop Prot 73:16–27
Jin J, Wang G, Liu X, Pan X, Herbert SJ (2005) Phosphorus application affects the soybean root response to water deficit at the initial flowering and full pod stages. Soil Sci Plant Nutr 51:953–960
Jin J, Wang G, Liu X, Pan X, Herbert SJ, Tang C (2006) Interaction between phosphorus nutrition and drought on grain yield, and assimilation of phosphorus and nitrogen in two soybean cultivars differing in protein concentration in grains. J Plant Nutr 29:1433–1449
Jin J, Lauricella D, Armstrong R, Sale P, Tang C (2015) Phosphorus application and elevated CO2 enhance drought tolerance in field pea grown in a phosphorus-deficient vertisol. Ann Bot 116:975–985
Jones CA, Jacobsen JS, Wraith JM (2005) Response of malt barley to phosphorus fertilization under drought conditions. J Plant Nutr 28:1605–1617
Jukanti AK, Gaur PM, Gowda CL, Chibbar RN (2012) Nutritional quality and health benefits of chickpea (Cicer arietinum L.): a review. Br J Nutr 108:S11–S26
Kahraman A, Harmankaya M, Ceyhan E (2015) Nutritional variation and drought tolerance in chickpeas (Cicer arietinum L.). J Elem 20:331–341
Kaida R, Sage-Ono K, Kamada H, Okuyama H, Syono K, Kaneko TS (2003) Isolation and characterization of four cell wall purple acid phosphatase genes from tobacco cells. Biochim Biophys Acta (BBA) Gene Struct Expr 1625:134–140
Kaida R, Serada S, Norioka N, Norioka S, Neumetzler L, Pauly M, Sampedro J, Zarra I, Hayashi T, Kaneko TS (2010) Potential role for purple acid phosphatase in the dephosphorylation of wall proteins in tobacco cells. Plant Physiol 153:603–610

Kambhampati MS, Vu VT (2013) EDTA enhanced phytoremediation of copper contaminated soils using chickpea (Cicer aeritinum L.). Bull Environ Contemp Toxicol 91:310–313

Kamiya T, Islam R, Duan G, Uraguchi S, Fujiwara T (2013) Phosphate deficiency signaling pathway is a target of arsenate and phosphate transporter OsPT1 is involved in as accumulation in shoots of rice. Soil Sci Plant Nutr 59:580–590

Khan GA, Bouraine S, Wege S, Li Y, de Carbonnel M, Berthomieu P, Poirier Y, Rouached H (2014) Coordination between zinc and phosphate homeostasis involves the transcription factor PHR1, the phosphate exporter PHO1, and its homologue PHO1;H3 in Arabidopsis. J Exp Bot 65:871–884

Khan AG, Vogiatzaki E, Glauser G, Poirier Y (2016) Phosphate deficiency induces the jasmonate pathway and enhances resistance to insect herbivory. Plant Physiol 171:632–644

Khirbat SK, Jalali BL (1999) Biochemical basis of resistance to chickpea Ascochyta blight. Legum Res 22:46–50

Krishna KSSR, George PJ (2017) Effect of levels of phosphorus and zinc on growth and yield of Kabuli chickpea (*Cicer kabulium* L.). J Pharmacognosy Phytochemistry 6:013–1016

Kuang R, Chan KH, Yeung E, Lim BL (2009) Molecular and biochemical characterization of *AtPAP15*, a purple acid phosphatase with phytase activity, in *Arabidopsis*. Plant Physiol 151:199–209

Kumar P, Sharma MK (2013) Nutrient deficiencies in pulse crops: chickpea (Cicer arietinum Linn.). In: Kumar P, Sharma MK (eds) Nutrient deficiencies of field crops: guide to diagnosis and management. CABI, Boston, MA, USA

Kuwahara FA, Souza GM, Guidorizi KA, Costa C, Meirelles PRDL (2016) Phosphorus as a mitigator of the effects of water stress on the growth and photosynthetic capacity of tropical C4 grasses. Acta Sci Agron 38:363–370

Larson SR, Young KA, Cook A, Blake TK, Raboy V (1998) Linkage mapping of two mutations that reduce phytic acid content of barley grain. Theor Appl Genet 97:141–146

Larson SR, Rutger JN, Young KA, Raboy V (2000) Isolation and genetic mapping of a non-lethal rice (Oryza sativa L.) low phytic acid 1 mutation. Crop Sci 40:1397–1405

Lei L, Chen L, Shi X, Li Y, Wang J, Chen D, Xie F, Li Y (2014) A nodule-specific lipid transfer protein AsE246 participates in transport of plant-synthesized lipids to symbiosome membrane and is essential for nodule organogenesis in Chinese milk vetch. Plant Physiol 164:1045–1058

Leidi EO, Rodríguez-Navarro DN (2000) Nitrogen and phosphorus availability limit N2-fixation in bean. New Phytol 147:337–346

Lemma W, Wassie H, Sheleme B (2013) Response of chickpea (Cicer Arietinum L.) to nitrogen and phosphorus fertilizer S in halaba and taba, Southern Ethiopia. Ethiopian. J Nat Resour 13:115–128

Li D, Zhu H, Liu K, Liu X, Leggewie G, Udvardi M, Wang D (2002) Purple acid phosphatases of Arabidopsis thaliana comparative analysis and differential regulation by phosphate deprivation. J Biol Chem 277:27772–27781

Li RJ, Lu WJ, Guo CJ, Li XJ, Gu JT, Kai XIAO (2012a) Molecular characterization and functional analysis of OsPHY1, a purple acid phosphatase (PAP)-type phytase gene in rice (Oryza sativa L.). J Integr Agric 11:1217–1226

Li C, Gui S, Yang T, Walk T, Wang X, Liao H (2012b) Identification of soybean purple acid phosphatase genes and their expression responses to phosphorus availability and symbiosis. Ann Bot 109:275–285

Liang C, Tian J, Lam HM, Lim BL, Yan X, Liao H (2010) Biochemical and molecular characterization of *PvPAP3*, a novel purple acid phosphatase isolated from common bean enhancing extracellular ATP utilization. Plant Physiol 152:854–865

Liang C, Sun L, Yao Z, Liao H, Tian J (2012) Comparative analysis of *PvPAP* gene family and their functions in response to phosphorus deficiency in common bean. PLoS One 7:e38106

Liao H, Wong FL, Phang TH, Cheung MY, Li WYF, Shao G, Yan X, Lam HM (2003) *GmPAP3*, a novel purple acid phosphatase-like gene in soybean induced by NaCl stress but not phosphorus deficiency. Gene 318:103–111

Liu CM, Muchhal US, Uthappa M, Kononowicz AK, Raghothama KG (1998) Tomato phosphate transporter genes are differentially regulated in plant tissue by phosphorus. Plant Physiol 116:91–99

Liu PD, Xue YB, Chen ZJ, Liu GD, Tian J (2016) Characterization of purple acid phosphatases involved in extracellular dNTP utilization in *Stylosanthes*. J Exp Bot 67:4141–4154

López-Bucio J, Cruz-Ramírez A, Herrera-Estrella L (2003) The role of nutrient availability in regulating root architecture. Curr Opin Plant Biol 6:280–287

Lynch J (1995) Root architecture and plant productivity. Plant Physiol 109:7–13

Lynch JP (2011) Root phenes for enhanced soil exploration and phosphorus acquisition: tools for future crops. Plant Physiol 156:1041–1049

Lynch JP, Brown KM (2011) Topsoil foraging – an architectural adaptation of plants to low phosphorus availability. Plant Soil 237:225

Maheshwari SK, Gupta JS, Jhooty JS (1981) Effect on various cultural practices on the incidence of the wilt and root rot of peas. Ind J Agric Res 15:149–151

Malik JA, Goel S, Sandhir R, Nayyar H (2011) Uptake and distribution of arsenic in chickpea: effects on seed yield and seed composition. Commun Soil Sci Plant Anal 42:1728–1738

Mantri NL, Ford R, Coram TE, Pang ECK (2007) Transcriptional profiling of chickpea genes differentially regulated in response to high-salinity, cold and drought. BMC Genomics 8:303

Marschner H, Cakmak I (1986) Mechanism of phosphorus induced zinc deficiency in cotton. II. Evidence for impaired shoot control of phosphorus uptake and translocation under zinc deficiency. Physiol Plant 68:491–496

Martins D, Macovei A, Leonetti P, Balestrazzi A, Araijo S (2017) The influence of phosphate deficiency on legume symbiotic N2 fixation. In: Sulieman S, Tran LSP (eds) Legume nitrogen fixation in soils with low phosphorus availability: adaptation and regulatory implication. Springer, Cham

Meharg AA, Naylor J, Macnair MR (1994) Phosphorus nutrition of arsenate-tolerant and nontolerant phenotypes of velvet grass. J Environ Qual 23:234–238

Mehra P, Giri J (2016) Rice and chickpea GDPDs are preferentially influenced by low phosphate and CaGDPD1 encodes an active glycerophosphodiester phosphodiesterase enzyme. Plant Cell Rep 35:1699–1717

Mehra P, Pandey BK, Giri J (2016) Comparative morphophysiological analyses and molecular profiling reveal Pi-efficient strategies of a traditional rice genotype. Front Plant Sci 6:1184

Mehra P, Pandey BK, Giri J (2017) Improvement in phosphate acquisition and utilization by a secretory purple acid phosphatase (OsPAP21b) in rice. Plant Biotechnol J 15:1054–1067

Mengel K, Kirkby EA (1978) Principles of plant nutrition. International Potash Institute, Worblaufen-Bern

Mora-Macías J, Ojeda-Rivera JO, Gutiérrez-Alanís D, Yong-Villalobos L, Oropeza-Aburto A, Raya-González J, Jiménez-Domínguez G, Chávez-Calvillo G, Rellán-Álvarez R, Herrera-Estrella L (2017) Malate-dependent Fe accumulation is a critical checkpoint in the root developmental response to low phosphate. Proc Natl Acad Sci USA 114:E3563–E3572

Muchhal US, Raghothama KG (1999) Transcriptional regulation of plant phosphate transporters. Proc Natl Acad Sci U S A 96:5868–5872

Muchhal US, Pardo JM, Raghothama KG (1996) Phosphate transporters from the higher plant Arabidopsis thaliana. Proc Natl Acad Sci U S A 93:10519–10523

Murakami H, Nobusawa T, Hori K, Shimojima M, Ohta H (2018) Betaine lipid is crucial for adapting to low temperature and phosphate deficiency in nannochloropsis. Plant Physiol 177:181–193

Nakamura Y (2013) Phosphate starvation and membrane lipid remodeling in seed plants. Prog Lipid Res 52:43–50

Nakamura Y, Awai K, Masuda T, Yoshioka Y, Takamiya KI, Ohta H (2005) A novel phosphatidylcholine-hydrolyzing phospholipase C induced by phosphate starvation in Arabidopsis. J Biol Chem 280:7469–7476

Naz H, Naz A, Ashraf S (2015) Impact of heavy metal toxicity to plant growth and nodulation in chickpea grown under heavy metal stress. Int J Res Emerg Sci Technol 2:248–260

Neumann G, Römheld V (1999) Root excretion of carboxylic acids and protons in phosphorus-deficient plants. Plant Soil 211:121–130

Neumann G, Massonneau A, Langlade N, Dinkelaker B, Hengeler C, Römheld V, Martinoia E (2000) Physiological aspects of cluster root function and development in phosphorus-deficient white lupin (*Lupinus albus* L.). Ann Bot 6:909–919

Nielsen KL, Eshel A, Lynch JP (2001) The effect of phosphorus availability on carbon economy of contrasting common bean (*Phaseolus vulgaris* L.) genotypes. J Exp Bot 52:329–339

Ohwaki Y, Hirata H (1992) Differences in carboxylic acid exudation among P-starved leguminous crops in relation to carboxylic acid contents in plant tissues and phospholipid level in roots. Soil Sci Plant Nutr 38:235–243

Pacak A, Barciszewska-Pacak M, Swida-Barteczka A, Kruszka K, Sega P, Milanowska K, Jakobsen I, Jarmolowski A, Szweykowska-Kulinska Z (2016) Heat stress affects Pi-related genes expression and inorganic phosphate deposition/accumulation in barley. Front Plant Sci 7:926

Pande S, Siddique KHM, Kishore GK, Bayaa B, Gaur PM, Gowda CLL, Bretag TW, Crouch JH (2005) Ascochyta blight of chickpea (*Cicer arietinum* L.): a review of biology, pathogenicity, and disease management. Aust J Agric Res 56:317–332

Pandey BK, Mehra P, Verma L, Bhadouria J, Giri J (2017) *OsHAD1*, a haloacid dehalogenase-like APase enhances phosphate accumulation. Plant Physiol 174:2316–2332

Pang J, Bansal R, Zhao H, Bohuon E, Lambers H, Ryan MH, Ranathunge K, Siddique KHM (2018) The carboxylate-releasing phosphorus-mobilizing strategy can be proxied by foliar manganese concentration in a large set of chickpea germplasm under low phosphorus supply. New Phytol 219:518. https://doi.org/10.1111/nph.15200

Pant BD, Burgos A, Pant P, Cuadros-Inostroza A, Willmitzer L, Scheible WR (2015) The transcription factor PHR1 regulates lipid remodeling and triacylglycerol accumulation in Arabidopsis thaliana during phosphorus starvation. J Exp Bot 66:1907–1918

Péret B, Desnos T, Jost R, Kanno S, Berkowitz O, Nussaume L (2014) Root architecture responses: in search of phosphate. Plant Physiol 166:1713–1723

Pilu R, Landoni M, Cassani E, Doria E, Nielsen E (2005) The maize *lpa241* mutation causes a remarkable variability of expression and some pleiotropic effects. Crop Sci 45:2096–2105

Plaxton WC, Tran HT (2011) Metabolic adaptations of phosphate starved plants. Plant Physiol 156:1006–1015

Raboy V (2001) Seeds for a better future: 'low phytate' grains help to overcome malnutrition and reduce pollution. Trends Plant Sci 6:458–462

Raboy V (2002) Progress in breeding low phytate crops. J Nutr 132:503S–505S

Raboy V, Gerbasi PF, Young KA, Stoneberg SD, Pickett SG, Bauman AT, Murthy PP, Sheridan WF, Ertl DS (2000) Origin and seed phenotype of maize low phytic acid 1-1 and low phytic acid 2-1. Plant Physiol 124:355–368

Raghothama K (1999) Phosphate acquisition. Annu Rev Plant Physiol Plant Mol Biol 50:665–693

Rai V, Sanagala R, Sinilal B, Yadav S, Sarkar AK, Dantu PK, Jain A (2015) Iron availability affects phosphate deficiency-mediated responses, and evidences of cross talk with auxin and zinc in arabidopsis. Plant Cell Physiol 56:1107–1123

Ram S, Gupta MP (1988) Effect of nitrogen, phosphorus and potassium on the population of insect pests of fodder mustard (*Brassica campestris* L.) and its seed yield in India. Trop Pest Manage 34:435–437

Ramamoorthy P, Lakshmanan K, Upadhyaya HD, Vadez V, Varshney RK (2016) Shoot traits and their relevance in terminal drought tolerance of chickpea (*Cicer arietinum* L.). Field Crop Res 197:10–27

Rasool S, Abdel Latef AA, Ahmad P (2015) Chickpea. In: Azooz MM, Ahmad P (eds) Legumes under environmental stress. Wiley, Chichester

Ravichandran S, Stone S, Benkel B, Zhang J, Berrue F, Prithiviraj B (2015) Optimal level of purple acid phosphatase5 is required for maintaining complete resistance to *Pseudomonas syringae*. Front Plant Sci 6:568

Reddy TY, Reddy GHS (2005) Principles of agronomy. Kalyani Publishers, New Delhi

Reddy CS, Kim KM, James D, Varakumar P, Reddy MK (2017) PgPAP18, a heat-inducible novel purple acid phosphatase 18-like gene (PgPAP18-like) from Pennisetum glaucum, may play a crucial role in environmental stress adaptation. Acta Physiol Plant 39:54

Ribet J, Drevon JJ (1995) Increase in permeability to oxygen and in oxygen uptake of soybean nodules under limiting phosphorus nutrition. Physiol Plant 94:298–304

Rodriguez D, Goudriaan J, Oyarzabal M, Pomar MC (1996) Phosphorus nutrition and water stress tolerance in wheat plants. J Plant Nutr 19:29–39

Rose TJ, Impa SM, Rose MT, Pariasca-Tanaka J, Mori A, Heuer S, Johnson-Beebout SE, Wissuwa M (2013) Enhancing phosphorus and zinc acquisition efficiency in rice: a critical review of root traits and their potential utility in rice breeding. Ann Bot 112:331–345

Rubio G, Hong L, Yan X, Lynch JP (2003) Topsoil foraging and its role in plant competitiveness for phosphorus in common bean. Crop Sci 43:598–607

Ryan J (1997) A global perspective on pigeonpea and chickpea sustainable production system: present status and future potential. In: Asthana A, Kanpur AM (eds) Recent advances in pulses research in India. Indian Society for Pulses Research and Development, Kanpur

Rasmussen SK, Josefsen L, Burhenne K, Sorensen MB (2003) Phytase from wheat, barley and rice belongs to the purple acid phosphatase family. In: Courtin C, Veraberbeke WS, Delcour JA (eds) Recent advances in enzymes in grain processing. Universiteit Leuven, Katholieke, pp 147–151

Sadji-Ait Kaci H, Chaker-Haddadj A, Aid F (2016) Interactive effects of salinity and two phosphorus fertilizers on growth and grain yield of *Cicer arietinum* L. Acta Agric Scand B Soil Plant Sci 67:208–216

Sadji-Ait Kaci H, Chaker- Haddadj A, Aid F (2018) Enhancing of symbiotic efficiency and salinity tolerance of chickpea by phosphorus supply. Acta Agric Scand B Soil Plant Sci 68:534–540

Sahi ST, Ghazanfar MU, Afzal M, Habib A, Ilyas MB (2007) Role of N, P and K contents in resistance against ascochyta blight of lentil (lens culinaris Medik). Pak J Bot 39:2175–2181

Sánchez-Calderón L, López-Bucio J, Chacón-López A, Cruz-Ramírez A, Nieto-Jacobo F, Dubrovsky JG, Herrera-Estrella L (2005) Phosphate starvation induces a determinate developmental program in the roots of *Arabidopsis thaliana*. Plant Cell Physiol 46:174–184

Sánchez-Rodriguez AR, Del Campillo MC, Torrent J, Jones DL (2014) Organic acids alleviate iron chlorosis in chickpea grown on two P-fertilized soils. J Soil Sci Plant Nutr 14:292–303

Sarker BC, Karmoker JL (2011) Effects of phosphorus deficiency on accumulation of biochemical compounds in lentil (Lens culinaris medik.). Bangladesh J Bot 40:23–27

Schulze J, Temple G, Temple SJ, Beschow H, Vance CP (2006) Nitrogen fixation by white lupin under phosphorus deficiency. Ann Bot 98:731

Shane MW, Lambers H (2005) Cluster roots: a curiosity in context. Plant Soil 274:101–125

Sharma HC, Gowda CLL, Stevenson PC, Ridsdill-Smith TJ, Clement SL, Ranga Rao GV, Romeis J, Miles M, El-Boushssini M (2007) Host plant resistance and insect pest management in chickpea. In: Yadav SS, Redden RJ, Chen W, Sharma B (eds) Chickpea breeding and management. CABI Publishers, Cambridge

Shi J, Wang H, Wu Y, Hazebroek J, Meeley RB, Ertl DS (2003) The maize low-phytic acid mutant *lpa2* is caused by mutation in an inositol phosphate kinase gene. Plant Physiol 131:507–515

Shi J, Wang H, Schellin K, Li B, Faller M, Stoop JM, Meeley RB, Ertl DS, Ranch JP, Glassman K (2007) Embryo-specific silencing of a transporter reduces phytic acid content of maize and soybean seeds. Nat Biotechnol 25:930

Shukla UC, Yadav OP (1982) Effect of phosphorus and zinc on nodulation and nitrogen fixation in chickpea (*Cicer arietinum* L.). Plant Soil 65:239

Shukla T, Kumar S, Khare R, Tripathi RD, Trivedi PK (2015) Natural variations in expression of regulatory and detoxification related genes under limiting phosphate and arsenate stress in *Arabidopsis thaliana*. Front Plant Sci 6:898

Siddique KH, Johansen C, Turner NC, Jeuffroy MH, Hashem A, Sakar D, Gan Y, Alghamdi SS (2012) Innovations in agronomy for food legumes.A review. Agron Sustain Dev 32:45–64

Singh MV (2008) Micronutrient deficiencies in crops and soils in India. In: Alloway BJ (ed) Micronutrient deficiencies in global crop production. Springer, Dordrecht

Singh DK, Sale PW, Pallaghy CK, McKenzie BM (2000) Phosphorus concentrations in the leaves of defoliated white clover affect abscisic acid formation and transpiration in drying soil. New Phytol 146:249–259

Singh AP, Pandey BK, Deveshwar P, Narnoliya L, Parida SK, Giri J (2015) JAZ repressors: potential involvement in nutrients deficiency response in rice and chickpea. Front Plant Sci 6:975

Singh R, Pratap T, Singh D, Singh G, Singh AK (2018) Effect of phosphorus, Sulphur and biofertilizers on growth attributes and yield of chickpea (*Cicer arietinum* L.). J Pharmacognosy Phytochemistry 7:3871–3875

Srinivasarao C, Ganeshamurthy AN, Ali M, Venkateswarlu B (2006) Phosphorus and micronutrient nutrition of chickpea genotypes in a multi-nutrient-deficient typic ustochrept. J Plant Nutr 29:747–763

Stevenson-Paulik J, Bastidas RJ, Chiou ST, Frye RA, York JD (2005) Generation of phytate-free seeds in Arabidopsis through disruption of inositol polyphosphate kinases. Proc Natl Acad Sci U S A 102:12612–12617

Sulieman S, Tran LSP (2015) Phosphorus homeostasis in legume nodules as an adaptive strategy to phosphorus deficiency. Plant Sci 239:36–43

Sun F, Suen PK, Zhang Y, Liang C, Carrie C, Whelan J, Ward JL, Hawkins ND, Jiang L, Lim BL (2012) A dual-targeted purple acid phosphatase in *Arabidopsis thaliana* moderates carbon metabolism and its overexpression leads to faster plant growth and higher seed yield. New Phytol 194:206–219

Svistoonoff S, Creff A, Reymond M, Sigoillot-Claude C, Ricaud L, Blanchet A, Nussaume L, Desnos T (2007) Root tip contact with low-phosphate media reprograms plant root architecture. Nat Genet 39:792–796

Tan GZ, Das Bhowmik SS, Hoang TM, Karbaschi MR, Johnson AA, Williams B, Mundree SG (2017) Finger on the pulse: pumping iron into chickpea. Front Plant Sci 8:1755

Tanner EVJ, Sheldrake MWA, Turner BL (2016) Changes in soil carbon and nutrients following 6 years of litter removal and addition in a tropical semi-evergreen rain forest. Biogeosciences 13:6183–6190

Tariq A, Pan K, Olatunji OA, Graciano C, Li Z, Sun F, Zhang L, Wu X, Chen W, Song D, Huang D, Xue T, Zhang A (2018) Phosphorous fertilization alleviates drought effects on Alnus cremastogyne by regulating its antioxidant and osmotic potential. Sci Rep 8:5644

Theodorou ME, Plaxton WC (1993) Metabolic adaptations of plant respiration to nutritional phosphate deprivation. Plant Physiol 101:339–344

Tripathi RD, Srivastava S, Mishra S, Singh N, Tuli R, Gupta DK, Maathuis FJM (2007) Arsenic hazards: strategies for tolerance and remediation by plants. Trends Biotechnol 25:158–165

Tripathi LK, Thomas T, Kumar S (2013) Impact of nitrogen and phosphorus on growth and yield of chickpea (*Cicer arietinum* L.). Asian Journal of Soil Science 8:260–263

Vadez V, Rashmi M, Sindhu K, Muralidharan M, Pushpavalli R, Turner NC, Krishnamurthy L, Gaur PM, Colmer TD (2012) Large number of flowers and tertiary branches, and higher reproductive success increase yields under salt stress in chickpea. Eur J Agron 41:42–51

Valentine AJ, Benedito VA, Kang Y (2011) Legume nitrogen fixation and soil abiotic stress: from physiology to genome and beyond. In: Foyer CH, Zhang H (eds) Annual plant reviews, vol 42. Wiley-Blackwell, Oxford, UK, pp 207–248

Vance CP, Uhde-Stone C, Allan DL (2003) Phosphorus acquisition and use: critical adaptations by plants for securing a nonrenewable resource. New Phytol 157:423–447

Vats P, Banerjee UC (2004) Production studies and catalytic properties of phytases (myo-inositol hexakisphosphate phosphohydrolases): an overview. Enzym Microb Technol 35:3–14

Velez PA, Talano MA, Paisio CE, Agostini E, González PS (2016) Synergistic effect of chickpea plants and Mesorhizobium as a natural system for chromium phytoremediation. Environ Technol 38:2164–2172

Veneklaas EJ, Stevens J, Cawthray GR, Turner S, Grigg AM, Lambers H (2003) Chickpea and white lupin rhizosphere carboxylates vary with soil properties and enhance phosphorus uptake. Plant Soil 248:187–197

Wang C, Wei Q, Zhang K, Wang L, Liu F, Zhao L, Tan Y, Di C, Yan H, Yu J, Sun C, Chen WJ, Xu W, Su Z (2013) Down-regulation of OsSPX1 causes high sensitivity to cold and oxidative stresses in rice seedlings. PLoS One 8:e81849

Wang J, Si Z, Li F, Xiong X, Lei L, Xie F, Chen D, Li Y, Li Y (2015) A purple acid phosphatase plays a role in nodule formation and nitrogen fixation in *Astragalus sinicus*. Plant Mol Biol 88:515–529

Wei F, Fanella B, Guo L, Wang X (2016) Membrane glycerolipidome of soybean root hairs and its response to nitrogen and phosphate availability. Sci Rep 6:36172

Wongkaew A, Srinives P, Nakasathien S (2013) Isolation and characterization of purple acid phosphatase gene during seedling development in mungbean. Biol Plant 57:267–273

Wouterlood M, Cawthray GR, Turner S, Lambers H, Veneklaas EJ (2004) Rhizosphere carboxylate concentrations of chickpea are affected by genotype and soil type. Plant Soil 261:1–10

Wouterlood M, Lambers H, Veneklaas EJ (2005) Plant phosphorus status has a limited influence on the concentration of phosphorus-mobilising carboxylates in the rhizosphere of chickpea. Funct Plant Biol 32:153–159

Xiao K, Harrison MJ, Wang ZY (2005) Transgenic expression of a novel *M. truncatula* phytase gene results in improved acquisition of organic phosphorus by *Arabidopsis*. Planta 222:27–36

Xiao K, Harrison M, Wang ZY (2006) Cloning and characterization of a novel purple acid phosphatase gene (MtPAP1) from Medicago truncatula Barrel Medic. J Integr Plant Biol 48:204–211

Yahiya M, Samiullah, Fatma A (1995) Influence of phosphorus on nitrogen fixation in chickpea cultivars. J Plant Nutr 18:719–727

Yamaji N, Takemoto Y, Miyaji T, Mitani-Ueno N, Yoshida KT, Ma JF (2016) Reducing phosphorus accumulation in rice grains with an impaired transporter in the node. Nature 541:92–95

Yang H, Knapp J, Koirala P, Rajagopal D, Peer WA, Silbart LK, Murphy A, Gaxiola RA (2007) Enhanced phosphorus nutrition in monocots and dicots over-expressing a phosphorus-responsive type I H^+-pyrophosphatase. Plant Biotechnol J 5:735–745

Yergeau E, Sommerville DW, Maheux E, Vujanovic V, Hamel C, Whalen JK, St-Arnaud M (2006) Relationships between fusarium population structure, soil nutrient status and disease incidence in field-grown asparagus. FEMS Microbiol Ecol 58:394–403

Zhao L, Liu F, Xu W, Di C, Zhou S, Xue Y, Yu J, Su Z (2009) Increased expression of OsSPX1 enhances cold/subfreezing tolerance in tobacco and *Arabidopsis thaliana*. Plant Biotechnol J 7:550–561

Zhao FJ, McGrath SP, Meharg AA (2010) Arsenic as a food chain contaminant: mechanisms of plant uptake and metabolism and mitigation strategies. Annu Rev Plant Biol 61:535–559

Zhao H, Sun R, Albrecht U, Padmanabhan C, Wang A, Coffey MD, Girke T, Wang Z, Close TJ, Roose M, Yokomi RK, Folimonova S, Vidalakis G, Rouse R, Bowman KD, Jin H (2013) Small RNA profiling reveals phosphorus deficiency as a contributing factor in symptom expression for citrus huanglongbing disease. Mol Plant 6:301–310

Zheng G, Tian B, Li W (2014) Membrane lipid remodelling of *Meconopsis racemosa* after its introduction into lowlands from an alpine environment. PLoS One 9:e106614

Zhu H, Qian W, Lu X, Li D, Liu X, Liu K, Wang D (2005) Expression patterns of purple acid phosphatase genes in Arabidopsis organs and functional analysis of *AtPAP23* predominantly transcribed in flower. Plant Mol Biol 59:581–594

Chapter 10
Genome Engineering Tools for Functional Genomics and Crop Improvement in Legumes

Rashi Khandelwal and Mukesh Jain

10.1 Introduction

Targeted genome editing with zinc-finger nuclease(s) (ZFNs), transcription activator-like effector nuclease(s) TALENs, and clustered regularly interspaced palindromic repeat (CRISPR)/CRISPR-associated 9 (Cas9) has revolutionized almost all biological research with their ability to specifically alter genes with site-specific modifications (Joung and Sander 2013; Zhang et al. 2013; Fichtner et al. 2014; Gaj 2014; Kumar and Jain 2015). All of these nucleases create double-stranded breaks (DSBs) at a specific position on the genome, and nonhomologous end joining (NHEJ) and homology-directed repair (HDR) mechanisms are used for repairing damaged DNA at those positions (Rodríguez-Leal et al. 2017).

Legumes represent an economically important family of flowering plants, which is the third largest family encompassing 19,000 known species (Graham and Vance 2003). Around 88% of them are involved in nodule formation in association with rhizobia (de Faria et al. 2010). Legumes are not much less to cereals in their importance to humans. Many legume species, such as *Arachis hypogaea* (peanut), *Medicago sativa* (alfalfa), *Glycine max* (soybean), and *Cicer arietinum* (chickpea), are grown as crop plants. Biological nitrogen fixation gives an edge to legumes over other crops because of their role in nitrogen-deprived soils and is of considerable ecological and agronomic interest. With the advancement of efficient and affordable genome sequencing technologies, a large number of legume genomes have been sequenced, and their various data are now publicly available. This information can be exploited with the use of site-specific nucleases for legume functional genomic studies and improvement.

R. Khandelwal · M. Jain (✉)
School of Computational and Integrative Sciences, Jawaharlal Nehru University, New Delhi, India
e-mail: mjain@jnu.ac.in

S. H. Wani, M. Jain (eds.), *Pulse Improvement*,
https://doi.org/10.1007/978-3-030-01743-9_10

10.2 Functional Studies in Legumes

Functional genomics is a branch of molecular biology that helps in gene discoveries and their functional analysis, including noncoding transcripts and other regulatory factors. The availability of complete genome sequence has accelerated the functional genomic studies in legumes (Dash et al. 2016). Functional genomic studies help in discovery of key target genes via uncovering the differential expression profiles of genes in various tissue/cell types and conditions at different developmental stages (Garg and Jain 2013; Pelletier et al. 2017; Singh et al. 2017). Mutational analyses have played an important role in functional genomics of legumes (Kang et al. 2016). Insertional mutagenesis, RNAi (RNA interference), fast neutron mutagenesis, and *Tnt1* retrotransposon mutagenesis are a few genetic tools, which play a central role in functional genomics analyses. For these technologies to realize their full potential, reliable and resourceful databases, like Legume Information System (LIS), LegumeIP, SoyBase, LegumeBase, MedicagoGenome, Chickpea Transcriptome Database (CTDB), KnowPulse, Lotus Base, etc., are extremely important (Garg et al. 2011; Li et al. 2015; Verma et al. 2015; Dash et al. 2016). Identification and assessment of networks and pathways responsible for improved protein and oil composition, and high yield and understanding links to major crop disease resistance are critical, which can be achieved through functional genomics studies. In soybean, fast neutron and X-ray exposure have been used for inducing mutations to screen for desired seed fatty acid composition and understand nodulation mechanism (Men et al. 2002; Anai et al. 2005). Transcriptome studies have helped in understanding the development of seed/pod development (Severin et al. 2010; Verdier et al. 2013a, b; Garg et al. 2017). High-quality soybean research is growing enormously with the help of functional genomics tools, such as microarrays, real-time PCR, yeast-two hybrid system, ^{1}H NMR, and RNA-seq (Brown and Hudson 2017; Copley et al. 2017; Kim et al. 2017a). The information derived from these studies makes it an important system for progressive science and crop improvement. Nodule-transcription profiling studies have been done to discover the genes responsible for nodule-rhizobia interaction. Soybean, apart from nitrogen symbiosis, has several unique biological features that make it an interesting system for research, which include paleopolyploid genome and well-characterized biotic and abiotic stress tolerance genes (Schmutz et al. 2010). The small genome, diploidy, autogamy, and short generation time make *M. truncatula* and *L. japonicus* also as models for legume research (Barker et al. 1990; Handberg and Stougaard 1992). *Tnt1* tagged lines have been used for comparative and functional genomics research in *M. truncatula* and *L. japonicus* (Revalska et al. 2011). In *M. truncatula*, many functional studies have been done on seed coat composition and flavonoid accumulation to study the effect of tannins in the defense mechanism and floral color development (Dixon et al. 2013). Other functional genomics studies in *M. truncatula* were attempted on desiccation tolerance in seeds (Kang et al. 2016). However, much remains to be discovered till now. For rapid expansion of knowledge and understanding of biochemical, molecular, and genetic pathways, there is a great need for advanced genome-editing tools with high precision and efficacy, such as ZFNs, TALENs, and CRISPR/Cas9, which can expedite legume research.

10.3 Importance of Genome-Editing Technologies

To understand regulatory networks controlling desirable plant traits and important agronomic characteristics, the genome resources available for legumes along with new technologies for site-specific modifications can be used (Curtin et al. 2015; Khatodia et al. 2016; Nithin et al. 2017). Genome-editing technologies, such as TALENs, ZFNs, and CRISPR/Cas9, prove to be powerful tools in understanding functional aspects of genes in legumes (Abdallah et al. 2015; Arora and Narula 2017). Several studies have used these technologies to demonstrate their efficiency in different legumes and/or functional characterization of important genes of interest (Table 10.1). These high-throughput technologies are able to alter specific genes without distorting the genetic background and produce non-transgenic mutant plants in subsequent generations (Zhang et al. 2017). It will be possible to obtain a wide variety of mutated alleles for any desired gene of interest with scientific improvisation of these tools.

ZFN is composed of two parts: (a) a DNA-binding component consisting of 4–6 zinc-finger proteins (ZFPs) and (b) a nuclease component from Fok1 restriction endonuclease, isolated from *Flavobacterium okeanokoites*. In each zinc-finger motif (ZFM), Zn^{2+} ion stabilizes the conserved Cys_2His_2 residues of folded 30 amino acids, and each ZFM interacts with three base pairs. Fok1 requires Mg^{2+} for carrying out non-specific DNA-cleavage activity (Li et al. 1992). A 6 bp separation between the two DNA-binding sites of Fok1 gives the most efficient cleavage (Bibikova et al. 2001). Subsequently, the resulting DSBs trigger DNA repair through either HR or NHEJ mechanism (Jasin and Rothstein 2013; Carroll and Beumer 2014; Davies et al. 2017; Pannunzio et al. 2017).

TALENs represent a pair of TALEs which are fused to the catalytic domain of Fok1. Two TALENs are targeted to bind to the opposite strands at any desired genomic location (Joung and Sander 2013). DNA cleavage occurs upon binding of both TALENs and interaction of the Fok1 domains that leads to DSB. They follow the same basic mechanism as ZFNs, wherein Fok1 dimer controls the DNA cleavage. The DNA-binding specificity of TALENs is based upon designer TALEs, which follow the DNA recognition code of natural TALEs from *Xanthomonas*. The central repeat domain in the TALE mediates target site recognition, and each repeat binds specifically to a single nucleotide (Deng et al. 2012; Mak et al. 2012). Rearrangement of these repeats allows the assembly of an engineered TALEN to recognize a specific target DNA sequence.

CRISPR consists of two parts, a guide RNA (sgRNA) and a nuclear-localized Cas 9 protein. Cas9, derived from *Streptococcus pyogenes*, is a large protein which consists of two nuclease domains. The first 20 nucleotides of sgRNA are the targeting

Table 10.1 Genome-editing studies in legumes

Legume	Method used	Gene name	Phenotype/result observed	Reference
Glycine max	ZFN	*RNA-dependent RNA polymerase 6a* (*RDR6a*), *RNA-dependent RNA polymerase 6b* (*RDR6b*)	Two ZFNs recognized specific target copy and distinguished 2 bp difference in both the genes	Curtin et al. (2011)
Glycine max	ZFN	*Dicer-like 4a* (*DCL4a*)	T_0 generation exhibited large bulbous internodes, mostly underdeveloped and aborted seeds. Two viable seeds showed no inheritability of mutation	Curtin et al. (2011)
Glycine max	ZFN	*Dicer-like 4b* (*DCL-4b*)	2 bp insertion, increased lateral shoot growth	Curtin et al. (2011)
Glycine max	TALEN	*Fatty acid desaturase 2-1A* (*FAD2-1A*), *fatty-acid desaturase 2-1B* (*FAD2-1B*)	Increase in oleic acid content, decrease in linolenic acid content	Haun et al. (2014)
Glycine max	CRISPR/Cas9	*Glutamine synthase* (*GS1*)	2 bp deletion	Michno et al. (2015)
Glycine max	CRISPR/Cas9	*Chalcone-flavanone isomerase* (*CHI20*)	1–32 bp deletions	Michno et al. (2015)
Glycine max (*GFP* line)	CRISPR/Cas9	*Green fluorescent protein* (*GFP*)	Loss of fluorescence	Jacobs et al. (2015)
Glycine max	CRISPR/Cas9	*Glucosyltransferase* (*Glyma07g14530*)	Indels, SNPs, and replacements	Jacobs et al. (2015)
Glycine max	CRISPR/Cas9	*Glyma01g38150*, *Glyma11g07220*(*Arabidopsis thaliana DDM1 orthologs*)	Indel frequency of 21% for chr1 and 8.9% for chr11	Jacobs et al. (2015)
Glycine max	CRISPR/Cas9	*Glyma04g36150*, *Glyma06g18790* (*Arabidopsis thaliana MET1 orthologs*)	>97% indel frequency	Jacobs et al. (2015)
Glycine max	CRISPR/Cas9	*miR1514, miR1509*	>95% indel frequency in all 4 miR1514 and 3 out of 4 miR1509	Jacobs et al. (2015)
Medicago truncatula (*GUS* line)	CRISPR/Cas9	*β-glucuronidase* (*GUS*)	Loss of GUS staining	Michno et al. (2015)

(continued)

Table 10.1 (continued)

Legume	Method used	Gene name	Phenotype/result observed	Reference
Glycine max	ZFN	*Dicer-like 1a* (*DCL1a*), *Dicer-like 1b* (*DCL1b*)	Reduced seed size, aborted seedling development, defective miRNA precursor transcript processing efficiency, deregulated miRNA gene expression	Curtin et al. (2015)
Glycine max (*FAD2-1A*, *FAD2-1B*) knockout line	TALEN	*Fatty acid desaturase 3A* (*FAD3A*)	Further decrease in linolenic acid content	Demorest et al. (2016)
Glycine max	TALEN	*Glycine max phytoene desaturase 11 (GmPDS11), Glycine max phytoene desaturase 18 (GmPDS18)*	17.5–21.1% single targeting efficiency, 6.25% mutation frequency for simultaneous editing. Albino and dwarf phenotype	Du et al. (2016)
Glycine max	CRSIPR/Cas9	*Glycine max phytoene desaturase 11 (GmPDS11), Glycine max phytoene desaturase 18 (GmPDS18)*	11.7–18.1% single targeting efficiency with AtU6–26 promoter, double-mutation frequency of 12.5% and 43.4–48.1% with AtU6–26 and GmU6-16 g-1 promoter, respectively. Albino and dwarf phenotype	Du et al. (2016)
Lotus japonicus	CRISPR/Cas9	*Lotus japonicus leghemoglobin 1* (*Ljlb1*), *Lotus japonicus leghemoglobin 2*(*Ljlb2*), *Lotus japonicus leghemoglobin 3* (*Ljlb3*)	Change of nodule color from pink to white, small nodules, resulting in reduced symbiotic nitrogen fixation efficiency, chlorosis of leaves	Wang et al. (2016)
Glycine max	CRISPR/Cas9	*Glycine max dicer-like3a* (*GmDcl3a*)	Failed transmission of mutation to T_1 generation	Curtin et al. (2017)
Glycine max	CRISPR/Cas9	*Glycine max double-stranded RNA-binding2a* (*GmDrb2a*), *Glycine max double-stranded RNA-binding2b* (*GmDrb2b*)	Sensitive to drought stress, dark green coloration	Curtin et al. (2017)

(continued)

Table 10.1 (continued)

Legume	Method used	Gene name	Phenotype/result observed	Reference
Glycine max	TALEN	*Glycine max dicer-like 1a* (*GmDcl1a*), *Glycine max dicer-like 3a* (*GmDcl3a*)	Herbicide resistant T_0 plants, screening revealed no mutation	Curtin et al. (2017)
Glycine max	TALEN	*Glycine max dicer-like2b* (*GmDcl2b*)	Herbicide resistant T_0 plants, T_1 generation sequencing confirmed segregation of 3 bp deletion with six heterozygous, four homozygous, and two wild-type plants, including two transgene-free heterozygous lines	Curtin et al. (2017)
Medicago truncatula	CRISPR/Cas9	*Medicago truncatula Hua enhancer1* (*MtHen1*)	Seed mottling upon imbibition	Curtin et al. (2017)
Arachis hypogea	TALEN	*Fatty acid desaturase 2A* (*FAD2A*), *fatty acid desaturase 2B* (*FAD2B*)	Increase in oleic acid content, decrease in linolenic acid content	Wen et al. (2018)

site, and a hairpin structure at the 3′-end interacts with the Cas9 protein. Cas9 and sgRNA interact to identify complementary DNA sequence to the sgRNA and generate a DNA DSB. For repair of DSB, either nonhomologous end joining (NHEJ) or homologous repair takes place, which results in insertions or deletions and can be used to modify gene expression (activate or silence) by altering promoter sequences, epigenetic modification (Hu et al. 2014), or microscopic visualization of specific genomic loci (Chen et al. 2013). CRISPR/Cas9 has come out as a focal platform for this application because of its high efficiency and comparatively easier construct designing approach.

10.4 Genome Editing in Legumes via Zinc-Finger Nucleases

Modular assembly platforms, oligomerized pool engineering (OPEN), and context-dependent assembly (CoDA) are generally used for designing ZFNs (Wright et al. 2006; Maeder et al. 2009; Joung et al. 2010; Kim et al. 2010; Sander et al. 2011). CoDA is a rapid and easy platform developed by the Zinc Finger Consortium most recently. A previously characterized ZFN that targets *green fluorescent protein (GFP)* in *Drosophila*, *C. elegans*, plants, and humans (Maeder et al. 2008) was introduced into soybean (harboring homozygous *GFP* transgene) by *Agrobacterium rhizogenes* hairy root transformation method, and conditions for targeted mutagenesis by ZFNs in soybean were optimized by Curtin et al. (2011). Mutational analysis in the *GFP* coding region showed deletions ranging from 27 to 71 bp in 5 of 13 clones sequenced. It proved that hairy root expression system is an excellent way to screen ZFN mutagenesis in soybean in vivo. Two ZFNs that independently target *RNA-dependent RNA polymerase 6* (*RDR6*) homeologs, *RDR6a* and *RDR6b*, with only 2 bp difference were constructed to check whether CoDA ZFNs can discriminate between them or not (Table 10.1). The results indicated that ZFNs were able to discriminate the two homeologs, as the targeted copy was mutagenized at a much higher frequency than the other copy (Curtin et al. 2011). Heritable targeted mutagenesis of soybean *Dicer-like 4*(*dcl4*) gene was achieved via whole-plant transformation with ZFN targeting both paralogous copies of the *DCL4* gene (*DCL4a* and *DCL4b*) into soybean. The double mutants (*dcl4a/dcl4a/dcl4b/dcl4b*) were also obtained along with reactivation of expression of the ZFNs in a *dcl4a/dcl4b* T_1 line. For this, Curtin et al. (2011) used ZFN genome which provides information on ZFN target sites, chromosomal location, relative position to transcription initiation sites(s), and even frequency of occurrence within the genome. For soybean, ZFN genome tool predicted that 36,714 of 55,582 (~66%) protein encoding transcripts can be targeted by CoDA ZFNs.

ZFNs have also been used for characterizing the role of genes associated with miRNA pathway (Curtin et al. 2015). miRNAs play a key role in gene expression regulation via the RNA-silencing-mediated mechanism (Catalanotto et al. 2016). *DICER-LIKE 1*(*DCL1*), a ribonuclease III enzyme, functions as a central machinery to process hairpin-like precursor transcripts into mature RNAs (Comella et al. 2008). To confirm the role of *DCL1* homologues *DCL1a* and *DCL1b* in miRNA pathway, Curtin et al. (2015) employed ZFNs and created single and double mutants. *DCL1* homologues not only showed similarity at nucleotide level but also demonstrated indistinguishable expression pattern. This demonstrated that if a mutation is induced in any one of the homologous genes, miRNA biogenesis remains predominantly unaffected (Curtin et al. 2015). Single mutants exhibited normal morphological and molecular phenotype, whereas double mutants exhibited abnormal phenotype displaying aborted or reduced seed size as well as vast reduction in mature miRNAs. This data demonstrated that *DCL1a* and *DCL1b* play similar and very important role in miRNA pathway to regulate plant development.

10.5 TALEN-Mediated Mutagenesis in Legumes

TALENs have been used for precise gene editing in diverse organisms. However, TALENs have not been much studied in allopolyploid plants. Recently, Wen et al. (2018) intended to create targeted mutagenesis in peanut by TALENs. Composition of oleic acids and linolenic acids in peanut seed is approximately 36–67% and 15–43%, respectively (Pandey et al. 2014). *Fatty acid desaturase 2* (*FAD2*) is one of the important enzymes in peanut that catalyzes the conversion of oleic acid (monounsaturated fatty acid) to linolenic acid (polyunsaturated fatty acid) by removing two hydrogen atoms and creates a carbon/carbon double bond (Okuley et al. 1994). Since linolenic acid is a polyunsaturated fatty acid, it has more chances of oxidation, leading to faster putrefaction and hence shorter shelf life (Akhtar et al. 2014). In normal peanut varieties, where oleic acid and linolenic acid are in regular concentration, *FAD2A* and *FAD2B* homologous genes are expressed as usual. But, it has been demonstrated in a study that *FAD2B* expression is considerably reduced in high-oleate mutants (Chi et al. 2011). Therefore, considering these facts and knowing the harmful effects of increased linoleic acids, Chi et al. (2011) tried to increase the levels of oleic acid by suppressing the expression of *FAD2A* and *FAD2B* genes in peanut lines. To simultaneously suppress the expression of homologous genes, *A. hypogaea fatty acid desaturase 2*, *AhFAD2A*, and *ahFAD2B*, sequence-specific TALENs were designed. This was done to fulfill the vision of improving peanut lines. Fast TALETM TALEN Assembly Kit developed by SIDANSAI Biotechnology, (Shanghai) Co., Ltd was used to construct TALEN repeat arrays. *Agrobacterium rhizogenes*-mediated hairy root transformation of TALENs showed 8.33–12.38% mutagenesis frequency. Stable genetic transformation was obtained by *Agrobacterium*-mediated transformation via direct embryogenesis. Genetically stable *FAD2* mutant lines for both homologous genes were obtained in the T1 generation. The real-time PCR results showed that the relative expression of *FAD2* in the TALEN-induced mutant lines is significantly reduced as compared with the wild-type plants. In another study, TALENs were used to target *FAD3A* in the existing *FAD2-1A* and *FAD2-1B* double knockout *G. max* lines (Demorest et al. 2016). TALENs have also been used to target *phytoene desaturase* (*PDS*) genes in soybean, *GmPDS11* and *GmPDS18* (Du et al. 2016). Three TALEN pairs were designed, D1 that targeted both the genes and S1 and S2 that targeted *GmPDS11* at exons 4 and 5, respectively. *A. rhizogenes* hairy root-based transformation resulted in mutagenesis frequency in range of 17.5–21.1%. Mutation frequency of 6.25% with TALENS-D1 suggested that simultaneous mutation of two homeoalleles is possible (Du et al. 2016).

10.6 Targeted Genome Editing in Legumes Using CRISPR/Cas9

Among legumes, CRISPR/Cas9 was initially utilized for soybean genome modification via hairy root transformation mediated by *A. rhizogenes*. Jacobs et al. (2015) used CRISPR/Cas9 system in soybean to knock out a *GFP* transgene. CRISPR/Cas9 system-based knockout of *GFP* in soybean was observed by the loss of fluorescence. Loss of fluorescence in 15 out of 17 5′-target events and four of the 22 3′-target events showed knockout based on CRISPR/Cas9 assembly. Since complete fluorescence was lost, the result indicated that both *GFP* alleles were modified. Also, CRISPR/Cas9 was used to target a single-copy soybean gene, *Glucosyltransferase* (*Glyma07g14530*), and mutations in the form of indels, SNPs, and replacements were obtained. Jacobs et al. (2015) utilized CRISPR/Cas9 system to target soybean homologous gene pair (Glyma01g38150 and Glyma11g07220) singly and simultaneously. Three sgRNAs were designed: 01GDDM1 to target Glyma01g38150, 11GDDM1 to target Glyma11g07220, and 01 g + 11Gddm1 to target both genes. Single-targeting sgRNAs generated indel frequencies of more than 70%. Double targeting sgRNA (01 g +11Gddm1) also generated mutations with indel frequency of about 21% in chr1 and 8.9% in chr11. Targeting Glyma04g36150 and Glyma06g18790 gene pair, which are orthologs of *Arabidopsis thaliana MET1* (*methyltransferase 1*) gene, showed greater than 97% indel frequency (Jacobs et al. 2015). In addition, two miRNAs, miR1514 and miR1509, were targeted via CRISPR/Cas9, and >95% indel frequency in all the four miR1509 and three out of four miR1514 was observed (Jacobs et al. 2015). Though transformed hairy root serves as a perfect transgenic model system for soybean, it could not produce the whole plant, and hence heritable mutations could not be obtained. To introduce CRISPR-based mutations in whole plants, Cas9 constructs were transformed via biolistic as well as *Agrobacterium*-mediated methods in somatic embryos. The continued activity of Cas9 in somatic embryos generated additional genetic alterations. Seven out of nine targeting vectors showed indel frequency greater than 70%, which is around ten times higher than obtained with TALENs in soybean hairy roots (Haun et al. 2014). Both stably transformed hairy roots and somatic embryos demonstrated the functionality of the CRISPR/Cas9 system. These results illustrated that CRISPR/Cas9 is an efficient way for genome editing in soybean. Hence, CRISPR/Cas9 with its editing quality can develop novel phenotypes in plants which are of agronomic importance (Table 10.1).

Michno et al. (2015) exploited the CRISPR/Cas9 design website and *G. max* codon-optimized Cas9 (GmCas9) protein to create somatic mutations in both *G. max* and *M. truncatula*. Their web server (http://stuparcrispr.cfans.umm.edu/CRISPR) quickly identifies CRISPR/Cas9 target sites using either given DNA sequence or a soybean gene model (soybean gene identifier) as input. To target *Glutamine synthase-GS1*–gene identifier: Glyma18g04660, MDC32/GUS/GmCas9 destination vector was designed, which incorporated GmCas9 and a reporter gene (*GUS*). To rapidly screen for the effectiveness of the expression vector, target sequence in the vector was transformed in soybean hairy roots. The staining of transformed hairy roots with 5-bromo-4-chloro-3-indolyl glucuronide was confirmed. In addition, CAPS assay, using *Bsm*A1 restriction enzyme on each of the three root samples (top, middle, and bottom), revealed the presence of mutation throughout the root. Thus, the results demonstrated CRISPR/GmCas9-induced mutations in *GS1* gene but did not prove that the GmCas9 is necessarily better than other Cas9 for targeted genome editing.

Curtin et al. (2017) reported the generation of targeted mutations in loci central to small RNA pathways in *M. truncatula* and *G. max* using ZFN, TALEN, and CRISPR/Cas9 platforms. They found that CRISPR/Cas9 system was able to make double mutants for *double-stranded RNA-binding2* paralog pair (*GmDrb2a-GmDrb2b*) in T_0 generation which was found to be transmissible in the T_1 progeny. A CRISPR/Cas9 system with Gmubi-promoter was used to target *Dicer-like 3a*

(*GmDcl3a*) gene in soybean. Its transformation resulted into two putative T_0 mutant plants, but it was not transmitted to T_1 generation (Curtin et al. 2017).

To generate mutations in *Hua enhancer 1(HEN1)* gene, two *Hen1a* and *Hen1b* homologue copies in soybean and single *Hen1* copy in *M. truncatula* were targeted. In soybean, one plant appeared to carry mutations in both *Hen1* genes but failed to survive in tissue culture. A second plant was recovered in the T_1 generation but showed seed mottling phenotype, and these seeds failed to germinate completely (Curtin et al. 2017). The evaluation of *M. truncatula* T_1 generation plants showed three mutations each with transmittable in-frame mutations that seem to retain gene function (no disruption). But, none of the T_1 plants displayed phenotypes that are seen in *hen1* mutation of *Arabidopsis*. Cumulatively, it was concluded that a heterozygous frame-shift knockout of *hen1* might be lethal for embryos in legumes.

With CRISPR/Cas9, large deletions were created in *leghemoglobin* (*Lglb1*, *Lglb2*, and *Lglb3*) genes of *L. japonicus.* Stably transformed plants showed white nodules instead of pink leghemoglobin nodules, resulting in reduced symbiotic nitrogen fixation (Wang et al. 2016). A study conducted by Du et al. (2016) used CRISPR/Cas9 for targeting two soybean genes, *GmPDS11* and *GmPDS18*. They designed four sgRNAs (D7, S11, S12, and S13) for it. D7 was used to target both genes at exon 2 and obtained double mutation frequency of 12.5%. S11 and S12 were used to target *GmPDS11* at exons 4 and 6, respectively, and obtained mutation frequency of 15.3% and 13.8%, respectively. S13 targeted *GMPDS18* at exon 5, resulting in mutation frequency of 11.75%. With AtU6–26 promoter, it showed mutagenesis frequency between 11.7% and 18.1%, and with GmU6-16g-1 promoter, 43.5%–48.1% mutagenesis frequency was observed (Du et al. 2016).

10.7 Limitations of Genome-Editing Technologies

Despite being efficient, reliable, and advantageous, genome-editing tools exhibit certain limitations. Assembly of zinc-finger domains, such that it efficiently binds to nucleotide arrays with high affinity, is difficult (Ramirez et al. 2008). Number of off-targets associated with ZFN-based editing is very high. Large size of TALENs makes them a poor gene-editing tool (Gupta and Musunuru 2014). In addition, construction of TALE-binding domains is time-consuming and laborious (Nemudryi et al. 2014). Despite being an important tool in genome engineering, non-specific binding can cause problems in the CRISPR/Cas9 system, also allowing Cas9 to cleave off-target sequences. Even if there are mutations in the seed or PAM sequence, Cas9 in CRISPR/Cas9 system is able to identify and cleave sequences (Fu et al. 2013; Mali et al. 2013; Pattanayak et al. 2013). CRISPR-Cpf1 showed promising results in diminishing off-target effects (Kim et al. 2017b). One of the major limitations of CRISPR/Cas9 system is that only 18–20 nt region of sgRNA represents spacer sequence which can lead to off-targets. Thus, deletions and insertions are only limited to 18–20 nucleotides. More concentration of sgRNA/Cas9 complexes results in off-target cuts at higher rate than low concentration of the complexes (Fujii et al. 2013; Hsu et al. 2013; Pattanayak et al. 2013). Specificity is not increased through a mere extension of sgRNA (Ran et al. 2013). By employing two separate but specific sgRNAs and only one Cas9 can increase specificity by 50–1500-fold as compared to single sgRNA usage (Ran et al. 2013). Sequence-based targeting and analysis of off-target effects cannot be done efficiently, when there is a lack of complete sequence data in any legume species. Further, transformation and regeneration in legumes are quite challenging, time-consuming, and labor intensive (Atif et al. 1962), which are other major roadblocks in implementing genome editing in legume crops. It is very important to understand the pros and cons of each technique to execute efficient genome editing in legumes.

10.8 Conclusions and Future Perspectives

There has been an immense progress in crop improvement since past decade owing to the development of targeted genome engineering tools. The forte of genome editing in legumes has been radically changed by these tools. With specificity and accuracy of these advanced tools, it is possible to generate mutations, characterize the role of unknown genes, and regulate transcription of desired genes. In comparison to the conventional breeding technologies, less screening and backcrossing are required (Abdallah et al. 2015). Multiplexing is also a big advantage with CRISPR/Cas9, multiple gene mutations with the use of only one Cas9 and multiple sgRNAs can be achieved in legumes in the same way as in *Arabidopsis* (Xing et al. 2014). It will be a great boon to breeding technologies for legume crop improvement. Off-target effects of CRISPR/Cas9 is a limit but can be improved via development of wide range of PAM sequences and sgRNAs for RNA-guided nucleases (Karvelis et al. 2015). There is a recent ongoing work which is intended to identify better promoters for the expression of the Cas9 in legumes. TALENs and CRISPR/Cas9 genome engineering technologies are equally efficient, where CRISPR/Cas9 is the choice for most applications and TALENs have a variety of targeting range that might be advantageous for targeting a highly specific locus (Joung and Sander 2013). Agronomic traits, such as nutritional value, herbicide and pesticide resistance, as well as strength to survive in adverse environmental conditions, can be improved in legumes with these tools (Kamburova et al. 2017). Furthermore, accumulation of toxic metabolites can be prevented via genome engineering using these technologies. Epigenetic reprogramming can also be done by targeting the proteins involved in histone modifications and DNA methylation (Enríquez 2016). These genome-editing tools can also be potentially used to identify the regulatory proteins associated with DNA. Identification of genomic targets for improving crop performance and understanding of fundamental biology to create most advanced genome alterations are critical to realize the full potential of genome engineering tools.

Acknowledgments Work on CRISPR-Cas9-based genome editing in the author's laboratory is funded by the Science and Engineering Research Board (File no. EMR/2016/001311), Department of Science & Technology, Government of India, New Delhi.

References

Abdallah NA, Prakash CS, McHughen AG (2015) Genome editing for crop improvement: challenges and opportunities. GM Crops Food 6:183–205. https://doi.org/10.1080/21645698.2015.1129937

Akhtar S, Khalid N, Ahmed I, Shahzad A, Suleria HAR (2014) Physicochemical characteristics, functional properties, and nutritional benefits of peanut oil: a review. Crit Rev Food Sci Nutr 54:1562–1575. https://doi.org/10.1080/10408398.2011.644353

Anai T, Yamada T, Kinoshita T, Rahman SM, Takagi Y (2005) Identification of corresponding genes for three low-α-linolenic acid mutants and elucidation of their contribution to fatty acid biosynthesis in soybean seed. Plant Sci 168:1615–1623. https://doi.org/10.1016/j.plantsci.2005.02.016

Arora L, Narula A (2017) Gene editing and crop improvement using CRISPR-Cas9 system. Front Plant Sci 8:1932. https://doi.org/10.3389/fpls.2017.01932

Atif RM, Svabova L, Ondrej V, Klenoticova H, Jacas L, Griga M, Ochatt SJ (1962) Progress in botany. Nature 194:1023–1023. https://doi.org/10.1038/1941023a0

Barker DG, Bianchi S, Blondon F, Dattée Y, Duc G, Essad S, Flament P, Gallusci P, Génier G, Guy P, Muel X, Tourneur J, Dénarié J, Huguet T (1990) *Medicago truncatula*, a model plant for studying the molecular genetics of theRhizobium-legume symbiosis. Plant Mol Biol Report 8:40–49. https://doi.org/10.1007/BF02668879

Bibikova M, Carroll D, Segal DJ, Trautman JK, Smith J, Kim YG, Chandrasegaran S (2001) Stimulation of homologous recombination through targeted cleavage by chimeric nucleases. Mol Cell Biol 21:289–297. https://doi.org/10.1128/MCB.21.1.289-297.2001

Brown AV, Hudson KA (2017) Transcriptional profiling of mechanically and genetically sink-limited soybeans. Plant Cell Environ 40:2307–2318. https://doi.org/10.1111/pce.13030

Carroll D, Beumer KJ (2014) Genome engineering with TALENs and ZFNs: repair pathways and donor design. Methods 69:137–141. https://doi.org/10.1016/j.ymeth.2014.03.026

Catalanotto C, Cogoni C, Zardo G (2016) MicroRNA in control of gene expression: an overview of nuclear functions. Int J Mol Sci 17. https://doi.org/10.3390/ijms17101712

Chen B, Gilbert LA, Cimini BA, Schnitzbauer J, Zhang W, Li GW, Park J, Blackburn EH, Weissman JS, Qi LS, Huang B (2013) Dynamic imaging of genomic loci in living human cells by an optimized CRISPR/Cas system. Cell 155:1479–1491. https://doi.org/10.1016/j.cell.2013.12.001

Chi X, Yang Q, Pan L, Chen M, He Y, Yang Z, Yu S (2011) Isolation and characterization of *fatty acid desaturase* genes from peanut (*Arachis hypogaea L.*). Plant Cell Rep 30:1393–1404. https://doi.org/10.1007/s00299-011-1048-4

Comella P, Pontvianne F, Lahmy S, Vignols F, Barbezier N, DeBures A, Jobet E, Brugidou E, Echeverria M, Sáez-Vásquez J (2008) Characterization of a ribonuclease III-like protein required for cleavage of the pre-rRNA in the 3′ETS in Arabidopsis. Nucleic Acids Res 36:1163–1175. https://doi.org/10.1093/nar/gkm1130

Copley TR, Aliferis KA, Kliebenstein DJ, Jabaji SH (2017) An integrated RNAseq-1H NMR metabolomics approach to understand soybean primary metabolism regulation in response to Rhizoctonia foliar blight disease. BMC Plant Biol 17:1–18. https://doi.org/10.1186/s12870-017-1020-8

Curtin SJ, Zhang F, Sander JD, Haun WJ, Starker C, Baltes NJ, Reyon D, Dahlborg EJ, Goodwin MJ, Coffman AP, Dobbs D, Joung JK, Voytas DF, Stupar RM (2011) Targeted mutagenesis of duplicated genes in soybean with zinc-finger nucleases. Plant Physiol 156:466–473. https://doi.org/10.1104/pp.111.172981

Curtin SJ, Michno J-M, Campbell BW, Gil-Humanes J, Mathioni SM, Hammond R, Gutierrez-Gonzalez JJ, Donohue RC, Kantar MB, Eamens AL, Meyers BC, Voytas DF, Stupar RM (2015) MicroRNA maturation and microRNA target gene expression regulation are severely disrupted in soybean dicer-like1 double mutants. G3: Genes Genomes Genetics (Bethesda) 6:423–433. https://doi.org/10.1534/g3.115.022137

Curtin SJ, Xiong Y, Michno JM, Campbell BW, Stec AO, Čermák T, Starker C, Voytas DF, Eamens AL, Stupar RM (2017) CRISPR/Cas9 and TALENs generate heritable mutations for genes involved in small RNA processing of *Glycine max* and *Medicago truncatula*. Plant Biotechnol J 16:1125–1137. https://doi.org/10.1111/pbi.12857

Dash S, Campbell JD, Cannon EKS, Cleary AM, Huang W, Kalberer SR, Karingula V, Rice AG, Singh J, Umale PE, Weeks NT, Wilkey AP, Farmer AD, Cannon SB (2016) Legume information system (LegumeInfo.org): a key component of a set of federated data resources for the legume family. Nucleic Acids Res 44:D1181–D1188. https://doi.org/10.1093/nar/gkv1159

Davies JP, Kumar S, Sastry-Dent L (2017) Use of zinc-finger nucleases for crop improvement, 1st edn. Elsevier Inc. Dow AgroSciences, Indianapolis, IN, United States. https://doi.org/10.1016/bs.pmbts.2017.03.006

de Faria SM, Diedhiou AG, de Lima HC, Ribeiro RD, Galiana A, Castilho AF, Henriques JC (2010) Evaluating the nodulation status of leguminous species from the Amazonian forest of Brazil. J Exp Bot 61:3119–3127. https://doi.org/10.1093/jxb/erq142

Demorest ZL, Coffman A, Baltes NJ, Stoddard TJ, Clasen BM, Luo S, Retterath A, Yabandith A, Gamo ME, Bissen J, Mathis L, Voytas DF, Zhang F (2016) Direct stacking of sequence-specific nuclease-induced mutations to produce high oleic and low linolenic soybean oil. BMC Plant Biol 16:1–8. https://doi.org/10.1186/s12870-016-0906-1

Deng D, Yan C, Pan X, Mahfouz M, Wang J, Zhu J, Shi Y, Yan N (2012) Structural basis for sequence-specific recognition of DNA by TAL effectors. Science 335:720–723. https://doi.org/10.1126/science.1215670

Dixon RA, Liu C, Jun JH (2013) Metabolic engineering of anthocyanins and condensed tannins in plants. Curr Opin Biotechnol 24:329–335. https://doi.org/10.1016/j.copbio.2012.07.004

Du H, Zeng X, Zhao M, Cui X, Wang Q, Yang H, Cheng H, Yu D (2016) Efficient targeted mutagenesis in soybean by TALENs and CRISPR/Cas9. J Biotechnol 217:90–97. https://doi.org/10.1016/j.jbiotec.2015.11.005

Enríquez P (2016) CRISPR-mediated epigenome editing. Yale J Biol Med 89:471–486

Fichtner F, Urrea Castellanos R, Ülker B (2014) Precision genetic modifications: a new era in molecular biology and crop improvement. Planta 239:921–939. https://doi.org/10.1007/s00425-014-2029-y

Fu Y, Foden JA, Khayter C, Maeder ML, Reyon D, Joung JK, Sander JD (2013) High-frequency off-target mutagenesis induced by CRISPR-Cas nucleases in human cells. Nat Biotechnol 31:822–826. https://doi.org/10.1038/nbt.2623

Fujii W, Kawasaki K, Sugiura K, Naito K (2013) Efficient generation of large-scale genome-modified mice using gRNA and CAS9 endonuclease. Nucleic Acids Res 41:e187. https://doi.org/10.1093/nar/gkt772

Gaj T (2014) ZFN, TALEN and CRISPR/Cas based methods for genome engineering. Trends Biotechnol 31:397–405. https://doi.org/10.1016/j.tibtech.2013.04.004.ZFN

Garg R, Jain M (2013) Transcriptome analyses in legumes: a resource for functional genomics. Plant Genome 6. https://doi.org/10.3835/plantgenome2013.04.0011

Garg R, Patel RK, Jhanwar S, Priya P, Bhattacharjee A, Yadav G, Bhatia S, Chattopadhyay D, Tyagi AK, Jain M (2011) Gene discovery and tissue-specific transcriptome analysis in chickpea with massively parallel pyrosequencing and web resource development. Plant Physiol 156:1661–1678. https://doi.org/10.1104/pp.111.178616

Garg R, Singh VK, Rajkumar MS, Kumar V, Jain M (2017) Global transcriptome and coexpression network analyses reveal cultivar-specific molecular signatures associated with seed development and seed size/weight determination in chickpea. Plant J 91:1088–1107. https://doi.org/10.1111/tpj.13621

Graham PH, Vance CP (2003) Legumes: importance and constraints to greater use. Plant Physiol 131:872–877. https://doi.org/10.1104/pp.017004

Gupta RM, Musunuru K (2014) Expanding the genetic editing tool kit: ZFNs, TALENs, and CRISPR-Cas9. J Clin Invest 124:4154–4161. https://doi.org/10.1172/JCI72992

Handberg K, Stougaard J (1992) *Lotus japonicus*, an autogamous, diploid legume species for classical and molecular genetics. Plant J 2:487–496. https://doi.org/10.1111/j.1365-313X.1992.00487.x

Haun W, Coffman A, Clasen BM, Demorest ZL, Lowy A, Ray E, Retterath A, Stoddard T, Juillerat A, Cedrone F, Mathis L, Voytas DF, Zhang F (2014) Improved soybean oil quality by targeted mutagenesis of the *fatty acid desaturase 2* gene family. Plant Biotechnol J 12:934–940. https://doi.org/10.1111/pbi.12201

Hsu PD, Scott DA, Weinstein JA, Ran FA, Konermann S, Agarwala V, Li Y, Fine EJ, Wu X, Shalem O, Cradick TJ, Marraffini LA, Bao G, Zhang F (2013) DNA targeting specificity of RNA-guided Cas9 nucleases. Nat Biotechnol 31:827–832. https://doi.org/10.1038/nbt.2647

Hu J, Lei Y, Wong WK, Liu S, Lee KC, He X, You W, Zhou R, Guo JT, Chen X, Peng X, Sun H, Huang H, Zhao H, Feng B (2014) Direct activation of human and mouse *Oct4* genes using engineered TALE and Cas9 transcription factors. Nucleic Acids Res 42:4375–4390. https://doi.org/10.1093/nar/gku109

Jacobs TB, LaFayette PR, Schmitz RJ, Parrott WA (2015) Targeted genome modifications in soybean with CRISPR/Cas9. BMC Biotechnol 15:1–10. https://doi.org/10.1186/s12896-015-0131-2

Jasin M, Rothstein R (2013) Repair of strand breaks by homologous recombination. Cold Spring Harb Perspect Biol 5:1–18. https://doi.org/10.1101/cshperspect.a012740

Joung JK, Voytas DF, Cathomen T (2010) Reply to "Successful genome editing with modularly assembled zinc finger nucleases". Nat Methods 7(1):91–92. https://doi.org/10.1038/nmeth0210-91b

Joung JK, Sander JD (2013) TALENs: a widely applicable technology for targeted genome editing. Nat Rev Mol Cell Biol 14:49–55. https://doi.org/10.1038/nrm3486

Kamburova VS, Nikitina EV, Shermatov SE, Buriev ZT (2017) Genome editing in plants: an overview of tools and applications. Hindawi Int J Agron 2017:1. https://doi.org/10.1155/2017/7315351

Kang Y, Li M, Sinharoy S, Verdier J (2016) A snapshot of functional genetic studies in *Medicago truncatula*. Front Plant Sci 7:1175. https://doi.org/10.3389/fpls.2016.01175

Karvelis T, Gasiunas G, Young J, Bigelyte G, Silanskas A, Cigan M, Siksnys V (2015) Rapid characterization of CRISPR-Cas9 protospacer adjacent motif sequence elements. Genome Biol 16:1–13. https://doi.org/10.1186/s13059-015-0818-7

Khatodia S, Bhatotia K, Passricha N, Khurana SMP, Tuteja N (2016) The CRISPR/Cas genome-editing tool: application in improvement of crops. Front Plant Sci 7:506. https://doi.org/10.3389/fpls.2016.00506

Kim J-S, Lee HJ, Carroll D (2010) Genome editing with modularly assembled zinc-finger nucleases. Nat Methods 7:91 author reply 91–2. https://doi.org/10.1038/nmeth0210-91a

Kim E, Hwang S, Lee I (2017a) SoyNet: a database of co-functional networks for soybean *Glycine max*. Nucleic Acids Res 45:D1082–D1089. https://doi.org/10.1093/nar/gkw704

Kim H, Kim ST, Ryu J, Kang BC, Kim JS, Kim SG (2017b) CRISPR/Cpf1-mediated DNA-free plant genome editing. Nat Commun 8:1–7. https://doi.org/10.1038/ncomms14406

Kumar V, Jain M (2015) The CRISPR-Cas system for plant genome editing: advances and opportunities. J Exp Bot 66:47–57. https://doi.org/10.1093/jxb/eru429

Li L, Wu LP, Chandrasegaran S (1992) Functional domains in Fok I restriction endonuclease. Proc Natl Acad Sci 89:4275–4279. https://doi.org/10.1073/pnas.89.10.4275

Li J, Wei H, Liu T, Dai X, Zhao PX (2015) LegumeIP: an integrative platform for comparative genomics and transcriptomics of model legumes. Biol Nitrogen Fixat 2(2):807–816. https://doi.org/10.1002/9781119053095.ch80

Maeder ML, Thibodeau-beganny S, Osiak A, Wright DA, Anthony RM, Eichtinger M, Jiang T, Foley JE, Winfrey RJ, Townsend JA, Unger-wallace E, Sander JD, Fu F, Pearlberg J, Göbel C, Dassie JP, Porteus MH, Sgroi DC, Iafrate AJ, Dobbs D, Mc Cray PB Jr, Cathomen T, Voytas DF, Joung JK (2008) NIH public access. Mol Ther 31:294–301. https://doi.org/10.1016/j.molcel.2008.06.016.Rapid.

Maeder ML, Thibodeau-Beganny S, Sander JD, Voytas DF, Joung JK (2009) Oligomerized pool engineering (OPEN): an "open-source" protocol for making customized zinc-finger arrays. Nat Protoc 4:1471–1501. https://doi.org/10.1038/nprot.2009.98

Mak AN, Bradley P, Cernadas R a, Bogdanove AJ, Stoddard BL (2012) The crystal structure of TAL effector PthXo1 bound to its DNA target. Science 335:716–719. https://doi.org/10.1126/science.1216211

Mali P, Aach J, Stranges PB, Esvelt KM, Kosuri S, Yang L, Church GM (2013) CAS9 transcriptional activators for target specificity screening and paired nickases for cooperative genome engineering. Nat Biotechnol 31(9):833–838. https://doi.org/10.1038/nbt.2675

Men AE, Laniya TS, Searle IR, Iturbe-Ormaetxe I, Gresshoff I, Jiang Q, Carroll BJ, Gresshoff PM (2002) Fast neutron mutagenesis of soybean (*Glycine soja L.*) produces a supernodulating mutant containing a large deletion in linkage group H. Genome Lett 1:147–155. https://doi.org/10.1166/gl.2002.017

Michno JM, Wang X, Liu J, Curtin SJ, Kono TJ, Stupar RM (2015) CRISPR/Cas mutagenesis of soybean and *Medicago truncatula* using a new web-tool and a modified Cas9 enzyme. GM Crops Food 6:243–252. https://doi.org/10.1080/21645698.2015.1106063

Nemudryi AA, Valetdinova KR, Medvedev SP, Zakian SM (2014) TALEN and CRISPR/Cas genome editing systems: tools of discovery. Acta Nat 6:19–40. https://doi.org/10.1007/s11103-014-0188-7

Nithin C, Thomas A, Basak J, Bahadur RP (2017) Genome-wide identification of miRNAs and lncRNAs in *Cajanus cajan*. BMC Genomics 18:878. https://doi.org/10.1186/s12864-017-4232-2

Okuley J, Lightner J, Feldmann K, Yadav N, Lark E, Browse J (1994) Arabidopsis *FAD2* gene encodes the enzyme that is essential for polyunsaturated lipid synthesis. Plant Cell 6:147–158. https://doi.org/10.1105/tpc.6.1.147

Pandey MK, Wang ML, Qiao L, Feng S, Khera P, Wang H, Tonnis B, Barkley NA, Wang J, Holbrook CC, Culbreath AK, Varshney RK, Guo B (2014) Identification of QTLs associated with oil content and mapping *FAD2* genes and their relative contribution to oil quality in peanut (*Arachis hypogaea L.*). BMC Genet 15:1–14. https://doi.org/10.1186/s12863-014-0133-4

Pannunzio NR, Watanabe G, Lieber MR (2017) Nonhomologous DNA end joining for repair of DNA double-Strand breaks. J Biol Chem. https://doi.org/10.1074/jbc.TM117.000374

Pattanayak V, Lin S, Guilinger JP, Ma E, Doudna JA, Liu DR (2013) High-throughput profiling of off-target DNA cleavage reveals RNA-programmed Cas9 nuclease specificity. Nat Biotechnol 31:839–843. https://doi.org/10.1038/nbt.2673

Pelletier JM, Kwong RW, Park S, Le BH, Baden R, Cagliari A, Hashimoto M, Munoz MD, Fischer RL, Goldberg RB, Harada JJ (2017) LEC1 sequentially regulates the transcription of genes involved in diverse developmental processes during seed development. Proc Natl Acad Sci 114:E6710. https://doi.org/10.1073/pnas.1707957114

Ramirez CL, Foley JE, Wright DA, Müller-Lerch F, Rahman SH, Cornu TI, Winfrey RJ, Sander JD, Fu F, Townsend JA, Cathomen T, Voytas DF, Joung JK (2008) Unexpected failure rates for modular assembly of engineered zinc fingers. Nat Methods 5:374–375. https://doi.org/10.1038/nmeth0508-374

Ran FA, Hsu PD, Wright J, Agarwala V, Scott DA, Zhang F (2013) Genome engineering using the CRISPR-Cas9 system. Nat Protoc 8:2281–2308. https://doi.org/10.1038/nprot.2013.143

Revalska M, Vassileva V, Goormachtig S, Van Hautegem T, Ratet P, Iantcheva A (2011) Recent progress in development of *Tnt1* functional genomics platform for *Medicago truncatula* and *Lotus japonicus* in Bulgaria. Curr Genomics 12:147–152. https://doi.org/10.2174/138920211795564313

Rodríguez-Leal D, Lemmon ZH, Man J, Bartlett ME, Lippman ZB (2017) Engineering quantitative trait variation for crop improvement by genome editing. Cell 171:470–480.e8. https://doi.org/10.1016/j.cell.2017.08.030

Sander JD, Dahlborg EJ, Goodwin MJ, Cade L, Zhang F, Cifuentes D, Curtin SJ, Blackburn JS, Thibodeau-Beganny S, Qi Y, Pierick CJ, Hoffman E, Maeder ML, Khayter C, Reyon D, Dobbs D, Langenau DM, Stupar RM, Giraldez AJ, Voytas DF, Peterson RT, Yeh J-RJ, Joung JK (2011) Selection-free zinc-finger-nuclease engineering by context-dependent assembly (CoDA). Nat Methods 8:67–69. https://doi.org/10.1038/nmeth.1542

Schmutz J, Cannon SB, Schlueter J, Ma J, Mitros T, Nelson W, Hyten DL, Song Q, Thelen JJ, Cheng J, Xu D, Hellsten U, May GD, Yu Y, Sakurai T, Umezawa T, Bhattacharyya MK, Sandhu D, Valliyodan B et al (2010) Genome sequence of the palaeopolyploid soybean. Nature 463:178–183. https://doi.org/10.1038/nature08670

Severin AJ, Woody JL, Bolon YT, Joseph B, Diers BW, Farmer AD, Muehlbauer GJ, Nelson RT, Grant D, Specht JE, Graham MA, Cannon SB, May GD, Vance CP, Shoemaker RC (2010) RNA-Seq atlas of *Glycine max:* a guide to the soybean transcriptome. BMC Plant Biol 10:160. https://doi.org/10.1186/1471-2229-10-160

Singh D, Singh CK, Taunk J, Tomar RSS, Chaturvedi AK, Gaikwad K, Pal M (2017) Transcriptome analysis of lentil (*Lens culinaris Medikus*) in response to seedling drought stress. BMC Genomics 18:1–20. https://doi.org/10.1186/s12864-017-3596-7.

Verdier J, Dessaint F, Schneider C, Abirached-Darmency M (2013a) A combined histology and transcriptome analysis unravels novel questions on *Medicago truncatula* seed coat. J Exp Bot 64:459–470. https://doi.org/10.1093/jxb/ers304

Verdier J, Torres-Jerez I, Wang M, Andriankaja A, Allen SN, He J, Tang Y, Murray JD, Udvardi MK (2013b) Establishment of the *Lotus japonicus* gene expression atlas (LjGEA) and its use to explore legume seed maturation. Plant J 74:351–362. https://doi.org/10.1111/tpj.12119

Verma M, Kumar V, Patel RK, Garg R, Jain M (2015) CTDB: an integrated chickpea transcriptome database for functional and applied genomics. PLoS One 10:1–10. https://doi.org/10.1371/journal.pone.0136880

Wang L, Wang L, Tan Q, Fan Q, Zhu H, Hong Z, Zhang Z, Duanmu D (2016) Efficient inactivation of symbiotic nitrogen fixation related genes in *Lotus japonicus* using CRISPR-Cas9. Front Plant Sci 7:1333. https://doi.org/10.3389/fpls.2016.01333

Wen S, Liu H, Li X, Chen X, Hong Y, Li H, Lu Q, Liang X (2018) TALEN-mediated targeted mutagenesis of *fatty acid desaturase 2* (*FAD2*) in peanut (*Arachis hypogaea L.*) promotes the accumulation of oleic acid. Plant Mol Biol 97:177–185. https://doi.org/10.1007/s11103-018-0731-z

Wright DA, Thibodeau-Beganny S, Sander JD, Winfrey RJ, Hirsh AS, Eichtinger M, Fu F, Porteus MH, Dobbs D, Voytas DF, Joung JK (2006) Standardized reagents and protocols for engineering zinc finger nucleases by modular assembly. Nat Protoc 1:1637–1652. https://doi.org/10.1038/nprot.2006.259

Xing HL, Dong L, Wang ZP, Zhang HY, Han CY, Liu B, Wang XC, Chen QJ (2014) A CRISPR/Cas9 toolkit for multiplex genome editing in plants. BMC Plant Biol 14:1–12. https://doi.org/10.1186/s12870-014-0327-y

Zhang Y, Zhang F, Li X, Baller JA, Qi Y, Starker CG, Bogdanove AJ, Voytas DF (2013) Transcription activator-like effector nucleases enable efficient plant genome engineering. Plant Physiol 161:20–27. https://doi.org/10.1104/pp.112.205179

Zhang K, Raboanatahiry N, Zhu B, Li M (2017) Progress in genome editing technology and its application in plants. Front Plant Sci 8:177. https://doi.org/10.3389/fpls.2017.00177

Index

S. H. Wani, M. Jain (eds.), *Pulse Improvement*,
https://doi.org/10.1007/978-3-030-01743-9

The manufacturer's authorised representative in the EU is Springer Nature Customer Service Centre GmbH, Europaplatz 3, 69115 Heidelberg, Germany. If you have any concerns regarding our products, please contact ProductSafety@springernature.com

Printed and bound by CPI Group (UK) Ltd, Croydon, CR0 4YY
30/09/2026
02225440-0004